Hydropower Nation

China has the largest electricity generation capacity in the world today. Its number of large dams is second to none. Xiangli Ding provides a historical understanding of China's ever-growing energy demands and how they have affected its rivers, wild species, and millions of residents. River management has been an essential state responsibility throughout Chinese history. In the industrial age, with the global proliferation of concrete dam technology, people started to demand more from rivers, particularly when required for electricity production. Yet hydropower projects are always more than a technological engineering enterprise, layered with political, social, and environmental meaning. Through an examination of specific hydroelectric power projects, the activities of engineers, and the experience of local communities and species, Ding offers a fresh perspective on twentieth-century China from environmental and technological perspectives.

Xiangli Ding is Assistant Professor of History at the Rhode Island School of Design.

Studies in Environment and History

Editors

J. R. McNeill, *Georgetown University*
Ling Zhang, *Boston College*

Editors Emeriti

Alfred W. Crosby, *University of Texas at Austin*
Edmund P. Russell, *Carnegie Mellon University*
Donald Worster, *University of Kansas*

Other Books in the Series

Matthew P. Johnson *Hydropower in Authoritarian Brazil: An Environmental History of Low-Carbon Energy, 1960s–90s*
Ellen Arnold *Medieval Riverscapes: Environment and Memory in Northwest Europe, c. 300–1100*
Richard C. Hoffmann *The Catch: An Environmental History of Medieval European Fisheries*
Samuel Dolbee *Locusts of Power: Borders, Empire, and Environment in the Modern Middle East*
Andy Bruno *Tunguska: A Siberian Mystery and Its Environmental Legacy*
Lionel Frost et al. *Cities in a Sunburnt Country: Water and the Making of Urban Australia*
Adam Sundberg *Natural Disaster at the Closing of the Dutch Golden Age: Floods, Worms, and Cattle Plague*
Germán Vergara *Fueling Mexico: Energy and Environment, 1850–1950*
Peder Anker *The Power of the Periphery: How Norway Became an Environmental Pioneer for the World*
David Moon *The American Steppes: The Unexpected Russian Roots of Great Plains Agriculture, 1870s–1930s*
James L. A. Webb, Jr. *The Guts of the Matter: A Global Environmental History of Human Waste and Infectious Intestinal Disease*
Maya K. Peterson *Pipe Dreams: Water and Empire in Central Asia's Aral Sea Basin*
Thomas M. Wickman *Snowshoe Country: An Environmental and Cultural History of Winter in the Early American Northeast*
Debjani Bhattacharyya *Empire and Ecology in the Bengal Delta: The Making of Calcutta*
Chris Courtney *The Nature of Disaster in China: The 1931 Yangzi River Flood*
Dagomar Degroot *The Frigid Golden Age: Climate Change, the Little Ice Age, and the Dutch Republic, 1560–1720*
Edmund Russell *Greyhound Nation: A Coevolutionary History of England, 1200–1900*
Timothy J. LeCain *The Matter of History: How Things Create the Past*
Ling Zhang *The River, the Plain, and the State: An Environmental Drama in Northern Song China, 1048–1128*
Abraham H. Gibson *Feral Animals in the American South: An Evolutionary History*
Andy Bruno *The Nature of Soviet Power: An Arctic Environmental History*

A list of other books in the series can be found at the end of the volume.

Hydropower Nation

Dams, Energy, and Political Changes in Twentieth-Century China

XIANGLI DING
Rhode Island School of Design

Shaftesbury Road, Cambridge CB2 8EA, United Kingdom

One Liberty Plaza, 20th Floor, New York, NY 10006, USA

477 Williamstown Road, Port Melbourne, VIC 3207, Australia

314–321, 3rd Floor, Plot 3, Splendor Forum, Jasola District Centre,
New Delhi – 110025, India

103 Penang Road, #05-06/07, Visioncrest Commercial, Singapore 238467

Cambridge University Press is part of Cambridge University Press & Assessment,
a department of the University of Cambridge.

We share the University's mission to contribute to society through the pursuit of
education, learning and research at the highest international levels of excellence.

www.cambridge.org
Information on this title: www.cambridge.org/9781009426565
DOI: 10.1017/9781009426589

© Xiangli Ding 2024

This publication is in copyright. Subject to statutory exception and to the provisions
of relevant collective licensing agreements, no reproduction of any part may take
place without the written permission of Cambridge University Press & Assessment.

When citing this work, please include a reference to the DOI 10.1017/9781009426589

First published 2024

A catalogue record for this publication is available from the British Library.

*A Cataloging-in-Publication data record for this book is available from the Library
of Congress*

ISBN 978-1-009-42656-5 Hardback

Cambridge University Press & Assessment has no responsibility for the persistence
or accuracy of URLs for external or third-party internet websites referred to in this
publication and does not guarantee that any content on such websites is, or will
remain, accurate or appropriate.

Contents

List of Figures	page ix
List of Maps	xi
List of Tables	xiii
Acknowledgments	xv
Introduction: A Flow of Water and Power	1
Water History	2
Hydropower Nation	6
Energy Politics	10
Environment	12
Sources and Organization	15

PART I STARTING FROM SCRATCH

1 An Inexhaustible Source of Power	25
The Rise of "White Coal"	28
Local Pioneers	34
Yaolong and Jihe	34
Fuyuan	39
A New Man-Made Disaster	41
Conclusion	46
2 Mobilizing Rivers	48
The Wartime Energy Crisis	51
The National Resources Commission	52
Powering Arsenals and Other Factories	55
Tensions with Local Communities	57
Learning from the TVA	60

	Envisioning "China's TVA"	74
	Conclusion	77

PART II THE SOCIALIST BOOST

3	The Making of Red Hydro Technostructure	83
	Li Rui and the Institutionalization and Propagation of Hydroelectricity	85
	The "Hydroelectricity First, Thermoelectricity Second" Policy	89
	Learning from the Soviet Union	92
	Experts from the Old Society	97
	Red and Expert	104
	Practicing Maoism in Hydro Infrastructure Construction	108
	Conclusion	112
4	The Great Leap of Small Hydro	115
	Making Hydropower a Mass Campaign	118
	Model County of Small Hydropower: Yongchun	124
	Small Is Not Always Beautiful	133
	The Politicization of Small Hydropower	140
	Conclusion	143

PART III A HUGE SETBACK: THE SANMENXIA DAM

5	Silt and Hydroelectricity	149
	Sanmenxia before the Dam	151
	Japanese Planning	153
	Early Soil and Water Conservation Efforts	157
	Soviet Assistance	159
	The Challenge of Silting and Debates over the Soviet Design	163
	Constructing Sanmenxia with Propaganda	170
	Reconstructions	173
	Conclusion	174
6	The Human Cost	177
	Mobilizing Reservoir Inhabitants to Move to Dunhuang	181
	Daily Life and Struggles in Dunhuang	189
	Fleeing Dunhuang	196
	Compensation for Property Loss	199
	Conclusion	202
7	The Environmental Saga	204
	The Geographic Setting	205
	A Land in Transformation	207
	Riverbank Collapse	208
	Groundwater and Land Deterioration	210
	Land Use Change	211

    The Rise of Wetland Conservation 215
    Whooper Swan: An Environmental Windfall 218
    The Crackdown on Illegal Hunting 220
    Feeding the Swans 223
    Building a Benign Environment 225
    Marketing "Swan City" 226
    Conclusion 228
  Epilogue 231

Bibliography 239
Index 265

Figures

1.1	Shilongba Hydropower Plant	page 37
3.1	Struggle session at the Liujiaxia Dam site	112
4.1	Estimated annual total number of hydropower stations in rural China, 1970–1986	134
4.2	Electrical generation capacity of rural hydropower	134
5.1	Sanmenxia Dam under construction	171
6.1	Timeline of Sanmenxia Project construction and reservoir resettlement in Henan	181
7.1	Sanmenxia Reservoir at low water level	212
7.2	Feeding swans at the Sanmenxia Wetland Preserve	224

Maps

1.1 Major research sites   16
5.1 Sanmenxia Reservoir area   167

Tables

3.1	Education experience of selected hydropower engineers	*page* 101
5.1	Main stages of the Sanmenxia project	151
7.1	Sanmenxia Reservoir water level and wetland size	218

Acknowledgments

Nearly a decade has elapsed since the initiation of this project and, in reflecting upon this journey, I extend my heartfelt appreciation to numerous individuals and institutions.

Foremost among these expressions of gratitude is reserved for my advisors at the University at Buffalo (UB). Roger Des Forges, in particular, has exemplified unwavering patience, encouragement, and support throughout this academic odyssey. His meticulous review and commentary on the inaugural draft of this project were instrumental in shaping its trajectory. A debt of gratitude is also owed to Kristin Stapleton, whose feedback has consistently proven to be constructive. The scholarly expertise of Mark Nathan, specifically in the realms of Korean history and modern East Asia, has significantly broadened the horizons of my research, fostering a transnational perspective on history. Further appreciation is extended to Liu Yan and Adam Rome. Liu Yan provided invaluable guidance in formulating research proposals, while Adam Rome introduced me to seminal works in the field of American environmental history, enriching my academic landscape.

Acknowledgment is also due to my mentors at Nanjing University, where Hu Cheng, my advisor at Nanda, served as a source of inspiration during my formative years. Under his guidance, I learned the values of independent and thoughtful scholarship. Special thanks are extended to Li Yu, Ma Junya, and other faculty members of the History Department at Nanda for their contributions to both scholarship and mentorship.

I express my sincere gratitude to the editors of the Environment and History series, John McNeill and Zhang Ling, for their exceptional work. The invaluable insights I gained from this series have significantly enriched my own work, and it is a profound honor to have my first book featured within it. I am particularly indebted to Zhang Ling. Despite not having the opportunity to work with her within formal institutional frameworks, her support and critiques have been instrumental in bringing this book to fruition. Additionally, I extend my appreciation to the anonymous reviewers whose feedback played a crucial role in refining the manuscript.

Over the past few years, engaging in panel discussions, Q&A sessions, and conversations at conferences and workshops has been a rewarding experience, thanks to the scholarly input and constructive suggestions from individuals such as Robert Marks, Micah Muscolino, David Pietz, Victor Seow, Arunabh Ghosh, Ying Jia Tan, Jordan Sand, Donald Worster, Niu Jianqiang, Zheng Xiaoyun, Cao Shuji, Li Yushang, Hou Shen, Xia Mingfang, Li Huaiyu, Sigrid Schmalzer, Fan Fa-ti, Feng Fan, Xu Chongbao, Paul Kreitman, Sakura Christmas, Ren Ke, Zhang Meng, Matthew Johnson, John Hayashi, Fan Xin, Gao Yan, Wang You, among many others. Their thoughtful comments and queries have undoubtedly shaped and enhanced various aspects of this project. Throughout my field trip and the intricate writing process, I owe a debt of gratitude to many friends who generously offered their assistance. Special thanks to Chen Zhigang, Wen Yunfeng, Shao Weinan, Shen Yubin, He Zhiming, Chen Jie, Li Hongbin, Shi Yun, and many librarians and archivists in mainland China, Taiwan, and the United States.

Gratitude also goes to my colleagues and students at the Rhode Island School of Design, for creating a collegial and supportive professional environment.

Working with Rachel Blaifeder, Lucy Rhymer, and Rosa Martin from Cambridge University Press has been an absolute pleasure. I am deeply thankful for their patience and professionalism throughout the entire process. I extend my gratitude to my copy-editors Cynthia Col and Malcolm Thompson. Their meticulous editing significantly enhanced the manuscript's writing style, taking it to another level. Despite their invaluable contributions, I must acknowledge that any remaining errors are solely my responsibility.

The generous financial support from the Plesur Dissertation Fellowship, Department of History at UB, facilitated a year-long archival research trip in mainland China during 2014–2015. Additionally, the Asian Studies research grant from the Asian Studies Program at UB enabled me to travel to College Park, PA, where I collected Chinese hydrology materials from the Republican era. The Mark Diamond research grant from the Graduate Student Association allowed me to journey to Taiwan and gather archival documents related to the National Resources Commission at Academia Sinica. Furthermore, the Professional Development Fund and Humanities Fund of the Rhode Island School of Design played a crucial role in supporting the completion of the manuscript's writing and editing.

Lastly, I express deep gratitude to my family, especially my wife Ruiqi and two daughters Keyu and Zijin, whose unwavering companionship is indispensable. This book is dedicated to them.

Introduction

A Flow of Water and Power

In 1996, along the Yangtze River in Zigui 秭归, Hubei Province, the villagers of Guilin were in the middle of packing their belongings. Their friends and relatives were helping them to remove tiles, doors, and windows from the soon to be deluged homes. Down the hill, people were loading farm tools, furniture, and other essentials onto boats anchored along the river's banks. Amid the din of firecrackers and farewells from their fellow villagers, the first group of Three Gorges migrants set off for their government-designated places of resettlement. Because of the low-altitude location of their residence, Zhang Bing'ai 张秉爱, together with her disabled husband, and two children, were supposed to be part of the first group of migrants. For years, thanks to the help of her maternal family, Bing'ai had been able to cope with the burdensome farm work during busy seasons. Therefore, she insisted on staying for "moving up" resettlement, requesting a flat area for residential use at a higher altitude above the state designated displacement line, which was at odds with the government's resettlement policy. "I am just attached to this land. With land, you can have everything." Bing'ai refused to cooperate with the government's resettlement plan, but it was to no avail. Her home, along with what remained of those of her fellow villagers, was soon underwater. Her family ended up living in a temporary hut not far from the rising water.[1]

Among the millions of reservoir migrants in China, Bing'ai's struggle was not rare. Her attachment to the land and bond with her relatives were

---

[1] Feng Yan 冯艳, (dir.), *Bing'Ai* 秉爱(Beijing: Beisen Films, 2007), DVD.

authentic and could be understood intuitively without much infiltration of state rhetoric. A simple yet fundamental question arises: Why did her family have to be displaced? The answer seems obvious; it was for the construction of the Three Gorges Dam. But why did a dam need to be built on the river at the cost of millions of people's homes and livelihoods? Who should be held accountable for Bing'ai and her family's difficult situation? The Three Gorges Dam and its story of displacement are not the focus of this book, yet it serves as a starting point to trace the historical origin of Bing'ai's struggle and that of many others. This book aims to address these questions from a historical perspective by examining the entangled relationships among human beings, nature, hydropower technology, and political change in modern China.

WATER HISTORY

For as long as humans have existed, rivers have been intricately woven into the fabric of the Earth's complex ecosystems. As vital components of the hydrological system, these waterways have sustained a great diversity of flora and fauna and played a crucial role in the formation of alluvial plains and deltas. Rivers have been sources of immense generosity to humanity, providing us with drinkable water, fish (a valuable source of protein), and other essential nutrients. However, these same rivers can also pose a threat to human societies. The widespread adoption of agriculture and sedentary lifestyles in low-lying areas has made populations vulnerable to the unpredictable forces of flooding during rainy seasons. In response, communities have organized to build dikes and to dig irrigation channels to protect themselves and to ensure the preservation of their prosperity.[2]

According to the first national water census released in 2011, China has 45,203 rivers with catchment areas larger than 50 square kilometers. A total of 268,476 sluices and 413,679 kilometers of embankments have been built to regulate the flow of water, and the country operates 46,758 hydropower stations with a combined installed capacity of 333 million kilowatts (kW).[3] Over a long period, the natural river system has been

---

[2] Chris Courtney, *The Nature of Disaster in China: The 1931 Yangzi River Flood* (Cambridge: Cambridge University Press, 2018).
[3] Zhonghua renmin gongheguo shuilibu 中华人民共和国水利部, Guojia tongjiju 国家统计局, *Diyici quanguo shuilipucha gongbao* 第一次全国水利普查公报 (Bulletin of First National census for water) (Beijing: Zhongguo shuilishuidian chubanshe, 2011).

transformed into a complex human-engineered network. This transformation is the result of centuries of human intervention.

As the environmental historian Donald Worster astutely observes in his study of rivers in the American West, "To write history without putting any water in it is to leave out a large part of the story. Human experience has not been so dry as that."[4] Indeed, water is ubiquitous in our world, essential for life itself. The management of water has played an essential role in Chinese history as well. From the mythical flood mitigation efforts of Yu the Great, the founder of the first Chinese dynasty, the Xia (2070–1600 BCE), to the Zhengguo Canal that facilitated the rise of the Qin empire (221–206 BCE), water management has been indelibly imprinted on Chinese history.

In the twentieth century, economists and historians alike have delved into the significance of water management in shaping Chinese political structures. Ch'ao-ting Chi 冀朝鼎, an economist who later became a communist, pointed out the crucial role water control played in political struggles throughout Chinese history.[5] Decades later, Karl Wittfogel proposed the concept of a "hydraulic society," arguing that the need for large-scale hydraulic works had led to centralized, despotic political structures in China and other "oriental" countries.[6] This theory of "oriental despotism" has been challenged, however, by studies that have sought to uncover the high degree of local autonomy in the management of water resources, leading Akira Morita to propose the concept of the "hydraulic community."[7] Of course, the management of major projects like the Yellow River required far more human labor and financial investment than local irrigation systems.[8]

---

[4] Donald Worster, *Rivers of Empire: Water, Aridity, and the Growth of the American West* (Oxford: Oxford University Press, 1985), 5.
[5] Ch'ao-ting Chi, *Key Economic Areas in Chinese History: As Revealed in the Development of Public Works for Water Control* (London: George Allen & Unwin Ltd., 1936).
[6] Karl A. Wittfogel, *Oriental Despotism: A Comparative Study of Total Power* (New Haven: Yale University Press, 1957).
[7] Frederick W. Mote, "The Growth of Chinese Despotism: A critique of Wittfogel's Theory of Oriental Despotism as Applied to China," *Oriens Extremus* 8, no. 1 (1961), 1–41; Akira Morita, *Qingdai shuili yu quyu shehui* 清代水利与区域社会, trans. Lei Guoshan (Jinan: Shandong huabao chubanshe, 2008); Pierre-Etienne Will, "State Intervention in the Administration of a Hydraulic Infrastructure: The Example of Hubei Province in Late Imperial Times," in *The Scope of State Power in China*, ed. Stuart Schram (Hong Kong: Chinese University Press, 1985), 295–347; Yan Gao, *Yangzi Waters: Transforming the Water Regime of the Jianghan Plain in Late Imperial China* (Leiden: Brill, 2022).
[8] Peter Purdue, *Exhausting the Earth: State and Peasant in Hunan, 1500–1850* (Cambridge, MA: Harvard University Asia Center, 1987), 171.

As these scholars have shown, the history of water management reveals the "inevitable imbrication of human and natural processes."⁹ From individual households to the central organs of the state, a wide variety of human organizations have played a role in river management, either voluntarily or involuntarily. The larger the project, the more human labor was required, and with this emerged the need for a more efficient administrative system. However, it is important to note that water management was just one of multiple factors which contributed to the formation of large human organizations. Wars, human reproduction, and other factors have also had a great influence on the process of social stratification and state formation.

China contains two of the world's largest rivers, the Yellow River in the north and the Yangtze River in the south. Their basins have long been the core economic zones of China, and stable and centralized states have sought to control these rivers or to provide relief to communities affected by river-related crises. Failures of the river management system can be a symptom of a struggling state, whether they are due to financial limitations on maintenance, poor engineering, or organized sabotage. This can result in catastrophic consequences for people who live along these rivers. Residents of the Hebei Plain in the eleventh century, for example, or Wuhan in the early 1930s, and Henan in the seventeenth century and the late 1930s, all suffered from drowning, food shortages, infectious diseases, and death because of failed river management efforts.¹⁰ Thus, the history of water makes clear the close relationship between human and natural processes and the crucial role of river management in shaping human societies.

Despite their strengths, the concepts of "hydraulic society" and "hydraulic community" are inadequate to an analysis of water

---

⁹ Peter Purdue, "Is there a Chinese View of Technology and Nature?," in *The Illusory Boundary: Environment and Technology in History*, ed. Martin Reuss and Stephen Cutcliffe (Charlottesville: University of Virginia Press, 2010), 102.

¹⁰ See Roger V. Des Forges, *Cultural Centrality and Political Change in Chinese History: Northeast Henan in the Fall of the Ming* (Stanford: Stanford University Press, 2003); Micah Muscolino, *The Ecology of War in China: Henan Province, the Yellow River, and beyond, 1938–1950* (Cambridge: Cambridge University Press, 2015); Ling Zhang, *The River, The Plain, and the State: An Environmental Drama in Northern Song China, 1048–1128* (Cambridge: Cambridge University Press, 2016); Courtney, *Nature of Disaster in China*; Ma Junya 马俊亚. Beixisheng de jubu: Huaibei shehui shengtai bianqian yanjiu, 1680–1949 被牺牲的局部:淮北社会生态变迁研究 (*The Sacrificed Region: A Study of the Social Ecological Changes in Huaibei*) (Beijing: Beijingdaxue chubanshe, 2011).

management in the twentieth century. The advent of electricity produced significant changes in water management, particularly in relation to the generation of power. The installation of the Fourneyron turbine at Niagara Falls in 1895 was a major turning point as it enabled water to be converted into electricity on a large scale. This paved the way for the development of hydroelectricity and the long-distance transmission of electricity.[11] By the beginning of the twentieth century, advances in hydraulics, turbine design, and alternating current technology made the large-scale exploitation of hydroelectricity possible.[12] During the two decades between the world wars, water was responsible for generating up to half of the electrical power in many industrial countries. Hydropower was called "white coal" in Europe due to the abundance of power generated from glacial streams in the Alps.[13] To address high rates of unemployment during the Great Depression, countries such as Nazi Germany and the United States built large concrete dams to generate electricity.[14] The United States was a pioneer in river-basin management and multipurpose dams, as exemplified by the Tennessee Valley Authority and the Boulder Dam on the Colorado River. The development of the concept of multipurpose exploitation, which embraced flood control, irrigation, navigation, and electricity generation, was seen as a way to maximize the benefits of rivers.

In the twentieth century, the capacity to produce electricity became a crucial indicator of a nation's industrial strength. Concrete dams, seen as symbols of technological superiority and productivity, emblematized the modernist state's perspective on water management. Letting rivers run freely without exploiting them was viewed as wasteful by many state leaders. The importance of water management in the generation of

---

[11] In fact, the first hydroelectric plant on Niagara Falls was built in the 1880s. However, it produced direct current electricity, which had a very short transmission distance. Until Nikola Tesla's development of a poly-phase alternating current system of generator, motor, and transformer in the late 1880s, long-distance transmission of electricity was impossible.

[12] See Louis C. Hunter and Lynwood Bryant, *A History of Industrial Power in the United States, 1780–1930*, Vol. 3: *The Transmission of Power* (Cambridge, MA: The MIT Press, 1991).

[13] See Marc D. Landry II, "Europe's Battery: The Making of the Alpine Energy Landscape, 1870–1955," PhD Dissertation, Georgetown University, 2013.

[14] David Blackbourn, *The Conquest of Nature: Water, Landscape, and the Making of Modern Germany* (New York: W. W. Norton & Company, 2006); Richard White, *The Organic Machine: The Remaking of the Columbia River* (New York: Hill and Wang, 1995).

electricity, coupled with the appeal it held for engineers and political leaders, made it a significant aspect of a modern society.[15]

HYDROPOWER NATION

In this book, I introduce the concept of "hydropower nation" as a way to analyze the multidimensional transformation of rivers in twentieth-century China. Hydropower, harnessed from the natural force of river flows, has been converted into electricity to meet a variety of human demands. This transformation, however, requires not only specific technological innovations but also political and social engineering. In the twentieth century, building a strong and prosperous nation was a shared goal of Chinese elites and the Communist Party after 1949. They believed that the nation and its prosperity were embodied in the country's mountains and rivers, rather than in an imagined community that existed in texts only.[16] Regardless of their size, hydropower projects were integrated into the discourse of national development and became a driving force in propelling and securing the power of the ruling party, despite ideological and geopolitical shifts. In this sense, hydropower projects and political regimes were interdependent and mutually reinforcing. In the twentieth century, the convergence of rivers, hydropower technology, technocratic systems, and nation-state building projects led to the formation of a hydropower nation. It was conducive to an inseparable relationship between river management, energy production, and statist goals. At odds with the narratives of productivism and developmentalism, the hydropower nation brought forth both construction and destruction. The visible and often sublime concrete infrastructures came into existence at the cost of profound social and ecological disruptions.

In the age of great acceleration, China's rivers have been undergoing transformative changes at an unprecedented pace, as the country has embarked on a mission of national reconstruction and industrialization. With an abundance of rivers – more than any other country in the world – China is rich in hydropower potential, due to its mountainous terrain and the changes of elevation from the Qinghai–Tibetan Plateau to the Pacific coast. In a world that still relies heavily on fossil fuels, China has the

---

[15] John R. McNeill, *Something New under the Sun: An Environmental History of the Twentieth-Century World* (New York: W. W. Norton & Company, 2000), 157.

[16] Benedict Anderson, *Imagined Communities: Reflections on the Origin and Spread of Nationalism* (rev. ed.) (London: Verso, 2016).

largest capacity for hydropower production, far outstripping the second place United States. And this capacity is still growing.

Christopher Sneddon has called the proliferation of concrete dam technology in the historical context of the Cold War a "concrete revolution."[17] Yet, how this revolution unfolded in modern China remains unclear. My aim in this book is to tell the story of the planning, construction, and operation of China's hydropower projects. I analyze their historical significance and examine the role they played in energy production and the making of a nation-state and a socialist state. State building is a central narrative in the field of modern Chinese history. Scholars have engaged with this narrative from institutional, political, military, and medical perspectives.[18] In recent years, energy production has emerged as another lens through which to view the struggle to build the Chinese state.[19] This study situates itself in this discussion of energy and focuses on hydropower to examine the historical changes that have occurred in modern China. From small-scale private ventures to large state-run operations, hydropower projects have been inextricably linked to the political developments that occurred in China and elsewhere. I explore the complex interplay between energy, politics, and state-building as they have been shaped by these hydropower projects.

In the 1950s, under the guidance of the visionary Communist Li Rui 李锐 and other like-minded revolutionaries and technocrats, the People's Republic of China adopted an energy policy that prioritized hydropower over thermal power to drive industrialization. Instead of developing a carbon-intensive energy system, the country sought to build one based on water. Although it was short-lived, this policy led to the construction of numerous dams. These dams not only transformed China's landscape but

---

[17] Christopher Sneddon, *Concrete Revolution: Large Dams, Cold War Geopolitics, and the US Bureau of Reclamation* (Chicago: University of Chicago Press, 2015).

[18] Prasenjit Duara, *Rescuing History from the Nation: Questioning Narratives of Modern China* (Chicago: University of Chicago Press, 1997); Hans van de Ven, *War and Nationalism in China, 1925–1945* (London: Routledge, 2003); Klaus Muhlhahn, *Making China Modern: From the Great Qing to Xi Jinping* (Cambridge, MA: Belknap Press, 2019); Mary Augusta Brazelton, *Mass Vaccination: Citizen's Bodies and State Power in Modern China* (Ithaca: Cornell University Press, 2019).

[19] Shellen X. Wu, *Empires of Coal: Fueling China's Entry into the Modern World Order, 1860–1920* (Stanford: Stanford University Press, 2015); Hou Li, *Building for Oil: Daqing and the Formation of the Chinese Socialist State* (Cambridge, MA: Harvard University Asia Center, 2021); Ying Jia Tan, *Recharging China in War and Revolution, 1882–1955* (Ithaca: Cornell University Press, 2021); Victor Seow, *Carbon Technocracy: Energy Regimes in Modern East Asia* (Chicago: University of Chicago Press, 2022).

also became deeply intertwined with the social and political fabric of socialist China. My goal in this book is to describe the relationship between the concrete revolution of hydropower projects and the larger socialist revolution, and to determine how each influenced the other.

Ling Zhang, in her seminal work on the shifting course of the Yellow River, challenges the logic of the "hydraulic mode of production" that underlies much previous scholarship on water management in Chinese history. By exploring the human toll involved in maintaining the Yellow River–Hebei complex from 1048 to 1128, including costs, suffering, and loss of life, Zhang advances a theory of the "hydraulic mode of consumption." This analysis highlights the high cost of preserving a particular water-management system. Zhang's perspective provides a deeper understanding of the relationship between humans and water in China and shows that this mode of consumption continues to be widespread to this day.[20] It is important to note that examinations of the hydraulic modes of production and consumption have primarily been concerned with the (preindustrial) period of imperial China. In an agrarian society, water management primarily focuses on irrigation and transportation for the purpose of production. With industrialization, hydropower emerges as another source of productivity. David Pietz has demonstrated the Chinese state's more or less continuous historical commitment to harnessing the potential of the Yellow River, once called "China's Sorrow" because of its frequent flooding, and to transforming it into a catalyst for both agricultural and industrial growth.[21] The emphasis on production in Pietz's analysis, however, still outweighs that placed on consumption, as can be seen in his adherence to the "hydraulic mode of production" paradigm.

In my examination of the rise of the hydropower nation, I address both sides of the dual nature of hydropower, both as a source of power, in the form of energy and political might, and as a means of destruction, through the displacement of people and environmental degradation. By considering perspectives centered on the state, on local communities, and on the environment, I seek to incorporate both the analysis of power and energy production through the notion of the "hydraulic mode of production" and the evaluation of the tremendous costs incurred in the

---

[20] Zhang, *River, Plain, and State*, 178–179; 288–290.
[21] David Pietz, *The Yellow River: The Problem of Water in Modern China* (Cambridge, MA: Harvard University Press, 2015).

process of this pursuit of power through the concept of "hydraulic mode of consumption."

Throughout the twentieth century, Chinese states attempted to make China's rivers more productive through hydrology and hydraulic engineering. Despite substantial investments in large concrete dams, however, these efforts often produced unanticipated problems (such as silting) that diminished the anticipated gains in productivity. When the financial, social, and environmental costs of these dams are taken fully into account, it becomes difficult to argue that they have brought lasting benefits to either the environment or humanity. The relationship between the "hydraulic mode of production" and the "hydraulic mode of consumption" is complex and interdependent. Hydraulic engineering, technocratic ideas, social mobilization, the displacement of populations, and environmental degradation all played crucial roles in the emergence of China as a hydropower nation. It is clear that the two modes cannot be easily divided. They must be considered together to fully understand the impact of these projects.

The creation of a hydropower nation is a complex and multifaceted process that involves a variety of technological, political, social, and environmental factors. This book will examine these elements in detail, exploring the relationships between state, rivers, and people as they have evolved in modern China. By examining both small and large hydropower projects, I aim to provide a nuanced understanding of the evolution of state–river–people relationships. Rather than categorizing these projects as either successes or failures for the political regimes that enacted them, they are viewed instead as products and reflections of the natural, technological, and political systems in which they occur. Such an approach can provide a more complete understanding of the challenges and opportunities inherent in the making of a hydropower nation.

The making of a hydropower nation in China was influenced by the ideology of "high modernism." This ideology, as described by James Scott, is characterized by a strong belief in the power of scientific and technical progress, increased production, the mastery of nature, and the rational design of social order.[22] Despite China's technological limitations in the twentieth century, the Communist regime, with Soviet help, still made tremendous efforts to implement its ambitious river engineering projects. This was a localized manifestation of global high modernism,

---

[22] James Scott, *Seeing Like a State: How Certain Schemes to Improve the Human Condition Have Failed* (New Haven: Yale University Press, 1999), 4.

where Maoist radicals combined technical and political ambitions in a "war against nature" to realize the environmental and social engineering goals of the revolutionary state.[23] The result was a major role for mass participation, in the form of laborers, the incorporation of indigenous technologies, and involuntary displacement, all of which became prominent features of the Chinese form of the hydropower nation.

ENERGY POLITICS

In the industrial age, the capacity of human beings to produce energy has vastly surpassed that of any previous form of society, as if the sun were the only limit. With the emergence of the climate crisis, energy has become a focal point in global debates and collaborations. The shift toward an energy-intensive society and an anthropocentric environment was largely guided by scientists, engineers, entrepreneurs, and capitalists. Historians have a responsibility to critically examine the ways in which the increasing demand for energy has transformed human–nature and human–human relationships.[24] The study of energy has interested scholars for several decades, with experts such as Vaclav Smil and Alfred Crosby exploring the major sources of energy in human history, from grains to fossil fuels.[25] Both scholars express deep skepticism about the sustainability of our current energy-intensive form of society.[26] As the evolution of water management has demonstrated, energy and political power are inextricably related.[27] The development of hydropower in China is intricately linked to its political ecology. While recent research in modern China studies has largely focused on fossil fuels, the rapid rise of hydroelectricity in the twentieth century and its continued growth demand

---

[23] Judith Shapiro, *Mao's War against Nature: Politics and the Environment in Revolutionary China* (Cambridge: Cambridge University Press, 2001).

[24] One excellent example is Victor Seow's work on the history of Funshun colliery. Seow's critique of productivism behind the rise of carbon technocracy and modern industrial state in East Asia is intriguing. See Seow, *Carbon Technocracy*.

[25] See Muscolino, *Ecology of War in China*.

[26] Vaclav Smil, *Energy in World History* (Boulder: Westview Press, 1994); Alfred W. Crosby, *Children of the Sun: A History of Humanity's Unappeasable Appetite for Energy* (New York: W.W. Norton & Company, 2006); Vaclav Smil also wrote an overview of energy in modern China, see Vaclav Smil, *Energy in China's Modernization: Advances and Limitations* (Armonk: M. E. Sharpe, 1988).

[27] Timothy Mitchell argues that coal-centered socio-technical systems opened space for democracy, while oil led to the closing off of that space. See Timothy Mitchell, *Carbon Democracy: Political Power in the Age of Oil* (New York: Verso, 2011).

attention.²⁸ This book is an effort to address this gap in the existing literature, providing a comprehensive examination of the growth of hydropower in China. Victor Seow has proposed the concept of carbon technocracy as a way of analyzing the entanglement of fossil fuels and the formation of technocratic systems in East Asia. The concept of a hydropower nation sheds light on the significance of hydroelectricity in shaping modern China. Rather than being mutually exclusive, these concepts complement each other in uncovering the driving forces behind modern industrial society's unending pursuit of energy.

The study of energy history involves three key aspects: production, transportation/transmission, and consumption. Michael Faraday's discovery of electricity in the early nineteenth century and the subsequent development of alternating current technology was a revolutionary change in this history.²⁹ It made electricity the dominant way of transmitting energy, which it remains to this day. Hydropower technology transforms the movement of water into electricity that can be transmitted over vast distances through electrical grids. However, this presents a challenge in studying the consumption of the electricity generated from large hydropower projects. With the integration of regional and national grids, various sources of electricity are combined for redistribution. In my study of the Sanmenxia 三门峡 dam on the Yellow River (in Part III), my focus will, of necessity, be on production rather than transmission and consumption due to limitations of access to primary sources.³⁰ The latter must be left to future studies. On the other hand, my chapters on the energy crisis in wartime Chongqing (in Chapter 2) and small hydropower projects from the 1950s to the 1970s (in Chapter 4) will delve into the complete cycle of production, transmission, and consumption.

---

[28] Tan, *Recharging China in War and Revolution*; Seow, *Carbon Technocracy*.
[29] In his study of the technology of electricity, Thomas Hughes divides its history into four stages: invention, technology transfer, system growth, and substantial momentum. See Thomas Hughes, *Networks of Power: Electrification in Western Society, 1880–1930* (Baltimore: John Hopkins University Press, 1983). On the consumption of electricity, see David E. Nye, *Electrifying America: Social Meanings of a Technology, 1880–1940* (Cambridge, MA: MIT Press, 1992) and David E. Nye, *Consuming Power: A Social History of American Energies* (Cambridge, MA: MIT Press, 1997); On transmission, see Christopher F. Jones, *Routes of Power: Energy and Modern America* (Cambridge, MA: Harvard University Press, 2014).
[30] Because of the controversial past of the Sanmenxia hydropower project, my applications to access many of the documents listed in the archival catalogue of the Yellow River Conservancy Commission Archives were denied by authorities.

This book will address the social disruptions and environmental injustices inflicted by hydropower projects on local communities. Based on his examination of the sites of energy production and the transmission of electricity in America, Christopher Jones has argued that transport infrastructure systems deprived most consumers of knowledge of the environmental harms caused by their use of energy.[31] Similarly, the geographer Darrin Magee has developed the concept of the "powershed" to explain large hydropower projects in rural western China that supply electricity to urban eastern areas and the unjust political economy involved.[32] C. S. Lewis, the historian and philosopher, wrote that "what we call man's power over nature turns out to be a power exercised by some men over other men with nature as its instrument."[33] Under the moral principle of "sacrificing the minority for the good of the majority," hundreds of thousands of people were forcibly displaced to make way for hydropower reservoirs. During the Mao era (1949–1976), as it worked to build a national industrial system, China remained predominantly agrarian. The Sanmenxia Dam and other hydropower projects required not only the removal of people but also the mobilization of their labor for land reclamation projects in regions far away from their hometowns. Inhabitants of riparian areas bore the brunt of the industrial and agrarian complexes being created through the construction of large-scale hydropower projects. The effort to become a hydropower nation involves not just the building and maintenance of infrastructure, but also the sacrifice and destruction of communities and their ways of life.

ENVIRONMENT

Since the 1990s, a group of North American historians of technology have called for the development of more sophisticated research on the relationship between technology and the environment.[34] In particular, Sara Pritchard argues for the need to examine the material and discursive

---

[31] Jones, *Routes of Power*, 12.
[32] Darrin Magee, "Powershed Politics: Yunnan Hydropower under Great Western Development." *The China Quarterly* no. 185 (2006), 23–41.
[33] C. S. Lewis, "The Abolition of Man," in *The Complete C. S. Lewis Signature Classics* (Grand Rapids: Zondervan, 2007), 689–730, 719.
[34] See Jeffery K. Stine and Joel A. Tarr, "At the Intersection of Histories: Technology and the Environment," *Technology and Culture* 39, no. 4 (1998), 601–640.

historical interactions between ecological and technological systems.[35] Despite the significant environmental and social changes brought about by the adoption of large-scale technical innovations, the interactions of environment and technology in China during the age of great acceleration have received little attention.[36] By situating this study at the intersection of environmental, technological, and political changes, I aim to shed light on the proliferation of hydropower technology and the transformation of rivers in the Chinese context.

The widespread belief in the superiority of technological solutions to the problem of development in China coupled with increasing demand for energy have intensified the exploitation of natural rivers. Of all water-management technologies, concrete dams have the most profound and disruptive impact on riverine ecosystems, leading to declines and even extinctions of fish populations and hydrological disruptions. Further, other human activities connected to development, such as deforestation in the highlands, have exacerbated related problems, such as soil erosion, which can have long-term consequences for river systems and cause problems in the operation of concrete dams.[37] For instance, soil erosion on the Loess Plateau upriver from the Sanmenxia dam lead to sever silting in its reservoir and consequently undermined the dam's hydropower generation capacity and its ability to operate as a flood control mechanism. The historian Paul Josephson has characterized large-scale technical systems as "brute force technologies" that have a profound impact on the natural environment. In his studies of large concrete dams in the Soviet Union and the United States, he highlights the permanent harm to the environment and the disruption to communities that these massive projects cause. He argues that decentralized, smaller technologies would result both in a more harmonious relationship between humans and the environment and in greater social equity.[38]

In moving beyond the conventional narrative about large hydropower dams, this study also aims to provide a more comprehensive understanding of what has been called the "concrete revolution." Through a detailed

---

[35] Sara B. Pritchard, *Confluence: The Nature of Technology and the Remaking of the Rhone* (Cambridge, MA: Harvard University Press, 2011).
[36] See Seow, *Carbon Technocracy*.
[37] Robert B. Marks, *Tigers, Rice, Silk and Silt: Environment and Economy in Late Imperial South China* (Cambridge: Cambridge University Press, 1998); Ruth Mostern, *The Yellow River: A Natural and Unnatural History* (New Haven: Yale University Press, 2021).
[38] Paul R. Josephson, *Industrialized Nature: Brute Force Technology and the Transformation of the Natural World* (Washington: Island Press, 2002).

examination of both small and large hydropower projects in China from the 1910s to the 1970s, the nature of this revolution can be better understood. In 1973, E. F. Schumacher introduced the idea of "small is beautiful" and advocated for small hydro as an appropriate technology that would be less destructive to the environment and that could improve people's living standards in a sustainable way.[39] This view is reflected in the current worldwide push for the decentralization of energy production and the transformation of energy sources in response to climate change, as smaller hydro projects are viewed as a more manageable and environmentally responsible solution.[40]

The ideas and movements that currently advocate for decentralized energy systems and the promotion of the "small is beautiful" idea share a certain spirit with the utopian vision of a "small state" and local autonomy described by the Chinese philosopher Laozi. This book contributes to the global discussion of appropriate technology by examining the role of small hydropower projects in the context of contemporary political and economic circumstances and local conditions in China. It argues that a hydropower nation is not constructed solely out of mega projects, but also by means of small hydropower projects. By examining the small hydropower campaign during the Maoist period, this study reveals that small hydro projects also harmed local riverine systems.

In *Silenced Rivers*, the environmentalist Patrick McCully provides substantial evidence for the negative environmental and social impacts of large dams and calls for them to be taken down.[41] Meanwhile, the construction of the Three Gorges Dam on the Yangtze River prompted political and environmental activists in China – such as Dai Qing – to become vocal opponents of the project.[42] Rather than simply advocating that large dams be decommissioned, I am more in line with Richard White, who argues that the intertwining of technology and the environment constitutes a kind of organic entity.[43] We must situate ourselves in the historically entrenched mutual imbrication of technology and nature

---

[39] E. F. Schumacher, *Small Is Beautiful: Economics as if People Mattered* (London: Harper & Row, 1975).

[40] Astrid Kander, Paolo Malanima, and Paul Warde, *Power to the People: Energy in Europe over the Last Five Centuries* (Princeton: Princeton University Press, 2013).

[41] Patrick McCully, *Silenced Rivers: The Ecology and Politics of Large Dams* (New York: Zed Books, 2001).

[42] Dai Qing (ed.), *The River Dragon Has Come! Three Gorges Dam and the Fate of China's Yangtze River and Its People* (New York: Routledge, 1998).

[43] White, *Organic Machine*.

in water engineering that has been described as "technological lock-in."[44] This way, we can better understand the losses and gains that those projects involved, and make informed decisions about the future of hydropower in China. The construction of concrete dams has had a profound and lasting impact on China's riverine ecology. In addition, the consequences have not been uniform across all species: Waterfowls, for example, have adapted well while certain species of fish face extinction. It is important not only to acknowledge the negative effects of damming rivers, but also to recognize and take account of the complex and diverse dynamics at work in human-altered aquatic ecosystems.

SOURCES AND ORGANIZATION

Given the overwhelming scope of hydropower projects in China, it would be impossible in the space of a single book to examine all of them, even if I limited myself to the large ones. On the other hand, studying a single project or river would be insufficient to illuminate hydroelectric power's relationships with broader historical trends and to provide a comprehensive understanding of China's evolution into a hydropower nation. So, this book adopts a case study approach. In doing so, it seeks to describe China's journey toward becoming a hydropower nation and to provide insights into the interplay of technology, politics, the environment, and society.

The book combines a case study of a mega dam on the Yellow River, the Sanmenxia hydroelectric power project, with an examination of small-scale efforts to harness hydropower in the southwest and other areas from the 1910s to the 1970s. The focus will be on Chongqing, Yongchun 永春 county in Fujian Province, and Sanmenxia in Henan, each of which was closely linked to historical processes of political change, technological innovation, and social transformation during the twentieth century (see Map I.1). Chongqing, which is located on the upper stream of the Yangtze River, experienced a surge in hydroelectric production in the 1940s driven by wartime energy demands. Tributaries

---

[44] For Mark Elvin, the concept of technological lock-in describes a situation in which an established (yet suboptimal) technology continues to dominate approaches to problem-solving due to its prior establishment and the advantages that continue to derive from it. See Mark Elvin, *The Retreat of the Elephants: An Environmental History of China* (New Haven: Yale University Press, 2004), 123; Ruth Mostern also elaborates on the political economy that prevented imperial Chinese states from effectively tackling the silting problems of the Yellow River. See Mostern, *Yellow River*.

MAP 1.1 Major research sites

of the Yangtze River near Chongqing became the objects of state-led mobilization in the Nationalist government's war of resistance against Japan (1937–1945). Thanks to its abundant precipitation and mountainous terrain, Yongchun, located in southeast China, was renowned as a model county during the Great Leap Forward (1958–1961) and the Cultural Revolution (1966–1976) for its small-scale hydroelectric development. Meanwhile, Sanmenxia was the site of the first large concrete dam on the Yellow River, built with the help of the Soviet Union in the 1950s. These projects illustrate the wartime mobilization, mass participation, and megalomania aspects of hydropower development in China respectively, but their centrality in this book is not meant to diminish the significance of other hydropower projects in the country. Rather, they serve as starting points for an examination of how the Chinese state became increasingly entangled with nature.

Introduction: A Flow of Water and Power

Historical writings about river management provide us with a deeper understanding of the role of engineers connected to the Confucian intellectual tradition in the imperial past, thanks to the abundance of personal writings and official records that have been preserved. Similarly, in the twentieth century, our perspective on river management is largely shaped by the contributions of professionally trained engineers, technocrats, and political leaders. This book is the result of research conducted in mainland China, Taiwan, and the United States. To delve into the early proliferation of modern hydropower knowledge, I have analyzed writings found in newspapers and magazines published during the late Qing (1842–1912) and the early Republican period (1912–1927). My research on the Nationalist government's hydropower planning and construction efforts in southwest China has relied on documents from the National Resources Commission and the Fuyuan Hydropower Company, retrieved from the Chongqing Municipal Archives and the Modern History Institute Archives of Academia Sinica in Taipei. Access to official archives regarding the Sanmenxia project and the small hydropower campaign during the Great Leap Forward and the Cultural Revolution, however, is limited. Consequently, I have relied on a combination of published materials – newspapers, pamphlets, local gazetteers, and memoirs written by hydropower specialists and party cadres, together with a limited selection of archival documents. I was fortunate to gain access to provincial and county archives in Henan and Gansu. This allowed me to study the displacement process that the formation of the Sanmenxia reservoir entailed. In the late 1980s, the Chinese Hydroelectric Engineering Association undertook a systematic collection of historical records, primarily in the form of personal memoirs by its members and others involved in its projects. This resulted in the publication of a 28-volume series called *Chinese Historical Materials on Water Power*.[45] These memoirs and documents provide a comprehensive overview of developments in the field in the twentieth century.

This book explores the evolution of knowledge about hydroelectric power, the transnational experiences of engineers, and the planning and construction of hydroelectric power plants in China across the twentieth century. Focusing on the concept of a "hydropower nation," it also explores the connections between hydroelectric power and key historical

---

[45] Zhongguo Shuilifadian Shiliao Zhengji Bianji Weiyuanhui 中国水力发电史料征集编辑委员会. *Zhongguo Shuilifadian Shiliao* 中国水力发电史料 (China Historical Materials of Water Power), 28 volumes, 1987–1997.

events such as the war of resistance against Japan, the Great Leap Forward, and the Cultural Revolution. By placing specific hydroelectric projects, the actions of individual engineers, and the reactions of local communities in the context of the larger narrative of Chinese state-building and industrialization, this study provides a deeper understanding of the intersection of the technological, political, social, and environmental histories of modern China.

This book is divided into three parts and seven chapters. Part I, "Starting from Scratch," encompasses two chapters. Chapter 1 explores the making of the initial connection between politics and hydropower technology. I derive the term "hydropower nation" from a discourse among Chinese elites that linked the country's abundant hydropower resources to their concerns about its future in a context of foreign aggression. This chapter investigates the introduction of knowledge about the conversion of river flows into electricity to China in the late Qing and early Republican periods. Despite the prominence of fossil fuel energy in the industrialized world, certain Chinese intellectuals advocated harnessing the country's abundant river resources to produce electricity as a means of achieving full national independence. Local elites took the lead in constructing the first set of hydropower stations in southwest China, and afterward an increasing number of Chinese elites recognized the potential of hydropower in the country. As a result, in the context of a long-term national crisis, hydropower came to be, for many people, synonymous with the strengthening of the Chinese nation.

Chapter 2 investigates the ascent of technocrats to key roles in defining the technological foundations of the hydropower nation. The wartime crisis and the active involvement of the state in hydropower development and the training of engineers deepened the connection between hydropower and the Chinese nation. Before the Second Sino-Japanese War (1937–1945), the National Resources Commission of the Nationalist government conducted studies on China's hydropower potential. During the war it began constructing hydropower projects in the southwest to mitigate the energy shortage the war produced. By uncovering the many interactions between American and Chinese institutions and individuals, this chapter explores the importance of transnational exchange in strengthening the technological foundations of the hydropower nation. It also delves into the early social and environmental impacts of the nascent hydropower nation. Despite being limited in scale, social and environmental disturbances in the 1940s foreshadowed the significant human toll and ecological changes that would occur in later decades in a fully realized hydropower nation.

In Part II, "The Socialist Boost," Chapters 3 and 4 describe the formation of the socialist hydropower infrastructure following the victory of the Communist revolution in 1949. The entry of technocrats into the nascent revolutionary state gave rise to a potent, yet fragile, hydropower nation. Chapter 3 traces the establishment and distribution of hydroelectricity in the early People's Republic. Inheriting technical personnel from the defeated Nationalist regime, the Communist government worked to create its own infrastructure. This chapter argues that the cooperation between party leaders and "experts from the old society" functioned relatively smoothly until the rise of radical Maoism in 1958. The harnessing of rivers was matched and enabled by the taming of individuals with technical expertise.

Chapter 4 delves into the "great leap" of small hydropower during the Maoist period. It analyzes the distinctive feature of the expansion of the hydropower nation as it was influenced by Maoist ideology: mass participation. Despite the common assumption that Communist China blindly pursued mega dams, Mao believed in "walking on two legs": Large Soviet-style dams on the one hand, and small indigenous hydro projects that could be built and operated by the masses on the other. With the goal of boosting agricultural productivity and rural electrification, the PRC state mobilized communes nationwide to harness local rivers for the generation of electricity. This chapter examines local experiences of small hydro campaigns, focusing on Yongchun in Fujian Province. Across the country, tens of thousands of small hydroelectric power stations were constructed within a few years. The lack of prior hydrological investigations and professional knowledge, however, meant that many of these stations were not able to deliver stable electrical output, while they also resulted in the fragmentation of local rivers.

Part III, "A Huge Setback," explores the story of the Sanmenxia Hydropower Project on the Yellow River. The chapters of this section provide a comprehensive examination of the project in order to highlight the complex political, social, and environmental issues that accompany the development of a hydropower nation. In contrast to the narrative of national crisis that dominated discussions prior to 1949, the drive to build a socialist state played a central role in shaping the emerging hydropower nation.

Chapter 5 describes the planning, design, and construction of the Sanmenxia Hydropower Project in the 1950s and the early 1960s. Building on David Pietz's analysis of the Soviet influence on the general plan for the Yellow River, this chapter examines the negotiations that took place between Soviet experts, their Chinese counterparts, and local stakeholders during the process of designing the Sanmenxia Dam.[46] While the Communist Party proclaimed in 1957 that the primary purpose of the project was flood control, it is clear from the sources examined in this chapter that the generation of hydroelectricity was the principal element of the dam's design.

Chapter 6 examines the devastating toll that the development of the hydropower nation took on people's lives. In Henan, Shanxi, and Shaanxi, over 430,000 people were forced to leave their homes and communities due to the construction of the Sanmenxia Dam. This chapter focuses on the resettlement of over 7,000 residents from Henan to Gansu, revealing how the state's pursuit of hydroelectricity not only altered the Yellow River's physical landscape but also caused irreparable harm to the affected communities.

Chapter 7 addresses the environmental impacts of the hydropower nation. Rising water levels in the dam's reservoir area brought about unforeseen ecological consequences. Alongside the tragic human displacement, various species of fish and waterfowl have experienced a variety of changes in the transformed river landscape. In recent decades, environmental hazards – such as riverbank collapse – have persisted in the reservoir area. On the other hand, the creation of a wetland preserve and the emergence of a thriving population of whooper swans in the area offer evidence of the development of a nuanced relationship between humanity and nature in the Chinese state's pursuit of what it calls an ecological civilization. These developments complicate the overwhelmingly dominant narrative of environmental decline in the field of environmental history.

The relationships between humans, technology, and the environment are complex and it is critical that we consider them carefully, especially when it comes to river and hydropower projects. Each project must be analyzed in its specific natural, social, and political context if we are to grasp fully the tradeoffs and impacts it entails. By exploring the human and environmental narratives behind the formation of the hydropower

---

[46] See Pietz, *Yellow River*.

nation, we can gain a deeper understanding of our connection to the environment – a connection that can be both intimate and distant. The triangulated dynamics of and tensions between the environment, humans, and technology, as well as the concept of a hydropower nation, are important themes in many discussions of river projects in the geological epoch increasingly referred to as the Anthropocene. By approaching this issue with a holistic and interdisciplinary perspective, we can work toward more sustainable and equitable energy solutions for the current age.

PART I

STARTING FROM SCRATCH

I

An Inexhaustible Source of Power

In 1894, as the First Sino-Japanese War loomed on the horizon, a young Sun Yat-sen, then just twenty-eight years of age, wrote a passionate petition to Li Hongzhang, the most prominent official associated with the Self-Strengthening movement in the final decades of the Qing era. In his letter, Sun implored Li to enact reforms in crucial areas such as education, agriculture, technology, and transportation. In particular, he emphasized the importance of technology, foretelling the arrival of the electrical age and its boundless potential. Sun wrote, "Recently, a revolutionary new method has emerged – the harnessing of the power of waterfalls to generate and store electricity, an inexhaustible source of power that can be used at any time and any place."[1] In a period when the country still relied overwhelmingly on the manual labor of its people, some enlightened Chinese figures, like Sun, began to envision a future in which alternative energy production would play a crucial role in shaping the nation's future.[2]

Sun Yat-sen's prophetic words, referring to a "new method of producing an inexhaustible source of power," have come to fruition as what we know today as hydroelectricity. The utilization of waterpower, however, is by no means a novel concept. In ancient Europe, for example, watermills powered by the natural force of flowing water had become an

---

[1] Sun Yat-sen, "Shang Li Hongzhang shu" 上李鸿章书 (June 1894), in *Sun Yat-sen quanji* 孙中山全集, vol. 1 (Beijing: Zhonghua shuju, 1981), 12.

[2] On the Qing empire's exploitation of natural resources, see Peter Lavelle, *The Profits of Nature: Colonial Development and the Quest for Resources in Nineteenth-Century China* (New York: Columbia University Press, 2020).

integral part of daily life. These watermills, which converted the power of water into mechanical force, relieved countless people of the grueling and monotonous task of grinding grains. The following epigram from the *Anthology* speaks to this revolutionary change:

> Spare your hands, which have been long familiar with the millstone, you maidens who used to crush the grain. Henceforth you shall sleep long, oblivious of the crowing cocks who greet the dawn. For what was your task, Demeter has now handed on to the Nymphs.[3]

William the Conqueror's 1086 Domesday Book survey reveals the presence of thousands of small water-driven mills in England and estimates that one mill existed for every fifty households.[4] However, the use of these mills was limited by environmental factors such as meteorology, topography, and geology. The mills primarily performed essential tasks such as producing grist, sawing, carding, and fulling, all crucial for the agricultural society of the time. For over a millennium and a half prior to the invention of the steam engine in 1769, water-powered machinery represented the most advanced and widely utilized mechanical technology in human society.[5] Even after the dawn of the first Industrial Revolution and its increasing reliance on fossil fuels, waterpower remained an important source of energy in Europe and the Americas. In the 1820s, the people of Lowell, Massachusetts, sought to overcome seasonal fluctuations in waterpower capacity by constructing a complex system of hydraulic works: Large masonry dams and protective walls, networks of distribution canals equipped with water gates and spillways, and upstream storage reservoirs. This was the first successful experiment in overcoming the physical and natural limitations of waterpower and stabilizing its output for the operation of mills. As a result, Lowell was transformed into the first "factory town" and the center of the textile industry in North America.[6]

---

[3] Nymphs is a metaphor for machine motor power. Marc Bloch, *Land and Work in Medieval Europe*, trans. J. E. Anderson (Berkeley: University of California Press, 1967), 145.

[4] Margaret T. Hodgen, "Domesday Water Mills," *Antiquity* 13 (1939), 261.

[5] See Lynn White, Jr., *Medieval Technology and Social Change* (Oxford: Oxford University Press, 1962); Jean Gimpel, *The Medieval Machine: The Industrial Revolution of the Middle Ages* (London: Penguin Books, 1976).

[6] Louis C. Hunter, *A History of Industrial Power in the United States, 1780–1930*, Vol. 1: Waterpower in the Century of the Stream Engine (Charlottesville: University Press of Virginia, 1979), 113. Also see Theodore Steinberg, *Nature Incorporated: Industrialization and the Waters of New England* (Cambridge: Cambridge University Press, 1991).

What was the experience of waterpower in Chinese history? From the Han dynasty (206 BCE–220 CE) onward, people harnessed the power of water for a multitude of purposes. The watermill played a crucial role in processing grain for the agricultural economy, also powering the development of tea and textile workshops, enabling the development of a commodity economy in certain regions.[7] However, the "new method of producing an inexhaustible source of power" that Sun Yat-sen spoke of in his letter and that is now known as hydroelectricity was first developed in Europe. My extensive examination of primary sources from the late nineteenth and early twentieth centuries reveals a nuanced perspective on the dissemination of hydroelectric knowledge in China. While the story of hydroelectricity may initially seem to fit within the conventional narrative of the transfer of Western science and technology to China, a

---

[7] On the development of watermills in Chinese history, see Joseph Needham, Ling Wang, and Gwei-djen Lu, *Science and Civilization in China*, vol. 4, *Physics and Physical Technology, Part II: Mechanical Engineering* (Cambridge: Cambridge University Press, 1965); Zhang Baichun 张柏春, "Zhongguo chuantong shuilun jiqi qudong jixie" 中国传统水轮及其驱动机械 (China's Traditional Waterwheel and Its Motor Mechanism), *Ziran kexue shi yanjiu* 自然科学史研究 (*Studies in the History of Natural Sciences*) 13, no. 2 (1994), 155–163; Tan Xuming 谭徐明, "Zhongguo shuili jixie de qiyuan, fazhan jiqi zhongxi bijiao" 中国水力机械的起源、发展及其中西比较研究 (A Study of the Origin and Development of Hydraulic Machinery in China and Its Comparison with the West), *Ziran kexue shi yanjiu* 自然科学史研究 (*Studies in the History of Natural Sciences*) 14, no. 1 (1995): 83–95; Li Bozhong 李伯重, "Chucai Jinyong: Zhongguo shuizuan dafangche yu Yingguo Arkwright shuilifangshaji" 楚才晋用：中国水转大纺车与英国阿克莱水力纺纱机 (The Talent of Chu Put to Use by Jin: China's Water-Powered Spinning Wheel and Britain's Arkwright Water Frame), *Lishi yanjiu* 历史研究 (*Historical Studies*) 1 (2002), 62–74; Liu Xiaoping 刘小平, "Tangdai siyuan de shuinianai jingying" 唐代寺院的水碾硙经营 (The Management of Buddhist Temples' Watermills during the Tang Dynasty), *Zhongguo nongshi* 中国农史 (*Chinese Agricultural History*) 4 (2005), 44–50; Wang Lihua 王利华, "Gudai huabei shuilijiagong xingshuai de shuihuanjing Beijing" 古代华北水力加工兴衰的水环境 (The Water Environment and the Evolution of Waterpower Use in Ancient North China), *Zhongguo jingji shi yanjiu* 中国经济史研究 (*Chinese Economic History Studies*) 1 (2005): 30–39; Nien Chen-ho 粘振和, "Lun Bei Song shuimo chafa" 论北宋水磨茶法 (The Study of the Monopoly System of Tea by Water-Powered Mills in the Northern Song Dynasty), 成大历史学报 (*Cheng Kung Journal of Historical Studies*) 47 (2014), 1–28; Fang Wanpeng 方万鹏, "Xiangdi zuomo: Mingqing yilai Hebei Jingjing de shuilijiagongye – jiyu huangjingshi shijiao de kaocha" 相地作磨：明清以来河北井陉的水力加工业 – 基于环境史视角的考察 (Install Watermills According to the Terrain: The Hydraulic Machining in Jingxing of Hebei Province since the Ming and Qing Dynasties: Based on the Perspective of Environmental History Studies), *Zhongguo nongshi* 中国农史 (*Chinese Agricultural History*) 3 (2014): 51–58; In the Japanese-language literature, see Nishiijima Sadao 西嶋定生, *Tyuugoku Keizaisi Kennkyuu* 中國經濟史研究 (Tokyo: Tokyo Daigaku shuppan-kai, 1966), chapter 4. This book has been translated into Chinese as *Zhongguo Jingjishi Yanjiu* (Beijing: Nongye chubanshe, 1984).

closer examination reveals more complex dynamics.[8] This chapter argues that the development of hydropower in China was not simply a straightforward replication of the Western experience but rather a process of adaptive appropriation. This involved the adaptation of hydroelectric principles to the features of local communities, environmental factors, and the prevailing political situation in the country.

This chapter undertakes a dual exploration of the early development of hydropower in China. On the one hand, it provides an intimate look at the experiences of three local elites who built small hydroelectric plants in southwest China. On the other, it describes the transformation of Chinese elites' understanding of China's hydropower potential, from initial skepticism to an eventual recognition of abundance. In the context of a persistent national crisis, many Chinese elites began to see hydropower as a key component in the project of building a strong, self-sufficient nation that was free from foreign aggression. The co-occurrence in the early twentieth century of the spread of hydroelectric knowledge and technology from the West and Chinese elites' search for means of national strengthening created a unique opportunity for the flourishing of both forces. The recognition of the potential of China's abundant hydropower resources promised a bright future for the nation. Thus, the seed of a hydropower nation was planted and awaited nurturing in the decades to come.

THE RISE OF "WHITE COAL"

During the Industrial Revolution, the rise of fossil fuels overshadowed the use of waterpower, which until then had been limited to locations near streams and rivers and was best suited to areas with waterfalls. Despite this shift, a lineage of hydraulic engineers continued to refine the waterwheel, the cornerstone of waterpower technology, and to increase its efficiency. In 1837, the French engineer Benoit Fourneyron, building on the research of Claude Burdin, invented the turbine, a revolutionary new type of waterwheel. Capable of producing 2,300 revolutions per minute with 80 percent efficiency and 60 horsepower, the turbine was a significant improvement in waterpower technology. The challenge of transmission remained. In 1895, at Niagara Falls, the Fourneyron turbine was put into use, converting waterpower into electricity on a massive scale for the

---

[8] See Benjamin A. Elman, *On Their Own Terms: Science in China, 1550–1900* (Cambridge, MA: Harvard University Press, 2005).

first time in history. This breakthrough, coupled with the development by Nikola Tesla in the late 1880s of a poly-phase alternating current system of generator, motor, and transformer, made long-distance transmission possible and ushered in the new era of hydroelectricity.

In the second half of the nineteenth century, major technological advancements in hydroelectricity occurred, primarily in Europe and North America, but then quickly spread to the East through transnational networks of missionaries.[9] In 1868, the American missionary W. A. P. Martin made a notable contribution to this transmission of knowledge. As a teacher and interpreter for the Translation Bureau in Beijing, Martin wrote *Introduction to Science* (*Gewu Rumen* 格物入门), a comprehensive textbook on Western science. In the section on water, Martin highlights a type of waterwheel referred to as the "watermill horizontal wheel" (*shuimo wolun* 水磨卧轮), which is in fact the Fourneyron turbine. He explains the principle of its operation in clear and concise terms: "Water is directed into the wheel's rim through a ten-foot-long vertical penstock. The wheel is fitted with curved buckets, which cause it to rotate as the velocity of the water drives it. The longer the penstock, the greater the power and speed of the wheel's revolution."[10] At the beginning of the twentieth century, the American missionary Young J. Allen began reporting on the hydropower projects at Niagara Falls in the *Chinese Globe Magazine*. Further, in 1898, the editorial board of *Scientific Review*, a journal published in Shanghai, in response to a question from a reader about the importance of waterpower relative to other sources of energy, acknowledged the benefits of waterpower but also acknowledged its limitations with respect to China. They noted, "There are only a few waterfalls in China, making waterpower less viable than wind power. It would be more advantageous if we could develop steam power, which surpasses waterpower in ease of use."[11] Despite China's long history of using watermills, the conventional wisdom of the time saw the country as lacking in hydropower potential.

By the early twentieth century, Europe and North America were undergoing two major scientific and technological advancements that would pave the way for the widespread use of hydroelectricity.

---

[9] On the evolution of hydropower technology in the West, see Norman Smith, *Man and Water: A History of Hydro-Technology* (London: Peter Davies, 1976).

[10] W. A. P. Martin 丁韪良, *Gewu Rumen* 格物入门 (*Introduction to Science*) (Beijing: Beijing tongwenguan, 1868), 27.

[11] *Gezhi xinbao* 格致新报 (*Scientific review*) (1898) 近代中国史料丛刊三编第24辑 (Taipei: Wenhai chubanshe, 1987).

In addition to the turbine, hydraulic engineering was developed further to stabilize and maximize the energy-producing capacity of water, while the invention of alternating current made long-distance transmission of electricity a reality.[12] These technological innovations made hydropower more accessible and sustainable than coal-powered electricity, leading to a boom in its use in Europe and North America in the years following World War I. According to a 1934 census of installed electricity capacity in several major countries, hydropower dominated the energy mix in countries such as Canada, Italy, Norway, Sweden, and Switzerland, accounting for over 90 percent of total capacity. France and Japan also had substantial shares of hydropower, at 70 and 85 percent, respectively, while in the United States, Germany, and the Soviet Union the shares were 45, 40, and 20 percent, respectively.[13] In certain areas, such as Lowell in North America and the Alpine region in Europe, hydropower had become the primary source of electricity for industrial production and daily illumination. The term "white coal" initially referred to the glacial streams flowing down from mountains, but had by this period become synonymous with hydropower, reflecting its increasing importance and its equivalence to coal, if not yet its superiority.[14]

As the demand for energy continued to grow, people started to acknowledge the limitations of coal as a source of power. The burning of coal produced ash that contaminated the air and, as a finite resource, it could not meet the increasing energy needs of society forever. In contrast, hydropower was seen as a clean and renewable alternative, offering a promising future for modern industrial civilization. In the early twentieth century, the feeling among many was that the trend of industrialization was moving toward hydro-electrification.[15]

As Shellen Wu illustrates in her work, the exploitation of natural resources, such as coal mines, has long been intertwined with issues of state power and sovereignty.[16] In the early decades of the twentieth

---

[12] See Hunter and Bryant, *History of Industrial Power in the United States*.

[13] Huang Yuxian 黄育贤, "Zhanhou kaifa woguo shuili ziyuan zhi guanjian" 战后开发我国水力资源之管见, *Jingji jianshe jikan* 经济建设季刊 1, no. 4 (1943), 42–47. Matthew Evenden, *Allied Power: Mobilizing Hydro-Electricity during Canada's Second World War* (Toronto: University of Toronto Press, 2015).

[14] For the study of the rise of "white coal" in Europe, see Landry II, "Europe's Battery."

[15] Lu Weizhen 陆为震 "Zhongguo weilai zhi shuili jianshe" 中国未来之水力建设 (China's Hydropower Construction in the Future), *Hankou Shangye Yuekan* 汉口商业月刊 1, no. 8 (1934), 21.

[16] Wu, *Empires of Coal*, 197.

century, China's resistance against imperialist invasions, particularly the incursions of the Japanese military, fueled a surge of nationalism. In their pursuit of economic development and political independence, Chinese elites saw it as their duty to harness the full potential of their nation's resources, including harnessing the energy of its rivers through the generation of hydropower. Their goal was not only to reconstruct and improve robust local economies but also to forge a strong, sovereign nation that could compete and succeed on the global stage.

In the early years of hydropower development, the presence of waterfalls was felt to be essential. As China lacked large waterfalls, many believed that its potential for hydropower was limited. However, with the advent of hydro-engineering, particularly with improvements in the construction of dams and penstock, perceptions of China's hydropower potential changed dramatically. Advances in hydrology and scientific studies of rivers further fueled this shift, as engineers and stakeholders grew increasingly optimistic about the prospects for hydropower in the country. According to the National Resources Commission of the Nationalist government in the late 1930s, China was estimated to have a potential hydropower capacity of over 40 million kilowatts, which would rank it as the third largest producer in the world, after only the United States and Canada.[17] This new recognition of China's hydropower potential served as a foundation for the confidence of Chinese civil engineers in the importance of hydropower in the country's state-building efforts. In Shanghai, for example, a quarter of the city's power plants were under foreign control and relied heavily on petroleum as their primary energy source.[18] However, this dependence on imported petroleum was viewed as a threat to China's national sovereignty.[19] The belief that China had enormous potential for hydropower – as opposed to oil fields, which were understood to be limited – took hold among the interested public. As Lu Shiqian 卢世铃, a college student studying civil engineering in

---

[17] Zheng Youkui 郑友揆, Cheng Linsun 程麟荪, and Zhang Chuanhong 张传洪, *Jiu Zhongguo de Ziyuanweiyuanhui, 1932–1949: Shishi yu pingjia* 旧中国的资源委员会, 1932–1949: 史实与评价 (The National Resource Commission in Old China, 1932–1949: Historical Fact and Evaluation) (Shanghai: Shanghai shehui kexueyuan chubanshe, 1991), 86.

[18] Gu Yuxiu 顾毓秀, "Dianqi yu jianshe" 电气与建设 (Electricity and Construction), *Xinmin* 新民 8 (1931), 10.

[19] American Standard Oil company was one of the largest exporters of petroleum products to China before the Second Sino-Japanese War, see Sherman Cochran, *Encountering Chinese Networks: Western, Japanese, and Chinese Corporations in China, 1880–1937* (Berkeley: University of California Press, 2000), chapter 2.

Shanghai, put it: "We should build hydroelectric plants anywhere we can to power our factories. With a cheap energy supply, our products will be able to compete with foreign goods and keep profits in the country. If we have sufficient hydropower energy, we will be able to stop importing oil from abroad, bringing us far greater benefits!"[20] Lu was not alone in his concern for the future of China's energy industry and its place in the world. This economic nationalism, rooted in the logic of the self-strengthening movement of the late Qing period, persisted through the twentieth century and was evident not only in the movements against imported goods but also in the pursuit of energy self-sufficiency, with a focus on hydropower development.[21]

Since the outbreak of the first Sino-Japanese War in 1895, Japan had occupied the attention of Chinese intellectuals. The engineer Shan Yubin 单毓斌 visited a power plant in Japan and wrote about his experience for his fellow Chinese, lamenting,

> Alas! Japanese entrepreneurs have invested heavily in developing hydroelectricity for the sake of their nation, leaving us far behind. Our country has its own mountains and rivers, including the renowned Qutang and Ba Gorges. If we bring together capitalists and engineers, our nation can greatly benefit from exploiting its hydropower potential.[22]

Further, Japanese hydropower construction in Manchuria and Taiwan embarrassed many Chinese intellectuals. In 1924, Japan completed the Riyuetan 日月潭 Hydropower Plant in Taiwan with an installed capacity of 100,000 kilowatts, making it the largest in Asia at the time.[23] Despite the rapid growth of Japan's hydropower industry, it did not possess the technology to produce large generators and turbines and thus had to

---

[20] Lu Shiqian 卢世钤, "Shuili liyong yu guomin jingji" 水力利用与国民经济 (Hydropower Exploitation and National Economy), *Zhonghua Yubao* 中华月报 5 (1935), 6.
[21] Pan Junxiang 潘君祥, *Zhongguo jindai guohuo yundong* 中国近代国货运动 (National Products Movements in Modern China) (Beijing: Zhongguo wenshi chubanshe, 1996); Karl Gerth, *China Made: Consumer Culture and the Creation of the Nation* (Cambridge, MA: Harvard University Asia Center, 2003).
[22] Shan Yubin 单毓斌, "Kaocha Riben Dongjing diandeng huishe guichuan shuilifadiansuo jilue" 考察日本东京电灯会社桂川水力发电所纪略 (Visit to Japan's Tokyo Light Company Guichuan Hydroelectricity Plant), *Dianqixiehui Zhazhi* 电气协会杂志 8 (1914), 54–63, 56.
[23] "Dongyang diyi fadian gongshe (shuli fadian)" 东洋第一发电工事（水力发电） (The Largest Hydroelectricity Plant in East Asia), *Dianqigongye Zhazhi* 电氣工业杂志 2, no. 2 (1924), 74.

import them from Germany and the United States.[24] This further fueled the sense of urgency among Chinese elites, who were concerned that their country's energy sector would lag far behind Japan's. It is not surprising, then, that many civil engineers would become advocates for the development of hydropower in the aftermath of the Japanese invasions.

However, the question of how China could most quickly catch up with Japan and the other industrial nations in energy production loomed large. Some Chinese people looked northward for a solution. The Soviet Union, with its experience of electrification and hydropower development, emerged as an attractive alternative model.[25] As people in China learned about the Bolshevik revolution, they learned not only Marxist ideology but also about the Soviet experience of electrification. Impressed by what they saw as the Soviet Union's rapid transformation from a weak and poor nation into a strong industrial power, some Chinese sought inspiration from it.

Chen Zudong 陈祖东 was a civil engineer who graduated from Tsinghua University in 1935. He was also a cousin of Chen Guofu 陈果夫, who served as the minister of the Organization Department of the Kuomintang. In 1932, he lamented,

China's abundance of population and natural resources, including the mighty Yangtze River, should have given it a distinct advantage in supporting industrialization. Yet, the lack of a unified plan for economic construction and the ongoing wars between warlords have left the Chinese people to suffer. In contrast, the leaders of the Soviet Union display a pragmatic attitude, leading to its rapid growth and rising influence among world powers. It is a source of shame for the Chinese people to lag so far behind despite their wealth of resources![26]

During his visit to the Soviet Union in 1939, Chen wrote journals to record his experiences and was deeply impressed by the Soviet Union's central planning and authoritarian capacity to force state policies through to completion, which he believed had led to the rapid growth of industrial

---

[24] On the building of large hydropower projects in the Japanese empire, see Aaron Moore, *Constructing East Asia: Technology, Ideology, and Empire in Japan's Wartime Era, 1931–1945* (Stanford: Stanford University Press, 2015).

[25] For studies on the electrification of Russia, see Anne D. Rassweiler, *The Generation of Power: The History of Dneprostroi* (Oxford: Oxford University Press, 1988). Jonathan Coopersmith, *The Electrification of Russia, 1880–1926* (Ithaca: Cornell University Press, 1992). On Chinese views of Soviet Russia, see John Knight, "Savior of the East: Chinese Imagination of Soviet Russia during the National Revolution, 1925–1927," *Twentieth Century China* 43, no. 2 (2018), 120–138.

[26] Chen Zudong, "Su'e jianshe zuida shuilifadianchang" 苏俄建设最大水力发电厂 (Soviet Union Builds the Largest Hydroelectricity Plant), *Su'e Pinglun* 苏俄评论 5 (1932), 616.

production in Russia. He suggested that China should adopt the Soviet model to develop its energy sector.[27]

Many in China were aware of the close relationship between energy production and a country's global standing as they observed the experiences of electrification in industrial countries such as Japan and the Soviet Union. They understood that an underdeveloped country like China needed a comprehensive national plan and a competent, if not authoritarian, government with strong executive power if it was to advance on this front. By the mid-twentieth century, hydropower, despite its unfamiliarity to many, had become a crucial indicator of national competence. As Chen Zudong put it, the construction of hydropower capacity should be embraced without hesitation as "the formula for China's national reconstruction."[28]

LOCAL PIONEERS

Despite this change in the consciousness of certain elites, before 1935 the use of hydropower in China was limited to a few scattered localities.[29] Western scientific theories of hydroelectricity were introduced by foreign missionaries and progressive Chinese intellectuals, but it was Chinese entrepreneurs who put the knowledge into practice. They viewed the construction of hydropower plants as a sound investment, a "once and for all" source of power that would have long-term or even permanent benefits. Upon learning about the construction of hydroelectricity plants at Niagara Falls in the United States and Canada, some Chinese businessmen expressed amazement at the apparently endless benefits that such a project could offer: "How could the benefits ever be exhausted?"[30]

YAOLONG 耀龙 AND JIHE 济和

In the early years of China's interactions with foreigners in the late Qing and early Republican periods, local businessmen were the primary supporters of hydropower development. As Elisabeth Koll has shown, the construction of railroads in China involved extensive collaboration

---

[27] Chen Zudong 陈祖东, "Cong dianlishuili shuodao Sulianjianguo yu Zhongguojianguo" 从电力水力说到苏联建国与中国建国 (From Electricity, Hydropower to the Construction of Soviet Union and China), *Xinjingji* 新经济 2, no. 4. (1939), 88.
[28] Chen, "Cong shuilidianli shuodao Sulianjianguo yu Zhongguojianguo," 91.
[29] For electric power output in China before 1937, see Tim Wright, "Electric Power Production in Pre-1937 China," *The China Quarterly* 126 (1991), 356–363.
[30] Shangyuan 商原, *Shangwubao* 商务报 31 (1908).

between foreigners and Chinese.³¹ Indeed, this transformative infrastructure project played an important role in the development of China's first hydropower project. In 1897, France was granted permission to build a railway connecting Haiphong in Vietnam and Kunming in Yunnan province, but this sparked an intense negative reaction in Kunming.³² In May 1900, a Catholic cathedral was burned down and the religious leaders of all the Christian denominations in the city were forced to flee.³³ Despite the turmoil and upheaval, a tiny Anglo-French business community persevered. To provide power for the railway station that was then under construction, a French survey team proposed the construction of a hydropower plant along the winding course of the Tanglang River 螳螂川 in Yunnan. This river was the sole outlet of the massive Dian Lake 滇池 and flowed for over 360 kilometers before merging with the Jinsha River 金沙江 to the north. The steep terrain and the lake's potential as a natural reservoir promised an ample supply of hydropower.³⁴

While the French celebrated their own railway as a project "worthy of the genius of the French,"³⁵ many Chinese businesspeople saw it as a threat to their commercial interests and to China's larger national interests. These businesspeople believed that the exploitation of hydropower was an "economic benefit and a political right of the Chinese nation and people."³⁶ With the aim of harnessing the energy of the Tanglang River, Liu Lingfang 刘苓舫, the director of the Bureau of Business Promotion in Yunnan, proposed the establishment of a plant to be "jointly managed by officials and businesspeople." Despite his efforts, however, few

---

³¹ Elisabeth Koll, *Railroads and the Transformation of China* (Cambridge, MA: Harvard University Press, 2019).
³² On the history of Vietnam-Yunnan railroad, see Mi Rucheng 宓汝成, *Diguo zhuyi yu Zhongguo tielu, 1847–1949* 帝国主义与中国铁路 (*Imperialism and China's Railway, 1847–1949*) (Beijing: Jingji guanli chubanshe, 2007); Wu Xingzhi 吴兴帜, *Yanshen de pingxingxian: Dian-Yue tielu yu bianmin shehui* 延伸的平行线：滇越铁路与边民社会 (*Extended Parallel Lines: The Dian-Vietnam Railway and the Borderland Community*) (Beijing: Beijing daxue chubanshe, 2012).
³³ Robert Nield, *China's Foreign Places: The Foreign Presence in China in the Treaty Port Era, 1840–1943* (Hong Kong: Hong Kong University Press, 2015), 288.
³⁴ Yunnan Sheng difangzhi bianzuan weiyuanhui 云南省地方志编纂委员会, *Yunnan Sheng zhi. juan 1, Di li zhi* 云南省志·卷一·地理志 (*Yunnan Province Magazine: Roll 1, Geography Section*) (Kunming: Yunnan renmin chubanshe, 1998), 300–301.
³⁵ Gabrielle Vassal, *In and Round Yunnan Fou* (London: W. Heinemann, 1922), 33.
³⁶ Zheng Qun 郑群, *Zhongguo Diyisuo Shuidianzhan Shilongba Chuanqi* 中国第一座水电站石龙坝传奇 (*The Legend of China's First Hydroelectric Plant: Shilongba*) (Kunming: Yunnan jianyu chubanshe, 2012), 16.

businessmen showed interest in the project.[37] Undeterred, the Kunming Chamber of Commerce stepped in and petitioned the local county magistrate to allow them to raise funds among themselves to build the hydropower plant. Under the leadership of Wang Xiaozhai 王筱斋, the owner of a private bank, a joint-stock corporation called the Kunming Yaolong Light Company was established to manage the construction and operation of the plant.[38] With the help of private investment and local initiative, the first hydroelectric plant in China, the Shilongba 石龙坝, was finally completed and brought into operation in Kunming in 1910 (Figure 1.1).[39]

Despite the nationalist motivations behind the project's inception, the Kunming Yaolong Light company had no choice but to seek technical support and machinery from foreign sources.[40] Through the mediation of Carlowitz & Company, a German trading firm with an office in Shanghai, the electrical equipment and turbines were provided by Siemens and Voith, respectively. The latter also dispatched two engineers to help with the design and construction. By 1912, the Yaolong Company had constructed a 1.478-kilometer canal and installed 480 kW of electrical equipment, ready to provide power to Kunming.[41] This was a pioneering project in China in another way, as it was the first to utilize high voltage alternating current (23,000 V) to transmit electricity over the relatively long distance of 35 kilometers to downtown Kunming.

---

[37] In the nineteenth century, the Qing state assumed very limited responsibility for infrastructure and industrial development. See Madeleine Zelin, *The Merchants of Zigong: Industrial Entrepreneurship in Early Modern China* (New York: Columbia University Press, 2006).

[38] Local gentry-merchants and chambers of commerce were active and played a leading role in local societies in the nineteenth and early twentieth century China, see Susan Mann, *Local Merchants and the Chinese Bureaucracy, 1750–1950* (Stanford: Stanford University Press, 1987); Prasenjit Duara, *Culture, Power, and the State: Rural North China, 1900–1940* (Stanford: Stanford University Press, 1991); Ma Min, *Guanshang zhi jian: shehui jubianzhong de jindai shenshang* 官商之间:社会巨变中的近代绅商 (*Between Official and Merchant: The Modern Gentry-Merchants amid Drastic Social Change*) (Tianjin: Tianjin renmin chubanshe, 1995); Elisabeth Koll, *From Cotton Mill to Business Empire: The Emergence of Regional Enterprises in Modern China* (Cambridge, MA: Harvard University Asia Center, 2003).

[39] See Arunabh Ghosh, "Multiple Makings at China's First Hydroelectric Power Station at Shilongba, 1908–1912," *History and Technology* 38 (2022), 167–185.

[40] On the Sino-German relationship and trade, see William Kirby, *Germany and Republican China* (Stanford: Stanford University Press, 1984).

[41] "Shangban Yunnan Yaolong diandeng gongsi Shilongba gongcheng jilue" 商办云南耀龙电灯公司石龙坝工程纪略 (Brief Records of the Commercial Yunnan Yaolong Company Shilongba Project), *Zhongguo Shuilifadian Shiliao* 中国水力发电史料 (*China Historical Materials on Waterpower*) 1 (1987), 73.

FIGURE 1.1 Shilongba Hydropower Plant (Photo by zhouyousifang/Moment via Getty Images)

The introduction of electric light in Kunming was initially met with skepticism and resistance. Despite the technical advancement that the Shilongba Hydropower Plant represented, the inconsistent and sometimes unstable supply of power was a major hindrance to the widespread adoption of electricity. Every night, just before 10:00 pm, light bulbs in the city would dim to a mere "glow of incense sticks," and residents would use their kerosene lamps for light.[42] Further, the plant's output exceeded the initial market demand. But the Yaolong Company did not give up on its mission to bring the benefits of electricity to the people of Kunming. They undertook a concerted effort to advertise the advantages of electric light and installed light bulbs for free to entice potential customers. These efforts paid off and the operation of the company soon became viable. Four months after its inauguration, more than 3,000 lamps had been sold. Seven local mills and factories began to use electricity as their primary source for driving equipment as well as illumination.[43] By 1923, the

---

[42] Frank Dikötter, *Exotic Commodities: Modern Objects and Everyday Life in China* (New York: Columbia University Press, 2006), 136.

[43] "Yunnan fu, Zhongguo de diyige shuidianzhan" 云南府, 中国的第一个水电站, 西门子杂志 (China's First Hydropower Plant), *Zhongguo Shuilifadian Shiliao* 5 (1989), 70–72. Before the employment of ammeters, electric companies sold electricity according to the wattage of light bulbs.

growing demand for electricity compelled the company to expand its installed capacity to meet the new needs of the community. The Shilongba Hydropower Plant thus played a pivotal role in the electrification of Kunming and the surrounding areas.

Following the success of the Yaolong plant, another hydropower project – named Jihe – was established in 1925 along the Longxi 龙溪 River in Luzhou 泸州, Sichuan. The Jihe project was designed by a talented civil engineer, Shui Xiheng 税西恒, who was born in Luzhou and had received training in mechanical engineering in Germany. After working briefly for Siemens, Shui returned to Sichuan and served as the director of the Southern Sichuan Construction Bureau.[44] With his professional training and local connections, Shui was able to convince local merchants and officials to invest in a hydroelectric plant that would harness the hydropower of the Longxi River. A curved masonry dam, standing 2.5 meters tall and stretching 80 meters across the river, was built to control water for the generation of electricity. The equipment required for the project, including a turbine and a 140-kW generator, was imported from Germany.[45] In the construction of the Shilongba and Jihe projects, a unique solution was found to overcome the problem of a shortage of cement, and a sticky rice slurry was used as a substitute.[46] In Nanping 南平, Fujian province, meanwhile, Ji Tinghong 纪亭洪, a member of the local gentry, used an indigenous wooden water wheel to drive a generator, thus providing an early example of blending traditional knowledge and local materials with advanced engineering technology.[47] These innovations and experiments, much like Eugenia Lean's study of "vernacular industrialism," show that the development of

---

[44] Liu Shengyuan 刘盛源, *Shui Xiheng zhuan* 税西恒传 (*Biography of Shui Xiheng*) (Beijing: Tuanjie chubanshe, 2016).

[45] Yang Yongnian 杨永年, "Jianguoqian Sichuan de shuidian jianshe" 建国前四川的水电建设 (Hydropower development in Sichuan before 1949), *Zhongguo shuilifadian shiliao* 2 (1987), 44–45.

[46] On the production of cement in China, see Albert Feuerwerker, "Industrial Enterprise in Twentieth-Century China: The Chee Hsin Cement Co," in *Approaches to Modern Chinese History*, ed. Albert Feuerwerker, Rhoads Murphey, and Mary C. Wright (Berkeley: University of California Press, 1967), 304–342; Wang Yanmou 王燕谋, *Zhongguo Shuini Fazhanshi* 中国水泥发展史 (*The History of Cement in China*) (Beijing: Zhongguo jiancaigongye chubanshe, 2005); Micah Muscolino, "Energy and Enterprise in Liu Hongsheng's Cement and Coal-Briquette Business, 1920–37," *Twentieth-Century China* 41, no. 2 (2016), 159–179; Humphrey Ko, *The Making of the Modern Chinese State: Cement, Legal Personality, and Industry* (Singapore: Palgrave Macmillan, 2016).

[47] Nanpingshi difangzhi bianzhuan weiyuanhui 南平市地方志编纂委员会, (eds.), *Nanping diquzhi* 南平地区志 (*Gazetteer of Nanping*) (Beijing: Fangzhi chuabanshe, 2004).

hydroelectricity in China was not solely a transfer of Western technology but rather a process that involved adaptation, local expertise, and practical experience.[48] These factors are often overlooked in current narratives of hydropower engineering.

FUYUAN 富源

During the Second Sino-Japanese war (1937–1945), the picturesque town of Beibei 北碚, located 60 kilometers north of the bustling city of Chongqing, became a kind of sanctuary for various institutions and offices of the Kuomintang government, universities, merchants, and others. Beibei was situated along the banks of the Jialing River, providing convenient transportation to Chongqing despite heavy bombing in the area by the Japanese. The river featured rapid currents and smooth, rounded rocks that stood above the water, which provided the inspiration for the town's name: "Bei" meaning north and "bei" meaning rock, or "Northern Rock." The town's three main streets were full of activity, especially on market days when farmers from the surrounding countryside arrived before dawn to sell their produce. The migration of the Nationalist government to the area in the early phases of the war had brought increased business and higher prices, and boats on the Jialing River brought additional goods to the town.[49] At night, the town was illuminated by flickering vegetable oil lamps, while restaurants and households awaited the arrival of electricity. When it finally arrived, according to one observer, "suddenly there was a universal 'Ah!' and claps of hands and believe it or not, it was the electricity!"[50] Even in wartime, Beibei was an important hub of commerce and community.

The demand for energy in Beibei, including electricity, skyrocketed with the arrival of industries from elsewhere and the growth of the refugee population. The town's municipal authority had already built a coal-fired power plant to provide electricity to the district, but its capacity was insufficient to meet rising demand brought on by the war. Furthermore, coal was a precious resource and was earmarked by the government for powering munitions factories, which were prioritized during the war.

---

[48] Eugenia Lean, *Vernacular Industrialism in China: Local Innovation and Translated Technologies in the Making of a Cosmetics Empire, 1900–1940* (New York: Columbia University Press, 2020).

[49] Adet Lin, Anor Lin, and Meimei Lin, *Dawn over Chungking* (New York: The John Day Company, 1941), 54.

[50] Lin et al., *Dawn over Chungking*, 32.

It was against this backdrop that Lu Zuofu 卢作孚, the director of the shipping company Minsheng 民生 (People's Livelihood), proposed the construction of a hydropower plant on the Liangtan 梁滩 River at Gaokengyan 高坑岩 to support the growing refugee community.

In fact, the idea of harnessing the energy of the Liangtan River for generating electricity had been in Lu Zuofu's mind since 1933. He saw the potential of the river and its waterfalls and, despite facing initial financial challenges, submitted an application to the local government and sent engineers to survey the river in that year.[51] He was not alone in that idea, as the renowned educator and rural reconstruction activist Yan Yangchu 晏阳初, better known in English as Jimmy Yen, also saw the potential of the waterfalls and established the China Rural Construction College near Beibei in 1940 because of its proximity to this resource.[52] It was not until the latter stages of the Second Sino-Japanese war in 1943, however, that the rising demand for energy in the refugee-rich town of Beibei, combined with support from the Water Conservancy Commission, the Communication Bank, Jincheng Bank, and other local entrepreneurs, finally provided Lu with the opportunity to make his idea a reality. He relaunched his hydropower plan and established the hydroelectric company Fuyuan, meaning "source of wealth."

The Liangtan River originated in the towering peaks of Bi Mountain in Sichuan province and eventually converged with the Jialing River near Beibei. This river had tremendous potential for the production of hydroelectricity. The rapids at Gaokengyan, where the river plummeted 40 feet, were an excellent location for a hydropower plant. While hydroelectricity is attractive in terms of its affordability once facilities have been built, it requires substantial upfront investment. Thus, the Fuyuan Company, like many other small-scale hydropower plants, used a joint-stock system to

---

[51] Guanyu qingjiang gaokengyan huafen jiexian beian bing fentou zai Shanghai dinggou jiqi zhi Tao Jianzhong de tongzhi 关于请将高坑岩划分界限备案并分头在上海订购机器致陶建中的通知 (A Notice to Tao Jianzhong for Registration of the Demarcation of Gakengyan and Ordering Appliances in Shanghai) (February 17, 1933), 02070006000510100013, Chongqing Municipal Archives (hereafter CMA), Chongqing. On Lu Zuofu's contributions to the modernization of Beibei, see Zhang Jin 张瑾, *Quanli, Chongtu yu Biange: 1926–1937 nian Chongqing Chengshi Xiandaihua Yanjiu* 权力、冲突与变革：1926–1937 年重庆城市现代化研究 (*Power, Conflict and Reform: A study of the Modernization of Chongqing, 1926–1937*) (Chongqing: Chongqing chubanshe, 2003), chapter 6.

[52] Yan Yangchu 晏阳初, *Pingmin jianyu yu xiangcun jianshe yundong* 平民教育与乡村建设运动 (*Civilian Education and Rural Construction Movement*) (Beijing: Shangwu yinshuguan, 2014), 287.

raise the necessary funds. With an initial share capital of 20 million yuan, consisting of 20,000 shares of 1,000 yuan each, the company attracted a diverse array of investors.⁵³ At their first meeting, the founders elected Lu Zuofu as chair of the board of directors. Lu's company, Minsheng, took on the task of manufacturing the hydro turbines required for the project itself, setting it apart from other companies like Yaolong and Jihe who imported their turbines from abroad. This was a testament to Lu's commitment to bringing the vision of this project to life. The establishment of Fuyuan was made possible by the combined efforts of a number of influential figures who brought with them not only financial resources but technical expertise as well. As a joint-stock enterprise, the company aimed to provide electricity for both domestic and commercial purposes, including household lighting and power for small and middle-sized factories. Reports estimated that when Fuyuan was built Beibei already had around 600 light bulbs, averaging 15 watts and powered by coal. However, with their confidence in the potential of hydroelectricity, the founders of Fuyuan believed they had the capacity to power another 3,000 bulbs.⁵⁴ The future of the company seemed promising.

A NEW MAN-MADE DISASTER

Chris Courtney has elegantly explained that floods are the result of a complex interplay of natural forces and human interference.⁵⁵ Throughout Chinese history, human actions have altered the landscape and waterways, and the creation of concrete dams and the implementation of hydropower technology have only enhanced our ability to manipulate rivers. The Liangtan River had tremendous hydropower potential but its flow was unpredictable, fluctuating greatly depending on the amount of rainfall in the area. To ensure a stable flow and the presence of enough water to generate electricity consistently throughout

---

⁵³ Fuyuan shuili fadian gongsi faqiren xingming jingli ji rengu shumu 富源水力发电公司发起人姓名经历及认股数目清册 (List of names, profiles and committed share amounts of Fuyuan hydroelectric company initial founders), 0060000200144000002, CMA, Chongqing.

⁵⁴ Fuyuan shuili fadian gufen youxian gongsi chuanli huiyilu ji diyi, er, sanci dongjian lianxi huiyi jilu 富源水力发电股份有限公司创立会决议录及第一、二、三次董监联席会议记录 (Fuyuan Hydroelectric Power Company, Ltd. founding meeting resolution records and the first, second and the third board of directors meeting records) (June 2, 1943), 0220000100000027000, CMA, Chongqing.

⁵⁵ Courtney, *Nature of Disaster in China*.

the year, a dam had to be constructed upstream from the power plant. Unfortunately, this led to conflict between the local community and the power company.

The Liangtan River had been used as a source of mechanical power for decades but not by the hydropower company. A number of watermills were in operation along the river, processing grain for local communities. The construction of the dam, however, caused the water level to rise and disrupted the operation of the watermills. Low-lying areas along the river were also flooded for the first time. To mitigate the negative impacts, the company built a lower dam with a sluice gate to regulate the water level in case of emergencies.[56] The company also implemented strict regulations for the use of the sluice gate. Despite these efforts, the reservoir still submerged farmlands, bridges, and roads. To address this, the company raised the height of bridges and roads to maintain accessibility, but it could not prevent the submerging of farmlands whose owners it had to compensate. To gain the support of local communities, the company reached out to local leaders and security groups. Jiang Gengqiao 蒋耕樵, the town chief of Xinglongxiang 兴隆乡, was supportive and saw supporting the nation's development efforts as a duty of local societies. He recommended two members of the local gentry, Lü Siqi 吕思齐 and Liu Yingzhou 刘瀛洲, to the company as advisors, as he believed they not only appreciated the value of the hydropower project but were also well-acquainted with the local situation.[57]

Lü Siqi, however, owned one of the watermills that had been impacted by the dam. His millhouse was in danger of collapsing because of the rising water. He wrote a letter to the company's manager, expressing concern about the financial loss the dam had caused him.[58] The company

---

[56] Guanyu jiansong Gaokengyan shuili fadianchang Liangtanqiao mofang, xushuiba gongcheng jianshe jingguo qingxing ji gongcheng yingxiang zhoubian bing jinxing bujiu deng de cheng, han, daidian, ling 关于检送高坑岩水力发电厂梁滩桥磨坊、蓄水坝工程建设经过情形及工程影响周边 并进行补救等的呈、函、代电、令 (Reports, letters, telegraphs and orders on Gaokengyan hydropower plant reservoir project's impact on Liangtanqiao watermills and its surrounding area) (August 6, 1943), 0220000100185000000100o, CMA, Chongqing.

[57] Xinglong was a township under Bishan county and was located on the upper stream of the Liangtan River. On rural social life and the changes it underwent during the war, see Isabel B. Crook, Christina K. Gilmartin, Yu Xiji, Gail Hershatter, and Emily Honig, *Prosperity's Predicament: Identity, Reform, and Resistance in Rural Wartime China* (New York: Rowman & Littlefield, 2013).

[58] Guanyu Baxian Xinglongxiang shangmin Lü Siqi, Wu Jicheng konggao Fuyuan shuilifadian gufen youxian gognsi lanjian fangwu, fanghai tongcheshuili bing banli peichang sunshi shiyi de cheng, han, daidian, xunling 关于巴县兴隆乡商民吕思齐、吴继澄控告富

promised to raise his watermills and compensate him for his losses, but months went by and nothing was done. Frustrated, Lü took further action. A heavy rainfall in 1946 resulted in accidental flooding of farmlands and watermills along the dammed river, further fueling local residents' dissatisfaction. Lü, as an influential member of the gentry, not only defended his own interests but also rallied local residents to protect their properties. He organized a petition delegation and brought their concerns to the public media, putting pressure on the hydropower company and local government to act.

In their petition, Lü and his followers referred to the flood as a manmade disaster and blamed it on the construction of the dam. They argued that the project posed a threat to public welfare, treated the local community unjustly, and undermined the efforts of resistance and reconstruction.[59] The hydropower company defended its project, claiming that the flood was an inevitable natural disaster that could have occurred regardless of the dam's construction. The company also emphasized in its response that the project was designed for the benefit of the public rather than for profit.[60]

The conflict surrounding the dam's construction highlighted the existence of differing perspectives on what "progress" entailed. Some saw the dam as a symbol of progress and believed it should be built, while others criticized its negative economic and ecological impacts on local society. It is important to understand that the development of hydroelectric projects is not a straightforward and simple process of progress, but a multifaceted one. In challenging the conventional modernist or developmentalist discourse that portrays the hydroelectricity project as solely positive, I aim to shed light on the diverse experiences of the different groups involved in and impacted by the project.

源水力发电股份有限公司滥建房屋、妨害筒车水利并办理赔偿损失事宜的呈、函、代电、训令 (Reports, letters, telegraphs and orders on merchants Lü Siqi, Wu Jicheng from Xinglong township, Ba county, who sued Fuyuan Hydroelectric Power Company, Ltd. for building infrastructure which harmed local water conservancy, and on the issue of compensation) (April 6, 1946), 02200001000660000001000, CMA, Chongqing.

[59] Baxian Liangtanhe liangan nongmin bei Fuyuan shuidian gognsi du chengzai qingyuantuan 巴县梁滩河两岸农民被富源水电公司堵口成灾请愿团 (Petition group of Ba county farmers along the Liangtan River whose lands were negatively affected by the disaster caused by the Fuyuan Hydroelectric Power Company) (April, 1946), 02200001000660000001000, CMA, Chongqing.

[60] Guanyu Baxian Xinglongxiang shangmin Lü Siqi, Wu Jicheng konggao Fuyaun shuilifadian gufenyouxian gongsi lanjian fangwu, fanghai tongcheng shuili bing banli peichang sunshi shiyi de cheng, han, daidian, xunling.

The conflict described here raises important questions about who has the right to use the river and what such a right entails. Traditionally, when disputes over access to water resources arose, individuals or families sought the help of respected members of the local gentry or government officials for mediation. These disputes were often resolved based on land rights and the chronological sequence of water use, but they sometimes escalated into inter-clan conflicts.[61] During the War of Resistance against Japan, the Kuomintang government prioritized the use of resources to support the war effort. To this end, in 1942, the government introduced the Shuili Jianshe Gangyao (Principles of Water Conservancy Construction), which emphasized the need to prioritize the demands of the war while ensuring the protection of water sources. The Principles stated that hydroelectric projects should be developed to the fullest extent possible, taking into account the specific industrial and social demands of the time.[62] It is clear that this conflict over the use of the river reflected the larger issue of who possessed the power to determine how natural resources would be used and developed. While the government's principles prioritized the needs of the war and reconstruction, it is important to consider the effects on local communities and the environment as well.

As a center of resettlement located near the wartime capital of China, Beibei faced a pressing need for a more stable energy supply. In light of this, elites residing in the area were highly supportive of the Liangtan River hydroelectric project, which appeared poised to receive a water-use permit from the local government. In 1943, the Ba County government dispatched engineers to assess the project and wrote an informal ruling

---

[61] On water management and water-rights disputes in Chinese history, see Purdue, *Exhausting the Earth*; R. Keith Schoppa, *Xiang Lake: Nine Centuries of Chinese Life* (New Haven: Yale University Press, 1989); Thomas M. Buoye, *Manslaughter, Markets, and Moral Economy: Violent Disputes over Property Rights in Eighteenth-Century China* (Cambridge: Cambridge University Press, 2000), chapter 3; Chao Xiaohong 钞晓鸿. "Guangai, huanjing yu shuili gongtongti – jiyu Qingdai Guanzhong zhongbu de fenxi" 灌溉、环境与水利共同体 (Irrigation, Environment and Hydraulic Community: An Analysis based on the Central Part of the Guanzhong Plain), *Zhongguo Shehui Kexue* 中国社会科学 4 (2006), 190–204; Zhang Junfeng 张俊峰, *Shuili Shehui de Leixing: Ming Qing yilai Hongdong Shuili yu Xiangcun Shehui Bianqian* 水利社会的类型：明清以来洪洞水利与乡村社会变迁 (*The Pattern of Hydraulic Society: Water Conservancy and Rural Social Changes in Hongdong since the Ming and Qing dynasties*) (Beijing: Beijing Daxue chubanshe, 2012).

[62] Tian Dongkui 田东奎, "Zhongguo jindai shuiquan jiufen jiejue jizhi yanjiu" 中国近代水权纠纷解决机制研究 (Study on the Mechanism of Solving Water Rights Dispute in Modern China), PhD Dissertation, Zhongguo Zhengfa University, 2006, 222.

that the company would secure its permit.[63] However, in 1947, due to the ongoing controversy surrounding the dam and the devastating flood of 1946, the company was yet to receive the permit.[64]

Despite the lack of a permit, the project was still completed, and electricity was generated for the town of Beibei starting in 1944. However, the lack of a permit was not the only challenge that the project faced. The natural flow of the river posed another challenge: The water level was subject to change, which affected the river's ability to power the turbines. The engineers had designed the dam with a sluice gate to regulate the water level and thereby ensure consistent power, but the company was unable to build a higher dam due to cost constraints and resistance from local residents. As a result, it was compelled to build a lower dam, which then led to periodic shortages of water for the turbines. During the driest months, the turbines could only run for ten hours per day, which resulted in a limited power supply for the town.[65]

The success of small hydroelectric projects in southwest China was conditioned by several factors. First, the region's topography, with its waterfall and steep river channels, provided ideal conditions for harnessing hydropower. Also, the availability of steel turbines and electric generators through international trade networks facilitated the implementation of these projects. As the central government in this period was weak and no comprehensive planning was possible, the driving force behind these initiatives was local business and political elites who sought to meet private and civic energy needs. In this context, the role of the state (represented by local officials) was limited to providing support and assistance.

---

[63] Guanyu Fuyuan shuili fadian gufenyouxian gongsi liyong Gaokengyan pubu jingying shuilifadian ji banli shuiquan dengji, kancha shiyi, jiansong shuiquan dengji shenqingshu deng de cheng,han,pi 关于富源水力发电股份有限公司利用高坑岩瀑布经营水力发电及办理水权登记、勘察事宜、检送水权登记申请书等的呈、函、批 (Reports, communications, and responses on the Fuyuan Hydroelectric Power Company using Gaokengyan waterfall to generate electricity and registration of water rights, investigation, and submission of application) (June, 1943), 0220000100011000002000, CMA, Chongqing.

[64] Fuyuan shuili fadian gufenyouxian gongsi di er jie di san, sidengci dongjian liangxihuiyi, gudong dahui jilu(fafang guxi,xiuzheng gongsi zhangcheng, zengjia ziben deng) 富源水力发电股份有限公司第2届第3、4次董监联席会议、股东大会记录 (发放股息、修正公司章程、增加资本等) (The third and fourth board meetings of the Fuyuan Hydroelectric Power Company, Ltd., on interest distribution, revising company bylaws, and increasing capital) (October 24, 1947), 022000010000300000001000, CMA, Chongqing.

[65] Ibid.

CONCLUSION

From the beginning, China's hydroelectric enterprises relied heavily on the advanced technologies developed in Europe and North America. The first hydropower plant in China, Yaolong, was designed by German engineers and imported its major pieces of equipment from Germany. Despite their limited scope, these projects still had a significant impact on local communities. The process of adapting and incorporating new technologies into existing systems is an inescapable element of how the relationships between humans and the environment are shaped.[66] In the case of the Liangtan River project in Beibei, the construction of the hydropower plant submerged watermills and farmlands, resulting in a loss of access to water for the riverside community. At first, only the more affluent households and businesses in the town benefited from the project, not the farmers whose properties adjoined the river or people living near the dam and the plant.[67] Although the flood in 1946 was not necessarily caused by the dam, it further impacted the livelihoods of the local residents, who saw the dam as the cause of a man-made disaster. These small-scale projects, which negatively affected a relatively small number of people, nonetheless portended the environmental and social disruptions that would be occasioned by the rapid expansion of hydropower in the second half of the twentieth century. Despite this, the social implications of these projects are usually overlooked in current discussions of the early history of hydropower in China.

With the beginning of the electrical age in the late nineteenth century, China was slow to experience the major technological innovations that were transforming life elsewhere. Despite this, people in China responded quickly to the energy transitions and technological advancements coming from Europe and North America. The emergence of small hydropower plants in southwest China shows that early Chinese responses were not led by the state, which was in disarray at the time. Rather, they were driven by the initiative of private individuals and local elites. These local hydropower projects were fueled by a combination of resistance to foreign influence and a concern for China's status and prosperity during a tumultuous period. Although China could not be described as

---

[66] This is partially inspired by Nye, *Electrifying America*.
[67] Nicole Barnes, *Intimate Communities: Wartime Healthcare and the Birth of Modern China, 1937–1945* (Berkeley: University of California Press, 2018).

a hydropower nation in this period, certain Chinese elites recognized the potential to harness the country's hydropower resources as a means of securing its future. In response to the deepening national crisis caused by Japanese military aggression and the development of concrete mega-dams, technocrats serving in the Kuomintang government sought to harness the rivers in China's hinterland for the dual purpose of military resistance and national reconstruction.

2

Mobilizing Rivers

Located at the intersection of the Yangtze and Jialing Rivers, the city of Chongqing is often enveloped in mist and shrouded in a gray veil, its rugged terrain rising up from the banks of the rivers. Before it became the wartime capital of the Kuomintang government, it was essentially a remote outpost in the southwest, a largely forgotten corner of the country. Its new status as the seat of government during the war, however, did little to improve its architectural or aesthetic appeal. The city remained a patchwork of makeshift structures, clusters of homes perched on bamboo stilts and clinging to the cliffs, linked by narrow walkways and makeshift bridges.[1] The city's hills were scarred by countless dugouts, made by hand and left dark and dank, their walls sweating in the rain. The scars of Japanese bombs were still visible, leaving parts of the city in ruins. Along the Yangtze River, rows of men toiled, their bodies bent forward on ropes as they pulled their junks upstream. The ropes were fastened to the tops of the masts, acting as levers as the men on shore leaned into them. A chant leader led each group, his voice rising in song, echoed by the rhythmic chorus of the workers as they swayed in unison.[2] On the riverbanks, pairs of workers carried sedan chairs, transporting newly arrived travelers with wealth and influence, privileged enough to be able to pay to avoid having to climb the stairs in the oppressive heat of the Chongqing summer. In the city's streets, among the ruins, slogans were painted on walls that

---

[1] Duncan McRoberts, *Pleading China* (Grand Rapids, MI: Zondervan Pub. House, 1946), 96.
[2] Daniel Nelson, *Journey to Chungking* (Minneapolis: Augsburg Publishing House, 1945), 124.

reminded the people of the ongoing struggle for national salvation: "Those who have money give money; those who have strength give strength." "Victory in resistance; success in reconstruction."[3] The work of the barge haulers and sedan carriers, as well as the slogans on the walls, testified to the centrality of manual labor, still a primary source of energy in China even as it fought a modern war.

"Modern warfare is a struggle over energy. The more energy a country can mobilize, the stronger it will be. Without enough energy, a country cannot be well defended," declared Huang Wenxi 黄文熙, a hydraulic engineer educated in the United States, in 1942.[4] People have recognized for a long time the importance of various forms of energy for warfare, but the new forms of warfare that emerged in the twentieth century brought a new appreciation of significance of nonrenewable sources of energy. The exploitation of the Fushun coal mine by Japan and its invasion of Southeast Asia with the aim of securing oilfields are stark examples of this trend.[5] As tensions between China and Japan escalated in the 1930s, the Nationalist government in Nanjing realized that war was imminent. But they were ill-prepared for the military confrontation that lay ahead. In 1936, China's electric capacity was just 631,000 kW, with over 90 percent of that located in the coastal regions of Shanghai, Jiangsu, and Zhejiang. Japan, on the other hand, had a capacity of 7,560,000 kW.[6] When the war broke out and the Nationalist armies were quickly defeated, this forced a retreat to the west and the loss of almost all power plants.[7] In late 1938, the Japanese army laid siege to Wuhan, a major

---

[3] Lin et al., *Dawn over Chungking*, 25.

[4] Huang Wenxi 黄文熙, "Shuili jianshe zouyi" 水力建设刍议 (Comments on Hydropower Construction), *Jingji jianshe jikan* 经济建设季刊 (*Economic Reconstruction Quarterly*) 1 (1942), 154.

[5] McNeill, *Something New under the Sun*.

[6] Xue Yi 薛毅, *Guomin zhengfu ziyuan weiyuanhui yanjiu* 国民政府资源委员会研究 (Study on the Nationalist National Resource Commission) (Beijing: Shehui kexue wenxian chubanshe, 2005), 225; Takeo Kikkawa, "The History of Japan's Electric Power Industry before World War II," *Hitotsubashi Journal of Commerce and Management* 46, no. 1 (2012), 10–16, 13.

[7] For studies on electricity industry of China before 1937, see Wang Jingya 王静雅, "Nanjing Guomin zhengfu jianshe weiyuanhui dianye guihua yu shijian yanjiu" 南京国民政府建设委员会电业规划与实践研究 (A study of the Nanjing Nationalist Construction Commission's Electricity Industry Planning and Practice), *Huabei dianli daxue xuebao* 华北电力大学学报 6 (2012), 19–26; Zhao Xingsheng 赵兴胜, "Zhang Jingjiang in 1928–1937" 1928–1937 年的张静江, *Jindaishi yanjiu* 近代史研究 1 (1997), 237–251; Wang Shuhuai 王树槐, "Zhongguo zaoqi de dianqi shiye 1882–1928" (The Early Electricity Industry in China, 1882–1928) 中国早期的电气事业, 1882–1928, in Zhongyangyanjiuyuan jinshisuo (ed.), *Zhongguo xiandaihua lunwenji* 中国现代化论文

center of energy production, leading the National Resources Commission to move equipment for producing electrical power from Hubei and Hunan provinces to Sichuan and Yunnan in a desperate effort to save the remaining energy industry and to support the war effort. However, compared to the major cities in the east, the energy industry in the southwest was underdeveloped and incapable of meeting the urgent demand for electricity.[8]

In the face of Japanese aggression during World War II, the Nationalist government came to understand the vital importance of harnessing the country's rich potential hydropower reserves. The National Resources Commission was at the forefront of this effort, playing a key role in the construction of the Longxi River hydropower project and the planning of the Three Gorges project. This transformation of scale, from the local initiatives discussed in Chapter 1 to state-led efforts, marked a new era in the development of China's hydropower industry. Much of the existing literature on this subject, however, is comprised of general, surface-level information about the achievements of the National Resources Commission.[9] The actual story is far more complex and multifaceted than such accounts suggest. The construction of hydroelectric plants was not simply an end in itself. It was also entangled with the environment, with international relations, and with local societies in southwest China.

Through an exploration of the Longxi River project and the exchange of engineers between China and the United States, this chapter analyzes the intricate connections between war, energy, the environment, and local

---

集 (*Essays on the Modernization of China*) (Taipei: Institute of Modern History of Academia Sinica, 1991), 443–472; Wright, "Electric Power Production in Pre-1937 China."

[8] See William C. Kirby, "The Chinese War Economy," in *China's Bitter Victory: The War with Japan, 1937–1945*, ed. James C. Hsiung and Steven I. Levine (London: M. E. Sharpe, 1992), 185–212. In the eight years of the war, the Guomindang government successfully rebuilt only 7,986 kW of capacity in the southwest. Sun Yusheng 孙玉声, "Kanzhan banianlai zhi dianlishiye" 抗战八年来之电力事业 (The Electricity Industry in the Eight-Years War of Resistance), *Ziyuan weiyuanhui jikan* 资源委员会季刊 6 (1946), 143. The Nationalist army's resistance at Wuhan frustrated Japan's attempt to secure a decisive victory. The army's efforts gained time for industries and government and social institutions to be moved. For details, see van de Ven, *War and Nationalism in China, 1925–1945*, chapter 6.

[9] For studies on the development of the electrical industry under the National Resources Commission, see Xue, *Guomin zhengfu ziyuan weiyuanhui yanjiu*, chapter 11; Zhongguo shuili fadianshi bianji weiyuanhui 中国水力发电史编辑委员会, *Zhongguo Shuilifadianshi* 中国水力发电史 (*The History of Hydroelectric Power in China*), Vol. 1 (Beijing: Zhongguo dianli chubanshe, 2005), chapter 1.

communities in wartime Nationalist China. It aims to shed light on the emergence of a cohort of professionally trained Chinese hydropower engineers who would serve as the technical backbone of an incipient hydropower nation. By untangling this complex web of interactions, this chapter offers a new understanding of the challenges and triumphs China experienced during one of its most difficult periods.

THE WARTIME ENERGY CRISIS

The Nationalist government attempted to harness alternative sources of energy in the southwest region, including the Tianfu coal mine, despite the fact that its lower quality coal commanded higher prices due to the difficulties in extraction.[10] According to Madeleine Zelin, the use of natural gas in Sichuan to boil salt brine dates back to the Song dynasty.[11] Aware of this, in the early 1930s the Nationalist government launched an investigation into natural gas resources in the area. However, due to technological limitations, large-scale exploitation was not feasible. To sustain the energy needs of the ammunition industry, the government had to restrict, and at times even cut off entirely, electrical services for civilian use, which exacerbated the already dire fuel shortage caused by Japanese bombing.

The government encouraged the conservation of electricity in various ways. They advertised in newspapers, reminding the public that saving electricity at home was a way of supporting the war effort on the front lines. To further drive home this message, the government deployed quantitative methods to show the direct relationship between energy conservation and the production of war materials. For instance, its notices calculated that a reduction in use of 1.5 kWh could result in the possibility of manufacturing one additional mortar shell or mine. Saving 156 kWh of electricity meant that one more machine gun could be produced. Citizens were encouraged to use electricity efficiently and to reduce consumption by 40 percent through the use of low-wattage light bulbs.[12] The government also encouraged early bedtimes and early rising to reduce nighttime electricity usage and even considered banning certain

---

[10] See Tim Wright, *Coal Mining in China's Economy and Society, 1895–1937* (Cambridge: Cambridge University Press, 1984).

[11] Zelin, *Merchants of Zigong*.

[12] "Yaolong dianli gongsi jieyueyongdian jinji tongzhi" 耀龙电力公司节约用电紧急通知 (Yaolong Electrical Company Emergent Notice on Electricity Saving) (November 14, 1943), file no. 133-7-45 at Yunnan Provincial Archives (hereafter YPA), Kunming.

forms of nonessential electricity consumption such as store advertisement lights, exhibition windows, and hair dryers and curlers.[13] However, these measures could only alleviate energy scarcity to a limited degree. Ian J. Miller observed that total mobilization required the prioritization of resources for the war effort, even at the expense of other important needs.[14] Developing new sources of energy would likely have been a more effective solution in the long-term.

As the nation faced the crisis of war, many people in China came to understand the significance of their natural resources, including hydropower. The Sichuan region, however, accounted for only 4 percent of China's total electrical power-generating capacity.[15] As Rana Mitter notes, this was "a pitifully small base" upon which to build a wartime industrial economy.[16] In stark contrast, China as a whole had an estimated 40 million kW of hydropower potential, placing it third in the world behind only the United States and Canada. Much of this potential was located in southwest China, particularly in the provinces of Sichuan and Yunnan. Despite this, according to official statistics, China (excluding the northeast, which was under Japanese control) had only 2,726 kW of installed hydroelectric capacity in 1932.[17] In 1939, General Chen Zhang 陈章 of the Nationalist army wrote about the untapped potential of China's rivers and proposed that if they were properly exploited, they could not only support the war effort against Japan but also pave the way for reconstruction in the future.[18]

THE NATIONAL RESOURCES COMMISSION

Long before the PRC government's western development strategy in the 1990s, the Kuomintang regime played a critical role in the development of

---

[13] "Lun jieyu yongdian" 论节约用电 (On Saving Electricity), file no. 133-7-74 at YPA, Kunming.
[14] Ian J. Miller, *The Nature of the Beasts: Empire and Exhibition at the Tokyo Imperial Zoo* (Berkeley: University of California Press, 2013), 96.
[15] Kirby, "Chinese War Economy," 190–191.
[16] Rana Mitter, *Forgotten Ally: China's World War II, 1937–1945* (Boston: Mariner Books, 2013), 182–183.
[17] Zheng et al., *Jiu Zhongguo de ziyuan weiyuanhui, 1932–1949*, 86. The total electrical-generating capacity of the National Resources Commission's power plants reached 26,975 kW by 1945.
[18] Chen Zhang, "Duiyu shuili fadian yingyoude renshi" 对于水力发电应有的认识 (What We Should Know about Hydroelectric Power), *Xinminzu* 新民族 3 no. 14 (1939), 4–7.

hydropower in southwest China.[19] While the hydropower industry in China certainly remained underdeveloped during the Kuomintang period, the efforts and limited achievements of its officials and engineers in the first half of the twentieth century should not be overlooked. William Kirby has highlighted the emergence of the developmental state and institutional evolution during the Nanjing decade, when the Kuomintang government established the National Defense Strategy Commission in 1932. This commission was later reorganized as the National Resources Commission, with the renowned geologist Weng Wenhao 翁文灏 as its chairman. It was tasked with surveying and exploiting China's natural resources, as well as planning and developing key industries for China's reconstruction.

One of the central tasks of the commission was to supply energy to support industrialization, and so the Office of the National Electric Industry was established to conduct surveys and to oversee the design and construction of electrical projects in Nationalist-controlled areas, including hydroelectric projects mainly located in the southwest.[20] The commission's efforts stand out as one of the most comprehensive attempts to apply science and engineering to the work of government and the ongoing tasks of reconstruction, earning it the nickname "the engineers' stronghold in the National Government."[21]

Despite facing difficult constraints in terms of financial and technological resources, the Nationalist government nevertheless made early strides in developing hydropower projects in the southwestern interior of China during the 1930s. The fertile Sichuan region, with its numerous waterways, had tremendous potential for hydroelectric power generation. However, resource limitations posed a challenge to fully exploring all the

---

[19] Magee, "Powershed Politics"; Bryan Tilt, *Dams and Development in China: The Moral Economy of Water and Power* (New York: Columbia University Press, 2015).

[20] On the expansion of the energy industry led by the National Resources Commission, see Morris L. Bian, *The Making of the State Enterprise System in Modern China: The Dynamics of Institutional Change* (Cambridge, MA: Harvard University Press, 2005), 59–61.

[21] William C. Kirby, "Engineering China: Birth of the Developmental State, 1928–1937," in *Becoming Chinese: Passages to Modernity and Beyond*, ed. Yeh Wen-hsin (Berkeley: University of California, 2000), 137–160, 150. The rise of the National Resources Commission also signaled the shift to a military-oriented industrialization model in China. See Margherita Zanasi, *Saving the Nation: Economic Modernity in Republican China* (Chicago: The University of Chicago Press, 2006), 192; Megan J. Greene, *The Origin of the Developmental State in Taiwan: Science Policy and the Quest for Modernization* (Cambridge, MA: Harvard University Press, 2008).

rivers in the region within any kind of reasonable timeframe. To address both economic and political demands, the National Resources Commission selected the Sichuan Basin, with Chengdu at its center, and the area around Chongqing as the two primary areas of focus for the development of hydropower.

In 1935, the National Resources Commission dispatched a survey team to Chongqing under the leadership of Huang Yuxian 黄育贤, an experienced hydraulic engineer who had graduated from Tsinghua University and received further training at the California Institute of Technology and Cornell University. Over the course of six months, the team meticulously surveyed several rivers in the Sichuan Basin and the area surrounding Chongqing, including the Min, Qingyi, Dadu, Mabian, Qu, Jialing, Taohuaxi, and Longxi rivers. The survey revealed that the Dadu River had the greatest hydropower potential, but its remote location was a problem in terms of energy consumption and industrial development. On the other hand, the Longxi River was more accessible and was determined to be ideal for a hydropower plant to meet the energy demands of relocated industries in Chongqing. Considering both accessibility and the challenges of power transmission, Huang and his team concluded that the Commission should prioritize the development of the Longxi River.[22]

The Longxi (meaning Dragon Creek) River runs through Changshou 长寿 county in Chongqing and is a tributary of the Yangtze River. With three significant waterfalls along its route at the Upper and Lower Qingyan caves, which were as high as 40 meters, it was a source of frustration for local communities, who saw it as unnavigable and useful only for irrigation and transporting timber and coal during the high water seasons.[23] However, for Huang Yuxian and his survey team, these waterfalls presented an opportunity for a significant hydropower project. In July 1937, the National Resources Commission established the Office of the Longxi River Hydropower Project, with Huang as the director. The office soon began to prepare the site, constructing roads, offices, and housing in readiness for the project's construction. The plan was to build a 30-meter-high dam upstream of Shizitan 狮子滩 and to equip the power plant with four generators, each with a capacity of 4,000 kW.

[22] "Ziyuan weiyuanhui Longxihe shuili fadianchang gongchengchu nianliuniandu gongzuo baogao" 资源委员会龙溪河水力发电厂工程 年度工作报告 (The National Resources Commission Longxi River Project Engineer Office's Report of 1937), 02230004000230000001, CMA, Chongqing.
[23] Lu Qixun 卢起勋 and Liu Junxi 刘君锡, *Changshou Xian zhi* 长寿县志: *16 juan* (Taipei: Chengwen chubanshe, 1976), 44.

The retreat of the Nationalist government to the hinterland of the southwest and the pressing need for energy to sustain its war of resistance elevated the importance of the Longxi River project. However, the conflict also placed severe limitations on the material supplies available for the National Resources Commission's efforts. Despite these difficulties, the dam was built on the river at Shizitan and the sluice connecting the reservoir and the power plant was ready to channel water and drive the turbines by 1939. The only thing missing was the generators, but the Japanese military advance disrupted the transportation of large equipment, including the hydropower generators, from abroad. As a result, the National Resources Commission had no choice but to abandon the nearly completed Shizitan project. Their efforts were redirected toward smaller projects on the Taohuaxi River, a tributary of the Longxi River. This was a disappointment to Huang Yuxian and the other engineers involved in the project, who had imagined a brighter future for the Longxi River project.

POWERING ARSENALS AND OTHER FACTORIES

These smaller hydropower plants were a valuable source of energy for China's war effort, both on the home front and the frontline. In December 1940, when the Taohuaxi plant began supplying electricity to Changshou, the plant had only 149 users and 2,712 kWh were sold in the first month. By August 1941, however, when all the generators were operational, the number of users had increased to 558, as many manufacturing firms from the east had relocated to the area by then. These included the Hengfeng and Yunli rice factories, the China Matches factory, and the China Industrial Oil Refinery. Monthly consumption of electricity rose to 250,000 kWh. By December 1943, these figures had increased to 681 customers and a monthly consumption of 300,000 kWh. In January 1944, the Xiadong plant started generating electricity, increasing the total capacity of all the plants to 1,550 kW. In March 1948, with the addition of two more generators, this reached 3,866 kW. At this point there were 1,288 lighting users, 75 factory-power users, and 7 industrial-heating users; monthly consumption reached 2,340,000 kWh – almost 10 times more than 3 years prior.[24]

---

[24] Zhu Chengzhang 朱成章, "Longxihe tiji shuidianzhan kaifa jishi" 龙溪河梯级水电站开发纪实 ("Records of the Longxi River Cascade Hydropower Exploitation"), *Zhongguo shuilifadian shiliao* 中国水力发电史料 1 (1987), 37.

In the 1940s, Changshou had the most affordable electricity in China, attracting many energy-intensive industries to the area. Small private enterprises, large-scale state factories, and arsenals related to national defense could also be found in Changshou. Electricity distribution was dominated by the smelting industry, at 48 percent, followed by the munitions industry, at 29 percent, light industry, at 16 percent, and household lighting, at 7 percent.[25] The largest single user of electricity was the 26th Arsenal of the National Defense Commission, which was granted top priority for access to electricity during shortages. Like other factories, and despite its special status, the arsenal was required to sign a commercial contract with the power plant.

The supply of energy in southwest China was challenged not only by the ongoing war with Japan but also by the region's natural environment. Rivers in China normally follow seasonal patterns of rising and falling, which entails a drop in water levels during the dry season, typically in winter and early spring. This decline in water levels caused a decrease in hydroelectric output, forcing power plants to ration electricity among major users and forcing enterprises to resort to using fossil fuels as a backup during these periods. In April 1945, for example, the plants' generator capacity dropped to 2,000 kW, far below the 3,000 kW specified in contracts with major users. Even factories related to national defense had to make do with reduced electrical power during these times.

During the war, the residents of Changshou took precautionary measures to safeguard their critical infrastructure, including their hydropower plants, from Japanese aerial attacks, principally the implementation of camouflage strategies aimed at rendering the buildings inconspicuous. Despite the frequent bombing raids, the power plants survived and remained operational. Following the Nationalists' defeat in the civil war to the Communists in 1949, however, the former warlord and Nationalist mayor of Chongqing, Yang Sen 杨森, decided to destroy the Taohuaxi and Xiadong hydropower plants before he fled to Taiwan.[26] Once again, then, the area around Changshou was without electricity.

---

[25] Dushi jihua weiyuanhui 都市计划委员会, *Guanyu shouji ziliao niding changshou longxihe yidai ge shuili fadian gongcheng jianmingbiao* 关于收集资料拟定长寿龙溪河一带各水力发电工程简明表、工程概况等的呈 (Memorial on Collecting Information about Changxi Longxi River Hydropower Projects, Brief Chart and General Situation), 0076-0001-00021-0000-009-001, CMA, Chongqing.

[26] Zhu, "Longxihe tiji shuidianzhan kaifa jishi," 37.

TENSIONS WITH LOCAL COMMUNITIES

Rana Mitter has pointed out that, while the Nationalist government's wartime mass mobilization was impressive in appearance, it was not always convincing in practice.[27] The "mobilization" of rivers for hydropower projects faced a complex set of challenges, both natural and social. The generation of hydropower requires a consistent and stable source of water, which is generally achieved through the construction of a reservoir. But the construction of reservoir infrastructure also submerges land in the surrounding riparian areas that could have been used for other things.

The landscape of the Chongqing region is characterized by rolling hills and verdant valleys, with very little flat terrain. Over time, rural communities in the area have shaped the hilly terrain into a mosaic of terraced rice paddies, the result of centuries of population growth and agricultural expansion.[28] As Isabel Crook shows in her ethnographic study of Xinglongchang, local farmers have developed three distinct types of irrigated fields: flood paddies (*goutian* 沟田), bordering paddies (*pangtian* 旁田), and hill paddies (*shantian* 山田). This diversity of fields, nestled in the waterways, testify to the ingenuity and resourcefulness of the farmers who have cultivated the soil for generations. Through Crook's work we gain a deeper understanding of the intricate networks of irrigation that sustain the livelihoods of these rural communities.

Flood paddies were easy to fill with rainwater and good for storing water. In a dry year they were the most reliable, but in a wet year they flooded easily. The prime fields were the bordering paddies that were easy to fill and drain because they bordered the flood paddies. Hill paddies, with far less rainwater draining into them, took a long time to fill and sometimes lost water to seepage. With plentiful rain, the farmer with hill paddies could reap a good harvest, but without it he would have to choose a less lucrative dry crop, therefore suffering a loss, since he still was obligated to pay his rent in rice.[29]

Aside from sporadic groves of bamboo and the occasional tree, by the early twentieth century nearly every inch of land was put to agricultural use in the farming communities around Chongqing.[30] Men scoured the hillsides in search of fuel for their cooking stoves, while children and the

---

[27] Mitter, *Forgotten Ally*, 178.
[28] On population settlement in the Sichuan area, see Robert Entenmann, "Migration and Settlement in Sichuan 1644–1796" (PhD Dissertation, Harvard University, 1982).
[29] Crook et al., *Prosperity's Predicament*, 47.
[30] Crook et al., *Prosperity's Predicament*, 48; John Lossing Buck, *Land Utilization in China* (New York: Paragon Book Reprint Corp, 1964), 78.

elderly gathered weeds from the sides of roads to feed their pigs. The family's livestock were also put to work, roaming the area in search of their own sustenance.[31]

The arrival of the Nationalist government in Chongqing and of trained hydropower engineers marked a turning point for the region's water-based landscape. The state-owned Taohuaxi and Xiadong projects, operated by the National Resources Commission, embarked on a mission to harness the waters to support the war of resistance as well as more general state-building efforts. The Commission's overwhelming power during the process of expropriating land for the projects effectively silenced the voices of local residents, leaving only numbers from official documents to shed light on the extent of land that was taken.

At the end of 1940, it was reported by the National Resources Commission that all compensation for the 430 mu of land expropriated for the Lower Qingyan Cave and Huilongzhai 回龙寨 projects had been distributed. Top-grade irrigated land was compensated at 180 yuan per mu, while second-grade land received 160 yuan and third-grade land 140 yuan. Nonirrigated lands were similarly divided into three classes, each mu being compensated at 130, 120, and 110 yuan, respectively.[32] Thus, the war and state-building efforts of the Nationalist government left an indelible imprint on the water-based landscape of the Chongqing region at the expense of the ancestral lands of its residents.

With the scarcity of available farmland, local residents believed that land was an intrinsically valuable asset, worth hanging on to even if a given plot was too small to function as a profitable farm.[33] With the exacerbation of inflation and steep rises in land and rice prices in southwestern China during the war, many landowners and tenants were deeply reluctant to transfer their properties to the National Resources Commission. In 1942, the Commission faced strong opposition from landowners in Fuyuan 复元 district, who were dissatisfied with the offered compensation. The sale price of irrigated land had skyrocketed from 180 yuan per mu in 1940 to around 2,000 yuan per mu in 1942, and it was still rising. So, landowners in Fuyuan demanded as much as 4,000

---

[31] Crook et al., *Prosperity's Predicament*, 49.
[32] "Ziyuan weiyuanhui longxihe shuili fadianchang gongchengchu 1939, 1940, 1941 niandu gongzuo baogao" 资源委员会龙溪河水力发电厂工程处 1939、1940、1941 年度工作报告 (Annual Reports of the National Resources Commission Longxi River Hydropower Engineering Office, 1939, 1940, 1941), 02230004000260000001 at the CMA, Chongqing.
[33] Crook et al., *Prosperity's Predicament*, 50.

yuan per mu for the land to be expropriated. But the Commission refused to pay more than the current market price and maintained further that they had the authority to utilize the expropriated lands even before full payment had been made. This announcement startled the landowners and stiffened their resolve to defend their properties. They gathered in protest and attempted to obstruct ongoing construction work. In response, the Commission summoned the local armed police to suppress the resistance.[34] This episode marks the first instance in China in which the state was mobilized to support the exploitation of hydroelectricity due to its connection with national defense.

Besides these initial land disputes during the construction of the hydropower project, tensions between the project managers and the local community persisted during its operation. In the summer of 1943, a severe rainstorm flooded the Taohua River in Duzhou district, destroying a bridge, thirteen houses, and adjacent crops. Many in the local community blamed the flood on the water gates that were part of the Taohuaxi hydroelectric project infrastructure, which were designed to contain water for the generation of electricity. However, because the mechanism was manually operated, the flood revealed that the initial plan had neglected to consider the risk of flooding to upstream communities. In an effort to prevent further damage, plant personnel hired ten people from the neighboring community to assist in lifting the water gates. Despite the rush and the urgency of the situation, only six out of thirty water gates could be successfully lifted, and this is what resulted in the flood. The following day, hundreds of frustrated flood victims, led by the district chief Huang Nanqiao 黄南樵, stormed the hydropower plant and demanded compensation and an upgrade of the water gates. They argued that state construction had to consider the needs of the people and that incidents like this, which harmed the community, ought never to be tolerated by the law.[35]

In its response, the hydropower company took steps to address the local community's concerns by compensating them for their losses of property. It also upgraded the water gates from a manual to a mechanical

---

[34] "Ziyuan weiyuanhui Longxihe shuili fadianchang" 资源委员会龙溪河水力发电厂 (Longxi River hydropower plant of National Resources Commission), 18-31-02-006-01, Modern History Institute Archives of the Academia Sinica (hereafter MHIAAS), Taipei.

[35] "Ziyuan weiyuanhui Longxihe shuili fadianchang" 资源委员会龙溪河水力发电厂 (Longxi River hydropower plant of the National Resources Commission), 18-31-02-005-02, MHIAAS, Taipei.

system of operation to ensure sustainability. This was crucial in ensuring the longevity and viability of the hydroelectricity enterprise while ensuring the wellbeing of the local community.

The flood in the Duzhou district highlights the lack of consideration for the environment and the local community during the construction of state hydroelectric projects in the 1940s. The National Resources Commission, tasked with directing the project, imposed its presence upon the local hydrosphere and did succeed in harnessing the power of the rivers for the state's benefit. But for those living in close proximity to the rivers, this intrusion was to be a double-edged sword. It left them vulnerable to disasters caused by the poor project management.

The 1940s saw the forging of a profound connection between the rivers, farmers, and the nation, as the state exerted its influence through the exploitation of hydropower, administrative reform, the expansion of public schooling, and most significantly, increased taxation and conscription.[36] These were new experiences for the people of southwest China, and they would shape their futures in ways they could never have imagined. Yet, despite the far-reaching impact of these state projects, they have often been overlooked in the larger narrative of China's experience of the war. Set aside the destruction seen in cities like Chongqing, the hydropower infrastructure built along these local rivers may seem unremarkable, but it laid the foundation for far more substantial changes to the nation's water landscape in the following years.

LEARNING FROM THE TVA

Despite the efforts of the National Resources Commission in the southwest region, the shortage of technical personnel and material constituted a significant obstacle to the large-scale expansion of hydropower. In response, the Commission turned to foreign assistance in its efforts to develop the capacity for comprehensive river engineering. This section examines the internationalization of hydropower engineering during World War II. I argue that the rise of hydropower in China was not simply a national effort but was also dependent on international cooperation.

In 1933, as part of Franklin Roosevelt's New Deal, the US Congress passed the Tennessee Valley Authority (TVA) Act, which aimed to address

---

[36] See Crook et al., *Prosperity's Predicament*; Diana Lary, *The Chinese People at War: Human Suffering and Social Transformation, 1937–1945* (Cambridge: Cambridge University Press, 2010).

the issues of flooding and poverty in the Tennessee Valley region. The establishment of the TVA involved the construction of a series of dams along the Tennessee River and its tributaries. While the official goals of the agency initially focused on flood control and river navigation, the focus had shifted by 1935 toward the generation and transmission of electricity.[37]

As noted by David Lilienthal, the chairman of the TVA, in 1944, "No major river in the world is so fully controlled as the Tennessee, no other river works so hard for the people, for the force that used to spend itself so violently is today turning giant water wheels. The turbines and generators in the TVA powerhouses have transformed it into electric energy. And this is the river's greatest yield."[38] Through comprehensive river-valley planning and construction, the once flood-threatened and poverty-stricken Tennessee Valley region was transformed into a thriving and productive one. As news spread around the world, the Tennessee Valley Authority (TVA) became a widely recognized model for other countries to follow. People from countries like Mexico, Chile, England, and Australia visited the valley with the hope of replicating the project's success in their own nations.[39] Lilienthal also noted in the same year that the agency was serving as a "training ground" for foreign technicians. Chinese visitors were particularly noteworthy for the frequency and scale of their visits in the 1940s. This section examines the wartime transnational exchange between China's National Resources Commission and the TVA by analyzing archival and published documents preserved in the United States and Taiwan.

In the 1930s, staff at the Rockefeller Foundation shared the latest information about the TVA with Chinese officials working on rural development, hoping to inspire them with the foundation's liberal vision of social and economic development.[40] In May 1940, Francis Hutchins,

---

[37] Carl Kitchens, "The Role of Publicly Provided Electricity in Economic Development: The Experience of the Tennessee Valley Authority, 1929–1955," *The Journal of Economic History* 74, no. 2 (2014), 389–419.

[38] David Lilienthal, *TVA: Democracy on the March* (New York: Harper & Brothers Publisher, 1944), 16.

[39] River management projects around the world followed the model of the TVA. Examples include the São Francisco River Project in Brazil, the Damodar Valley Authority in India, the Mekong River Commission in Southeast Asia, and the Snowy Mountain Authority in Australia. See Rohan D'Souza, "Damming the Mahanadi River: The Emergence of Multi-Purpose River Valley Development in India (1943–46)," *Indian Economic Social History Review* 40, no. 1 (2003), 81–105.

[40] David Ekbladh, "Meeting the Challenge from Totalitarianism: The Tennessee Valley Authority as a Global Model for Liberal Development, 1933–1945," *The International History Review* 32, no. 1 (2010), 47–67, 60.

the president of Berea College in Kentucky, extended an invitation to Dr. Hu Shi, China's ambassador to the United States, to visit Berea and the nearby Tennessee Valley Authority.

David Lilienthal himself was thrilled to welcome the Chinese ambassador and provided a TVA plane so Hu could view the valley from above. The ambassador was impressed by the TVA's accomplishments and, on his train journey back to Washington, DC he learned of the costs of the project and wrote to Lilienthal: "I was amazed to find that they totaled only $340,000,000, which is less than one-third of what President Roosevelt asked for the day before for emergency defense. How much good mankind could do with the hundreds of millions of dollars that are being spent every day on futile destruction!"[41] The war between China and Japan had inflicted devastating damage on China since 1937, seriously hindering its progress in industrialization. However, Hu had no reservations about his admiration of the TVA's accomplishments. Indeed, the visit had a lasting impact on him, impressing him with the power of social and economic development and inspiring a sense of hope for a better future.

Hu and his compatriots were inspired by this visit to forge a partnership between China and the United States, with the goal of harnessing the power of China's underdeveloped rivers. By the spring of 1941, with the passage of the Lend-Lease Act, collaboration between the TVA and the Chinese government was officially underway. In that year, the TVA dispatched three technical experts from its Health and Safety Division to China to assist with malaria control in connection with the construction of the Burma–Yunnan Railroad. This marked the beginning of a productive partnership between the TVA and the Nationalist government.[42]

In late 1941, following the Japanese attack on Pearl Harbor, the United States declared war on Japan and formally joined forces with the Republic of China. This strengthened technological collaboration between the two countries, as most Americans came to see China as a

---

[41] Letter from Hu Shih to David Lilienthal, dated May 20, 1940, in the Records of the Tennessee Valley Authority, Record Group 142 (hereafter RG 142), General Manager, Board of Directors Admin Files, box no. 345, folder 184C (China) Foreign, stored at The National Archives at Atlanta, GA.

[42] Correspondence between David Lilienthal and Ambassador Hu Shih, September 11, 1941, in RG 142 TVA, General Manager, Board of Directors Admin Files, box no. 345, folder 184C (China) Foreign, stored at the National Archives at Atlanta, GA.

On international medical assistance and progress being made in China, see Brazelton, *Mass Vaccination*; Barnes, *Intimate Communities*.

vital ally in the war effort.⁴³ In response, the National Resources Commission, facing a shortage of engineers and technicians with advanced training for its industrial reconstruction plans, developed a plan to send young and talented engineers and technicians to the United States for practical training. Chen Liangfu 陈良辅, an electrical engineer, was appointed the Commission's representative of the Nationalist government and the supervisor of the trainees in the United States. In the first cohort, thirty-one trainees left China in late 1942, with four students who were already in the United States joining the group there. Six of these trainees were sent to the Tennessee Valley Authority.⁴⁴

The experience of these six trainees at the TVA was an unforgettable one, filled, according to their own accounts, with both professional and personal growth. Among the trainees was Jiang Guiyuan 蒋贵元, an electrical engineer from Jiangdu 江都, Jiangsu province, with a background in civil engineering and extensive experience at the National Defense Strategy Commission and at a number of hydropower plants. His task was to spend six months designing and twelve months building hydropower plants at the TVA. His fellow trainees there, Shi Hongxi, Wang Pingyang, Xie Peihe, and Sun Yunxuan, each brought their own expertise in electrical instruments, power grids, and thermal power.⁴⁵ Even though strict restrictions on non-US citizens' access to the facility during wartime were in place, the US State Department nonetheless recognized the value of training these young, talented engineers and technicians. The Chinese government was also mindful of security and financial concerns: it guaranteed the loyalty of the trainees to the

---

⁴³ Michael Schaller, "FDR and the 'China Question'," in *FDR's World: War, Peace, and Legacies*, ed. David Woolner, Warren Kimball, and David Reynolds (New York: Palgrave MacMillan, 2008), 145–174, 145.

⁴⁴ Letter from Charles Thomson, chief of the Division of Cultural Relations of the Department of State, to David Lilienthal, dated October 26, 1942 in RG 142 TVA, General Manager, Board of Directors Admin Files, box no. 345, folder 184C (China) Foreign, stored at The National Archives at Atlanta, GA. Those six trainees were P. Y. Wang, Y. S. Sun, K. Y. Chiang, P. H. Hsieh, Q. T. Chang, and H. H. Sze. Also see Xue Yi, "Kangzhanshiqi de sanyi xueshe" 抗战时期的三一学社 ("The Sanyi association during the War of Resistance against Japan") *Kangri Zhanzheng yanjiu* 抗日战争研究 2 (2003), 87–107. On the general role of the National Resources Commission, see Greene, *Origin of the Developmental State in Taiwan*.

⁴⁵ Cheng Yufeng 程玉凤 and Cheng Yuhuang 程玉凰 (eds.), *Ziyuan weiyuanhui jishurenyuan fumei shishi shiliao, 1942* 资源委员会技术人员赴美实习史料, 1942 (*Archives on The National Resources Commission Technicians' training in the United States, 1942*) (Taipei: Academia Historica, 1988), 203. (Hereafter, *Fumei shixi shiliao*.)

war effort and covered all of their expenses during their time in the United States.⁴⁶

Jiangsu-born Zhang Guangdou 张光斗, a highly accomplished civil engineer, was selected to receive training in the field of hydropower. After obtaining his bachelor's degree in civil engineering from Jiaotong University in Shanghai, he honed his skills by obtaining a master's degree from both the University of California–Berkeley and Harvard. Despite having the opportunity to continue his education and to obtain a PhD at Harvard, Zhang returned to China in 1937 upon hearing of the beginning of the war between China and Japan. During the war he worked on the Longxi River and Wanxian hydropower projects before he went back to the United States in 1942.⁴⁷

At first, Zhang was going to receive training at the Bureau of Reclamation, another US federal institution responsible for large dam construction. After learning of the advanced hydroelectric construction techniques of the Tennessee Valley Authority, though, and at the suggestion of Chen Liangfu, Zhang changed his plans. Instead of spending his time at the Bureau of Reclamation, he began his training at the TVA in July 1943: Six months of project design and twelve months of construction, followed by a three-month tour of major hydroelectric projects around the United States.⁴⁸

As part of their practical training, all trainees were required by the National Resources Commission to submit monthly technical reports documenting their progress. Zhang Guangdou's time at the Tennessee Valley Authority was focused on gaining a comprehensive understanding of design details, construction, plant layout, and the methodologies used in building hydroelectric plants. In his technical reports, Zhang repeatedly emphasized the vastness and complexity of the science of hydroelectric engineering, noting that fully grasping its intricacies would require the collective effort of a group of individuals, not just one person.

The complexity and profundity of engineering theory and practice reveal how little we know and how much we should know in carrying out actual development. Given the limitations of the capacity and ability of a single individual, a system of division of labor would seem to be indispensable. In each field of engineering, there should be a group of men, organized and coordinated so that

---

[46] Letter from Charles Thomson, chief of the Division of Cultural Relations, Department of State to David Lilienthal, dated October 26, 1942, in RG 142 TVA General Manager, Board of Directors Admin Files, box no. 345, folder 184C (China) Foreign.
[47] *Fumei shixi shiliao*, 26. [48] *Fumei shixi shiliao*, 725.

their studies can cover the whole field and each is able to specialize in one branch. Only cooperation, specialization, and continuous research can succeed in industrial development and in raising our technical standards. Having an individual attempt to undertake an entire field of engineering is very inefficient, if not harmful.[49]

Through his training at the Tennessee Valley Authority, Zhang also came to understand the importance of practical experience in the construction of hydroelectric projects. He felt that China needed the guidance of foreign experts in its early years of large dam construction, so that a new generation of Chinese engineers could be created. In his technical reports, Zhang meticulously studied the design and construction methods of various hydroelectric projects in the Tennessee Valley, noting similarities between American and Chinese projects where they existed.[50] Zhang also kept in regular touch with his supervisor, Chen Liangfu, during his time in America. In their correspondence, they explored opportunities for institutional collaboration between the National Resources Commission and the TVA. As Zhang immersed himself in his training at Knoxville, he observed the TVA's collaboration with engineers from the Soviet Union, who were receiving technical assistance in designing hydroelectric and fossil fuel power plants. He learned that a group of six Russians were working as correspondents at the authority and that all designs were transmitted wirelessly to the Soviet Union, with the costs of this activity covered by the Lend-Lease Act.

Zhang also contributed to the ongoing debate about the development of China's electrical industry. In the 1930s, the Nationalist government attempted to nationalize the electrical industry, but this move faced strong opposition from the private sector.[51] As a hydroelectric engineer, Zhang had strong opinions on this and advocated for a government-led approach to developing the electrical industry, drawing inspiration from the success of the TVA. The technical and financial hurdles facing the Nationalist government in exploiting its hydropower resources in the southwest were significant, and Zhang argued for the benefits of collaboration and institutional support from the state.

Upon learning of the technical and financial assistance that the TVA was providing to the Soviet Union, Zhang shared this information with Chen Liangfu and the National Resources Commission. He worked hard

---

[49] *Fumei shixi shiliao*, 728–731.   [50] *Fumei shixi shiliao*, 742.
[51] On the Nationalist government's effort to nationalize the electrical industry, see Tan, *Recharging China in War and Revolution*.

to convince Chinese authorities to seek similar assistance from the TVA. He felt that the TVA, with its well-trained corps of technical personnel, would be willing to engage in overseas projects since the number of projects in the United States was declining. He pointed out that the National Resources Commission had already conducted hydrological and geological investigations of rivers in the southwest and had enough data to begin the planning and construction of hydropower plants. But the Commission was hampered by its limited engineering capacity and experience. With the close relationship between China and the United States as wartime allies, Zhang saw an opportunity for the Chinese government to accelerate its domestic reconstruction and to train more Chinese hydro-technical personnel.[52] Chen took Zhang's advice and wrote to Lilienthal to inquire about direct technical assistance from the TVA for the design and construction of projects in China. Lilienthal replied that, while the TVA was willing to provide assistance, authorization from the State Department or the Lend-Lease Administration was required. He also expressed concerns about the possible expenses of the project and the availability of staff.

Under the direction of Chen Liangfu, the Commission and its trainees had planned to gather semi-annually in various locations in the United States to discuss their training programs and to tour leading American industries. As Chen mentioned in another letter to Lilienthal, "All of our members in this country have long looked forward to visiting your TVA Project as waterpower development will be one of the top priorities we face in China after the war."[53] In July 1943, the Chinese trainees chose Knoxville, Tennessee – the headquarters of the TVA – as the location for their semi-annual convention. It was more than just the Commission trainees who were interested in the TVA; in December 1943, following the convention in Knoxville, S. D. Ren, the Vice President of the Universal Trading Corporation, and his associates also visited the Tennessee Valley. The Universal Trading Corporation was a branch of the Chinese government acting as its official purchasing agency in the United States. Along with procuring supplies for China's war effort, the company's engineering department also conducted studies for postwar reconstruction projects in China.

---

[52] *Fumei shixi shiliao*, 277–278.
[53] Letter from Chen Liangfu, Representative of Foreign Trade Office of National Resources Commission, to Gordon Clapp, general manager of TVA, dated June 1943, in RG 142 TVA, Records of the General Manager's Office Administrative Files, box no. 45.

During their ten-day stay in the valley, the Chinese delegation observed the TVA's various activities, including dam construction, power generation, water control, the improvement of navigation, soil conservation, food preservation, reclamation of waste land, the manufacture of fertilizer, rural electrification, and malaria control. They visited farms and spoke with local farmers, observing how TVA projects directly improved their quality of life. Impressed by the TVA's slogan "Built for the People of the United States," which was inscribed on all the concrete dams, Ren praised the TVA as a grand experiment in democracy and efficiency. He viewed the TVA's physical constructions as grand monuments of service to the people that would serve as models for generations to come and be replicated around the world.[54]

In early 1944, *Universal Engineering Digest*, a journal established by Chinese engineers abroad, released a special issue on the TVA. The issue provided an overview of the delegation's observations.[55] James Pope, the director of the TVA, praised the authors for the insightful coverage of the TVA that the issue provided. He commented, "I am impressed with how well you have captured the spirit and essence of this program ... It is a clear indication of the future development of resources in China that more Chinese engineers and students have visited the Authority than from any other country outside the US."[56] The visit of the Chinese delegation and the subsequent publication of their reports in *Universal Engineering*

---

[54] Letter from S. D. Ren to David Lilienthal, dated December 24, 1943, in RG 142 TVA General Manager, Board of Directors Admin Files, box no. 345, folder 184C (China) Foreign.

[55] A Chinese Society of Engineers had been formed in 1912, with 148 members under the leadership of Zhan Tianyou. The society's work in establishing the field in China, standardizing engineering education, and promoting a nationalistic agenda was augmented by the activities of the Chinese Engineering Society (Zhongguo Gongcheng Xuehui 中国工程学会), which was founded at Cornell University in 1919 by Chinese students pursuing advanced engineering studies in the United States. The merger of these two groups in 1931 formed a new Chinese Society of Engineers (Zhongguo Gongchengshi Xuehui 中国工程师学会) with some 2,300 members. The drive toward professional autonomy and self-regulation that had marked the earlier engineering association gradually gave way to greater cooperation with, and reliance on, the state that now provided funding for education and certified engineers. Over time their work would be incorporated in and become indistinguishable from the work of the new National Resources Commission, "the engineers' stronghold in the National Government." See Kirby, "Engineering China," 149–150.

[56] Letter from James Pope, the director of the TVA, to S. D. Ren, dated May 6, 1944, in RG 142 TVA General Manager, Board of Directors Admin Files, box no. 345, folder 184C (China) Foreign.

Digest highlighted the interest and potential of the Chinese in learning from the successes of the TVA.

From May 29 to June 2, 1944, another delegation of Chinese officials, led by Chiang Tingfu 蔣廷黻, the Chief Political Secretary of the Executive Ministry and the Chinese delegate to the United Nations Relief and Rehabilitation Administration, visited the TVA. One member of the delegation, Chang Kia-Ngua 张嘉璈, an adviser to the President of the Executive Ministry, was deeply inspired by the visit and expressed his desire to share the experience with his fellow countrymen in a letter of thanks to David Lilienthal.[57] The same year, Lilienthal's book *TVA: Democracy on the March* was published and Lilienthal sent a copy of the book, along with a letter, to Generalissimo Chiang Kai-shek through Vice-President Henry Wallace. In the book, Lilienthal writes about the universal language of the TVA, including its successes in soil fertility, forest management, electricity, phosphate, factories, minerals, and rivers.[58] He expresses pride in receiving visitors from China, including the young Chinese engineers who were receiving training at the TVA.[59]

The visits and training of Chinese at the TVA during World War II and later during the Cold War facilitated the spread of hydro-technological knowledge to China, as well as the idea of modern river-basin development. Although the TVA was highly politicized in the United States and on the global stage, the visits helped to create an unprecedented bond between the two countries.[60] As Chris Courtney and others have pointed out, America's willingness to assist China was not entirely altruistic and its assistance was not always impartial. The TVA's zeal for international

---

[57] Letter from Chang Kia-Ngau to David Lilienthal, dated June 16, 1944, in RG 142 TVA General Manager, Board of Directors Admin Files, box no. 345, folder 184C (China) Foreign. The United Nations Relief and Rehabilitation Administration precedes the formation of the UN in 1945.

[58] Lilienthal, *TVA*, 204.

[59] Letter from David Lilienthal to T. F. Tsiang, Office of Council Member for China, United Nations Relief and Rehabilitation Administration, dated May 9, 1944, in RG 142 TVA General Manager, Board of Directors Admin Files, box no. 345, folder 184C (China) Foreign.

[60] Eric Dinmore, "Concrete Result? The TVA and the Appeal of Large Dams in Occupation-Era Japan," *The Journal of Japanese Studies* 39, no. 1 (2013), 1–38; Arun Kumar Nayak, "The Mahanadi Multipurpose River Valley Development Plan in India," *World Affairs: The Journal of International Issues* 20, no. 4 (2016), 76–93; Fernando Purcell, "Dams and Hydroelectricity: Circulation of Knowledge and Technological Imaginaries in South America, 1945–1970," in *Itineraries of Expertise: Science, Technology, and the Environment in Latin America's Long Cold War*, ed. Andra Chastain and Timothy Lorek (Pittsburgh: University of Pittsburgh Press, 2020).

assistance was contingent on its own institutional interests, which sometimes overshadowed its goals of helping other countries.[61]

In July 1943, Zhang Guangdou continued his training at the TVA. At the same time, the Chief Engineer of the Bureau of Reclamation, John L. Savage, a world-renowned expert in large dam construction, was also visiting Knoxville. Zhang already knew Savage, having completed a three-month internship at the Bureau of Reclamation in 1936, where they had been introduced by his advisor. During their meeting in 1943, Zhang learned of Savage's upcoming trip to India to consult for a hydropower project and took the opportunity to invite Savage to visit China. Without hesitation, Savage expressed his interest in the idea and soon secured approval for the trip from the State Department.[62] Zhang promptly wrote to Chen Liangfu to encourage him to coordinate with the US government to arrange Savage's visit to southwest China for the purpose of hydropower planning.

Christopher Sneddon has explored the spread of high-dam technology and expertise driven by the US Bureau of Reclamation in the post-World War II era in the United States and abroad. However, Sneddon's account of the "concrete revolution" is limited by its US-centric perspective, ignoring the active involvement and the many contributions of Chinese engineers and officials in the success of concrete dam construction in the twentieth century. Throughout his career, Savage built a reputation as a master of concrete dam construction, responsible for iconic structures like the Hoover Dam, the Grand Coulee Dam, Parker Dam, and Shasta Dam. After the completion of the Hoover Dam, the first large dam equipped with a hydroelectric plant, America declared victory over the Colorado River. In an anonymous 1945 article "The Dams that Jack Builds," Savage was highly praised and celebrated for his contributions to the field.[63]

Although brief, Savage's visit to China marked a significant moment in the transnational transmission of hydropower knowledge between the United States and China during wartime. Initially, Savage was invited to inspect and design small to mid-sized hydropower projects near Chongqing. However, because of the Japanese military advance toward Yichang, Savage's scheduled visit to the nearby Three Gorges seemed impossible. During his stay, G. R. Paschal, an economic advisor to Donald Nelson, Chairman of the War Production Board and President Roosevelt's personal representative in China, proposed that the United States assist China in building a large dam on the Yangtze River to

---

[61] Courtney, *Nature of Disaster in China*, 175.   [62] *Fumei shixi shiliao*, 284.
[63] Sneddon, *Concrete Revolution*, 35–36.

harness its vast hydropower potential.⁶⁴ Inspired by this idea, Savage decided to undertake a field trip to the gorges, despite the Japanese military threat.⁶⁵ Chinese intelligence had also intercepted Japanese aerial surveys of the Three Gorges for dam design purposes, making Savage's survey even more important.⁶⁶ It is worth noting here that a group of Chinese engineers had actually conducted the first scientific survey of the Yangtze Gorges in 1932, with funding from the National Defense Commission.⁶⁷

Savage's preliminary report on the Yangtze River project proposed a 225-meter tall concrete straight-gravity dam, capable of producing 10,560 megawatts of hydroelectricity annually. He described the project as a "CLASSIC" (capitalized in the original) that would bring industrial development, employment, and a higher standard of living to China, transforming it from a weak to a strong nation.⁶⁸ At the time, China was in desperate need of a vast amount of electric power to rebuild its industries after the war. The initial requirement was estimated to be two to four million kilowatts, but this would rise to ten to twenty million kilowatts or more as industrialization proceeded and standards of living improved. To meet this challenge, Savage proposed that the United States provide both financial and technical assistance to China to build the

---

⁶⁴ Huaiyun Xu 徐怀云, "Yangzi jiang sanxia gaoba sheji jishi" 扬子江三峡高坝设计纪实 (Yangzi River Three Gorges High Dam Design Document), in *Zhongguo Chang Jiang Sanxia Gongcheng lishi wenxian huibian, 1918–1949* 中国长江三峡工程历史文献汇编, 1918–1949 (*Collected Records of the Three Gorges Project, 1918–1949*), ed., Zhongguo Changjiang Sanxia gongcheng lishi wenxian huibian bianweihui (Beijing: Zhongguo Sanxia chubanshe, 2010), 147.

⁶⁵ Wang Guanglun 王光纶 (ed.), *Qingxishanhe: Zhang Guangdou zhuan* 情系山河: 张光斗传 (*Biography of Zhang Guangdou*) (Beijing: Zhongguo kexuejishu chubanshe, 2014), 64.

⁶⁶ After the Japanese occupation of Yichang in 1940, the Japanese engineer Otani Kozui led a comprehensive survey and produced a design for a high dam at the Three Gorges. It was known as the Otani plan. See Deirdre Chetham, *Before the Deluge: The Vanishing World of the Yangtze's Three Gorges* (New York: Palgrave Macmillan, 2002), 116.

⁶⁷ Kirby, "Engineering China," 151. Decades later, the construction of Gezhouba and the Three Gorges Dam proved that the Chinese engineers' selection of possible dam sites in 1932 was superior to Savage's choice in terms of geology. The Nationalist government probably failed to provide the 1932 report to Savage during his visit in 1944. See Yun Zhen 恽震, "Guanyu Sanxia shuili shoucikance" 关于三峡水力首次勘测 (On the first survey of the hydropower of the Three Gorges), in *Zhongguo Chang Jiang Sanxia Gongcheng lishi wenxian huibian, 1918–1949* 中国长江三峡工程历史文献汇编, 1918–1949 (*Collected Record of the Three Gorges project, 1918–1949*), ed. Zhongguo Changjiang Sanxiagongcheng lishi wenxian huibian bianweihui (Beijing: Zhongguo Sanxia chubanshe, 2010), 144.

⁶⁸ Sneddon, *Concrete Revolution*, 40–41.

Yangtze Gorge hydropower project. The electricity thus generated could be used to produce large quantities of fertilizer, which could then be used to repay the American financial assistance.

Savage presented a compelling argument for the project's significance from both geopolitical and economic standpoints. He emphasized that China, after the devastating war, faced bankruptcy and lacked the financial resources to complete the project on its own. By offering assistance, he felt that the Chinese government would become more appreciative of the United States, thereby strengthening the alliance between the two countries and promoting peace in the Far East. Savage also saw a tremendous economic opportunity for the United States in helping China. A prosperous China would serve as a vast market for American exports and, although providing the aid might require short-term financial sacrifices, the long-term benefits would far outweigh them. He wrote, "A powerful and prosperous China, grateful to the United States for aid given in her rehabilitation, can be both a political and an economic asset." He warned that if the United States failed to provide aid, China's potential industrial resources would eventually be developed by some other country, possibly leading to economic domination by unfriendly nations. He argued that America should be involved in the process of rebuilding China. In his own words, "the first orders to be placed in building up the new China will be in effect the planting of seed, and for this reason we should see that the seed is American, so that the later harvest shall be American also."[69]

From an economic standpoint, Savage acknowledged that China may not have had much to offer in the short-term. However, he believed that a proposed power plant in the country could be used to produce nitrogen fertilizer, a product that was in high demand in American agriculture. Nitrogen production requires a large amount of energy, as it involves extracting nitrogen from air at high temperatures.[70] During the war, the use of nitrates in the production of bombs had taken precedence over the demands of agriculture. According to Savage's report, the annual fertilizer bill in the United States was substantial, reaching $354 million in 1942. Despite this enormous expenditure, American farms still faced shortages of fertilizer, resulting in decreased fertility and lower crop yields.

---

[69] "Hydroelectric Power for China and a Suggested Mode of Repayment to the United States Thereof," in RG 142 TVA General Manager, Board of Directors Admin Files, box no. 345, folder 184C (China) Foreign.

[70] Vaclav Smil, *Enriching the Earth: Fritz Haber, Carl Bosch, and the Transformation of World Food Production* (Cambridge, MA: MIT Press, 2001).

Savage also recognized the environmental and soil conservation benefits of increased fertilizer use. Proper fertilization, he argued, would reduce soil erosion and improve the rural living conditions that left millions of Americans in poverty. High rates of fertilization would also reduce production costs per acre, allowing for better use of farmland and the conservation of forests.[71] However, Savage also recognized that importing fertilizer from China would likely be opposed by American fertilizer producers. In his report, he highlighted the high costs and inadequate levels of production of existing plants in the United States, in an attempt to show that importing cheaper fertilizer from China would be beneficial to American farmers. In sum, Savage proposed that the US government should either directly finance or guarantee private financing for a large hydroelectric power plant on the Yangtze River in China, as he believed it would ultimately bring economic benefits to American agriculture.

Clearly a single engineer, no matter how accomplished, cannot bring a project as complex as the Yangtze Gorge to fruition on his or her own. To achieve his goal, Savage suggested that the Chinese government enter into a partnership with the United States government, in particular involving the Bureau of Reclamation and the Tennessee Valley Authority. Under this agreement, American specialists would aid in project design and in the training of Chinese engineers for the Yangtze Gorge Project. Savage presented his proposal in a report that was forwarded by the State Department to the Tennessee Valley Authority for review. While Savage was enthusiastic about the project's potential, board members at the TVA were less so.

The board of the TVA expressed skepticism toward the idea of financing the enterprise in exchange for nitrogen fertilizer exports. They pointed out that the data Savage presented on fertilizer was inaccurate, as the expected level of nitrogen imports from China (1,300,000 tons) would far surpass the annual agricultural use of nitrogen in the United States (631,000 tons). Further, the production of fertilizer in the United States had significantly increased during World War II and they felt that the wartime surplus would suffice to meet the demand for nitrogen in the country.[72] Thus, the board argued that the proposal should be rejected on both commercial and conservationist bases.

---

[71] "Hydroelectric Power for China and a Suggested Mode of Repayment to the United States Thereof," in RG 142 TVA General Manager, Board of Directors Admin Files, box no. 345, folder 184C (China) Foreign.

[72] "Informal note by Mr. Clapp, Neil Bass, January 11, 1945," in RG 142 TVA Records of the Chairman and The Members of the Board of Directors, box no. 58.

The board did, however, express interest in learning more about the potential for producing cheap fertilizer through exploiting Yangtze hydropower in China. They recognized the high demand for fertilizer in China and other neighboring countries in the Far East and viewed it as a potential market for American investors, as long as the cost was low enough that local farmers could afford to participate.[73] This attitude toward the Yangtze project and China was not uncommon at the time. The American ambassador in China, Clarence Gauss, for example, believed that the Chinese were open to receiving assistance from the United States, but not to giving it. He warned that excessive generosity could harm US interests in China.[74]

The TVA initially hesitated, then, but Savage's efforts, with the support of the federal government, did result in the formal investigation and design of the Yangtze hydropower project by the Bureau of Reclamation together with the Chinese National Resources Commission. In 1945, fifty Chinese technicians traveled to the United States to work on the design, while more than 500 American engineers and around 80,000 unemployed workers were going to be sent to China to complete the ground work.[75] However, planning was suspended in 1947 due to budget constraints and developments in the Chinese Civil War.[76] In his book *Recharging China in War and Revolution*, Tan Ying Jia highlights the drain on resources that the Yangtze River project may potentially have caused for smaller hydropower initiatives in the postwar years.[77] However, it is essential to consider the National Resources Commission's collaboration with the TVA in the larger context of the global growth of hydropower. While it is true that the Yangtze River project and Savage's involvement inspired many Chinese engineers and officials, it is important to acknowledge the American self-interest involved.

---

[73] Letter from E. A. Locke, Jr., on Executive Assistance to Donald Nelson written to David Lilienthal, dated February 14, 1945, in RG 142 TVA General Manager, Board of Directors Admin Files, box no. 345, folder 184C (China) Foreign.

[74] Sally Burt, *At the President's Pleasure: FDR's Leadership of Wartime Sino-US Relations* (Leiden: Brill, 2015), 86.

[75] Zhongguo Changjiang Sanxia gongcheng lishi wenxian huibian bianweihui (ed.), *Zhongguo Changjiang Sanxia gongcheng lishi wenxian huibian, 1918–1949* 中国长江三峡工程历史文献汇编, 1918–1949 (*Collected Record of the Three Gorges Project, 1918–1949*) (Beijing: Zhongguo Sanxia chubanshe, 2010), 135.

[76] For more detail on the early efforts of the Three Gorges project, see Liangwu Yin, "The Long Quest for Greatness: China's Decision to Launch the Three Gorges Project" (PhD Dissertation, Washington University, St. Louis, 1996).

[77] Tan, *Recharging China in War and Revolution*, chapter 5.

Despite the project remaining in the early stages of geological and hydrological investigation and preliminary design, the training and research materials provided to Chinese technical personnel by their American counterparts were invaluable assets in China's development of capacity in dam construction and river valley management. This collaboration illuminates the transnational dimensions of the making of a hydropower nation in twentieth-century China.[78]

ENVISIONING "CHINA'S TVA"

Not everyone was thrilled by Savage's proposal to build a massive dam on the Yangtze River. Xu Ying 徐盈, a Communist journalist, expressed his skepticism in this way:

> Construction and destruction are two sides of the same coin. Every gain comes with a cost ... How much land will be flooded? How many people will be displaced? How can we balance industrial development with the livelihoods of our people? ... We should learn from others' experiences, but we must not feel inferior. Our starting point must be the reality of China. There is no need to bear such an unbearable cost just to solve one problem – the fertilizer shortage.[79]

These concerns demanded serious consideration and scrutiny.

Similarly, Keh Chi-yang, a geographer who had obtained a PhD from the University of Michigan, conducted his own comprehensive study of the TVA and its potential impact on China's water-control program. He wrote, "The strength of a nation is directly proportional to the amount of its resources under development and utilization."[80] During the chaotic 1940s, Keh firmly believed in the idea of harnessing all resources, including water, land, and other natural resources, to aid China's postwar reconstruction. The full development of all of a nation's resources was a crucial but difficult task, especially for a country with as vast a territory as China's. Keh argued that the best approach was to reorganize political units based on river basins or physiographic regions. In his view, the TVA was the ideal model for China's task of total resource development.

---

[78] See David Pietz, *Engineering the State: The Huai River and Reconstruction in Nationalist China, 1927–1937* (New York: Routledge, 2002); Pietz, *Yellow River*.

[79] Zhongguo Changjiang Sanxia gongcheng lishi wenxian huibian bianweihui (ed.), *Zhongguo Changjiang Sanxia gongcheng lishi wenxian huibian, 1918–1949*, 132.

[80] Keh Chi-yang, "The TVA Program and the Water Control Program for China," 1. In RG 142 TVA, Records of the Chairman and The Members of the Board of Directors, 1939–1957, box no. 58.

Keh noted the TVA's success in developing both the water and the land resources of the Tennessee Valley. He described how the water control program in the river channel was managed through the construction of dams for flood control, navigation, and the generation of power, and that land resources were developed through various practices like reforestation, crop rotation, grass cover, and terracing, as well as fertilizing soil and liming in order to conserve water and soil.

Keh highlighted in particular the TVA's innovative approach, which he referred to as "multipurpose hydro-technical innovation," specifically the concrete dam, as a key factor in its success: "The simultaneous achievement of flood control, navigation, and power production through a series of multiple-purpose dams is one of the defining characteristics of the TVA program." In other words, total resource development required the use of multipurpose hydro-technical innovations like the concrete dam.

Despite noting relevant similarities between China and the United States, Keh Chi-yang highlighted cultural and demographic differences between the two nations as possible obstacles to replicating the success of the TVA in China. He noted that China faced heavy population pressure, with only seven acres of land for each person according to estimates, while the United States had a population density of 15 acres per person, less than half that of China. These differences would, Keh thought, present major challenges in terms of reservoir clearance and resettlement in large hydropower projects, particularly in terms of the economic, social, and public health consequences of inundating productive alluvial lowlands for reservoirs. Keh explained that such an activity would impact China's food production, involve the removal and relocation of families, cemeteries, institutions, and transportation and communication facilities, and exacerbate hazards like malaria.[81]

On the basis of his experience with the Tennessee Valley Authority, Keh believed that the loss of agricultural land could be offset by using fertilizers and soil conservation practices in upland areas. As had been demonstrated by farmers in the Tennessee Valley, the use of fertilizers led to higher crop yields per acre. Farmers were also assisted in adapting to upland and hillside farming through the implementation of soil conservation practices like contour farming, terracing, growing legumes and grass, and the proper use of fertilizers. To address the loss of arable land, Keh proposed the reclamation of uplands in western and northwestern

---

[81] Keh, "TVA Program," 22.

China's arid and semi-arid regions by government programs encouraging affected farmers and their families to migrate to these areas for reclamation projects. This transregional resettlement of population was not a new idea. The Nationalist government had already planned to develop its underdeveloped northwestern region in this way during the War of Resistance. However, defeat in the civil war meant that the opportunity to put these plans into action was lost. A decade later, the Communist government implemented this proposal for transregional demographic engineering, although it was not directly influenced by Keh.[82]

Keh's study clearly identified the challenges of upland reclamation in the west and northwest. He expected that the affected families would resist relocation because of the scarcity of water for irrigation and the lack of adequate farm machinery, factors that would render reclamation efforts extremely difficult. Beyond these environmental and materialistic obstacles, Keh highlighted two further cultural factors that would influence the attitudes of Chinese farmers toward resettlement. First, they possessed a strong attachment to their ancestral land and were closely tied to large family systems.[83] Second, opposition to the removal of cemeteries was likely to be strong because to the cultural significance of burial sites. Geomancers, consulted to determine the best location for graves based on the influence of wind and water, were often employed by families to choose the final site. Once the location had been selected, families usually kept the graves in place and did not allow them to be moved.[84]

Despite these challenges, Keh concluded with a strong argument for the need to develop China's resources fully. The power of a nation, he argued, is directly linked to the development and utilization of its human and natural resources. He stated that if either of these resources did not function properly, the power of the nation would suffer. He believed that if China was to achieve total resource development, nationwide planning for unified development would be required. The total development of China's resources would not only bring direct benefits to the nation, but would also increase its purchasing power, leading to prosperity.[85] It is unclear how widely Keh's ideas were circulated in the 1940s, but the

---

[82] See Diana Lary, *Chinese Migrations: The Movement of People, Goods, and Ideas over Four Millennia* (Lanham: Rowman & Littlefield, 2012); Gregory Rohlf, *Building New China, Colonizing Kokonor: Resettlement to Qinghai in the 1950s* (Lanham: Lexington Books, 2016).

[83] Keh, "TVA Program," 32.   [84] Keh, "TVA Program," 34.

[85] Keh, "TVA Program," 45.

concept of total resource development – as demonstrated by the TVA – had taken root in the Chinese engineering community. It was not until the 1950s that the Communist government, with the support of the Soviet Union, had both the political will and technical ability to transform China's rivers in the name of total resource development.

CONCLUSION

Richard Tucker and Edmund Russell have argued that demand for resources has had a greater impact on shaping the relationship between humans and nature than the weapons and battlefields of wars.[86] This statement is exemplified by the role of hydroelectric technology in the processes of industrialization and militarization in the twentieth century. All of the hydropower projects in the southwest were involved to some extent in the War of Resistance against Japan. The plants on the Longxi River, for instance, were primarily used for national defense, supplying energy to the munitions and smelting industries in Changshou. On the other hand, the Liangtan River project, which was commercially operated, sought to provide electricity to households, social institutions, and stores in the resettled community in Beibei, which was comprised of populations displaced by the war.[87]

To a certain extent, the war accelerated the growth of China's hydroelectric industry in the southwest, as the demand for energy was high.[88] In line with Grace Shen's analysis of geology and war in *Unearthing the Nation: Modern Geology and Nationalism in Republican China*, the wartime experience spurred the National Resources Commission to greater efforts.[89] Despite these efforts, though, it should be noted that the Nationalist-controlled areas generated only around 8 percent of the electricity produced in the occupied areas.[90] Although the capacity of

---

[86] Richard P. Tucker and Edmund Russell (eds.), *Natural Enemy, Natural Ally: Toward an Environmental History of War* (Corvallis: Oregon State University Press, 2004), 4.
[87] Other hydropower plants built by the Nationalists during the war include the Nanqiao 南桥 Hydropower Plant in Kaiyuan 开远, Yunnan province and the Tianmen River (天门河) hydropower plant in Tongzi 桐梓, Guizhou province.
[88] Morris Bian argues that the crisis of the Second Sino-Japanese War led to the development of the state-owned ordnance industry and other heavy industries. See Bian, *Making of the State Enterprise System in Modern China*.
[89] Grace Yen Shen, *Unearthing the Nation: Modern Geology and Nationalism in Republican China* (Chicago: The University of Chicago Press, 2014).
[90] Kirby, "Chinese War Economy," 192–193.

these projects seems modest by today's standards, they did mark the beginning of state-led hydropower engineering in modern China.

The War of Resistance against Japan required an immense amount of energy. Unlike the famine and environmental devastation caused by military activities in Henan studied by Micah Muscolino, small hydropower projects had a less intrusive and destructive impact on the environment. As Muscolino writes, "Militaries must constantly find new sources of useful energy and develop more effective mechanisms for handling large energy flow."[91] The use of surface water has been a major element of warfare since ancient times. However, not all of its effects on the environment were simply destructive. The Nationalists breached a dike to flood the Yellow River as a strategy to slow down the Japanese army in 1938. When they established the hydropower projects in the southwest that I have discussed in this chapter, they were trying to harness water in a more positive manner. Still, it would be inaccurate to assert that humans controlled natural rivers effectively. In these early scenarios, all projects had to accommodate the dry season. As Richard White observes, these rivers were part of energy systems that retained their natural qualities even after they were altered by human intervention.[92] When the rivers or reservoirs didn't contain enough water to turn the turbines, users of electricity had to adapt their daily routines to the unpredictable flow of the rivers.

William Kirby has highlighted the crucial role played by Chinese engineers in the survival of Nationalist China during its eight-year war against a technologically more advanced enemy.[93] However, these engineers did not operate in a vacuum. World War II greatly affected the transfer and exchange of hydropower technology between the United States and Nationalist China. One result of the wartime alliance was that China became the foreign country with the highest number of trainees and visitors to the Tennessee Valley in the 1940s. Both nations' goals in the postwar period were centered on reconstruction and peace and the collaboration of the National Resources Commission with the TVA and the Bureau of Reclamation reflected the global nature of China's experience of World War II. The American journalist Willard Espy noted in a report on the role of the TVA in the world that these projects were not just engineering feats, they were also aimed at reducing conflict in a variety regions, including Palestine, Greece, India, and China. Espy proclaimed

---

[91] Muscolino, *Ecology of War in China*, 9.   [92] See White, *Organic Machine*.
[93] Kirby, "Engineering China," 152.

that these projects were "dreams to put democracy on the march – these are dams to hold back the floods of war."[94] However, as Hans Van de Ven points out, the internationalization of the War of Resistance did not entirely bring the desired military benefits to the Nationalists.[95] Despite the best efforts of people like Zhang Guangdou, John L. Savage, David Lilienthal, and many engineers, the Chinese Civil War that followed in the late 1940s disrupted the Nationalist government's postwar reconstruction plans. Some engineers fled to Taiwan. Others chose to stay and work with the Communists.[96] Despite profound domestic changes and geopolitical shifts, the planning and design work, construction experience, personnel training, and technology transfers that occurred during the war established a technocratic foundation for the country's subsequent emergence as a fledgling hydropower nation. However, the social and environmental impact of the Longxi River project, among others, also revealed the darker side of this development. While the scale of infrastructure projects, social and environmental impacts, and political ambitions were limited in the Republican period, they would escalate to a whole new level in the People's Republic.

---

[94] Willard R. Espy, "Dams for the Floods of War," *New York Times Magazine*, October 27, 1946, 12–13.

[95] Hans van de Ven, *China at War: Triumph and Tragedy in the Emergence of the New China* (Cambridge, MA: Harvard University Press, 2018), 178.

[96] Of the six trainees sent to the Tennessee Valley Authority, five – Jiang Guiyuan, Zhang Guangdou, Shi Hongxi, Wang Pingyang, and Xie Peihe – stayed in mainland China. The Nationalist government sent Sun Yunxuan to take over the electrical industry left by the Japanese in Taiwan in 1945. Sun stayed in Taiwan as a leading engineer and government official until his death.

PART II

THE SOCIALIST BOOST

3

The Making of Red Hydro Technostructure

In 1954, a pamphlet was published for the Chinese public, beckoning them to an exploration of the grandeur of hydropower. Gone were the days of seeing rivers as ominous and dangerous beasts, wreaking havoc with their floods. The Communist Revolution, with the generous aid of the Soviet Union, brought forth a new appreciation of these waterways as valuable national treasures. This paradigm shift, the pamphlet proclaimed, was nothing short of a revolution in river management. Under the watchful eye of Chairman Mao and the Communist Party, the rivers of China were about to undergo a profound metamorphosis. No longer would they be left to wander wild and free, their power and potential left untapped. They would instead be tamed, harnessed, and put to work for the benefit of the Chinese people. The Party's plan was simple yet profound: to construct dams and reservoirs, thereby giving the rivers purpose and direction. These waterways would be brought under control, given a leader and an organizational form, and commanded to perform tasks of great importance. This pamphlet served as a clarion call, enjoining the Chinese people to join in a grand endeavor to harness the power of their rivers and usher in a new era of progress and prosperity.[1]

Despite misconceptions about the limited hydropower construction that had taken place during the Republican period, or perhaps because of deliberate neglect of those achievements, the idea of building concrete dams and transforming the might of rivers into electricity was still a new one to most Chinese people in 1949. Following victory in the civil war,

---

[1] Chang Jue 常珏, *Shuili Fadian* 水力发电 (*Hydropower*) (Beijing: Tongshuduwu chubanshe, 1954), 2–3.

Mao and the Communist Party set their sights on a grand vision of industrialization and improving the lives of the people. As they pursued their goal of socialism, every element of society – including China's rivers – needed to be brought into order and made to perform their duties with purpose. During the War of Resistance against Japan, hydropower development was seen as vital to the very survival of the nation. When the concrete revolution merged with the socialist revolution, hydropower was thrust to the forefront and would come to serve as both a symbol of the new socialist polity and a justification of its ideology. Although China's installed hydropower capacity during the Maoist period remained modest compared to countries like the Soviet Union and the United States, the union of socialism and water engineering created a hydropower nation characterized by enormous concrete feats of engineering and mass participation.

Despite the sweeping changes introduced by the Communist regime in 1949, the new state inherited much of the Nationalist regime's expertise and infrastructure in hydropower. As they set their sights on building a powerful technostructure, the cornerstone of a hydropower nation, people within the Communist state embarked on a series of institutional and ideological efforts to develop this important resource.[2] With the help of the Soviet Union, these Communists sought to institutionalize hydropower within the framework of the socialist system. During these years, engineers experienced a unique set of challenges and also opportunities as they navigated the Maoist era and its shifting demands. As David Pietz notes, the significance of comprehensive planning and development in the management of the Yellow River cannot be overstated.[3] However, it was the actions of specific individuals that brought these ideas to life. The party needed to bring engineers, with their expertise and know-how, under its control before it could move forward with its ambitious plans to tame the rivers.

This chapter examines three categories of people who played a role in this effort: party cadres, experts from the old society, and the "red engineers."[4] Through the stories of individuals like Li Rui, a revolutionary cadre, and Huang Yuxian and Zhang Guangdou, experts who

---

[2] Kendall Bailes, *Technology and Society under Lenin and Stalin* (Princeton: Princeton University Press, 1978), 15. According to Bailes, the "technostructure" includes applied scientists, engineers, agronomists, and technicians – all those who possess the specialized knowledge of technology necessary for advanced material production in a society.

[3] Pietz, *Yellow River*.

[4] I borrow the term "red engineer" from Joel Andreas, *Rise of the Red Engineers: The Cultural Revolution and the Origins of China's New Class* (Stanford: Stanford University

worked for the Nationalists before 1949 (Zhang later became a red engineer himself), we gain a deeper understanding of the challenges and triumphs of this pivotal time in China's history. In revisiting the revolution in education at Tsinghua University during the Cultural Revolution, Joel Andreas has shed light on the rise of "red engineers" in China.[5] This chapter provides a glimpse into the complex interactions between these groups, whose characteristics were defined and continually redefined in the political movements of the Maoist era. As I will show, the fluidity of these identities highlights the importance of both specific human interactions and Maoist politics in the making of China as a hydropower nation.

As discussed in Chapter 2, the National Resources Commission represented the rise of technocrats in the Nationalist party-state system, whose approach to hydropower was guided by developmentalist and technological priorities in the context of national survival and reconstruction. With the support of the socialist party-state and its technostructure, supported and led by top party leaders like Premier Zhou Enlai, the foundations of the hydropower nation became more firmly established. However, by the late 1950s, revolutionary politics had become so pervasive in Chinese society that the question of hydropower was inevitably politicized, which destabilized and undermined the very system that had brought it to its high point. Despite the limitations of the sources available to investigate this process, this preliminary sketch of hydropower development in Mao's China aims to illuminate the interplay between Maoist politics and the experiences of engineers in their efforts to create a hydropower nation.

LI RUI AND THE INSTITUTIONALIZATION AND PROPAGATION OF HYDROELECTRICITY

Arunabh Ghosh's examination of the role of statistics in the socialist economy in the 1950s underscores the significance of the formation of state institutions in our understanding of the early history of the People's Republic of China (PRC).[6] In contrast to previous scholarship, which primarily focuses on "campaign time," Ghosh highlights the need for a more comprehensive approach to understanding the formation of the

---

Press, 2009). In Andreas' work, it refers to professionally trained engineers who complied with the ideology and policies of the Chinese Communist Party.
[5] Andreas, *Rise of the Red Engineers*.
[6] Arunabh Ghosh, *Making It Count: Statistics and Statecraft in the Early People's Republic of China* (Princeton: Princeton University Press, 2020).

socialist state and its institutions.[7] In 1950, a hydroelectric planning group was established under the Fuel Industry Ministry, replacing the Hydroelectricity Survey Bureau of the Nationalist government. This planning group was later given formal status as the Hydroelectric Engineering Bureau. At that point it employed only thirty-three hydroelectric technicians, most of whom had been retained from the National Resources Commission. Huang Yuxian, the former director of the Nationalist government body responsible for hydropower, was appointed as its director. Initially, the bureau was limited to conducting surveys and providing technical consultation because ongoing political instability prevented it from undertaking or supervising projects. These tasks were instead carried out by local Communist authorities or under military supervision.

In the autumn of 1952, after serving as the director of propaganda in Hunan province since 1949, Li Rui made his way to Beijing. His transfer was triggered by the recent reappointment of his supervisor, Huang Kecheng 黄克诚, a distinguished former general of the Red Army and party secretary of Hunan. Li Rui's passion for economic construction, nurtured by his studies in mechanical engineering at Wuhan University in the 1930s, led him to request a transfer from the propaganda office to the sector of government that handled industrial construction. Because of his personal relationship with Chen Yun 陈云, the influential Vice Premier responsible for state finance and economic planning, Li's request was promptly approved, and he was quickly appointed the new director of the Hydroelectric Engineering Bureau.[8]

Upon his appointment, Li Rui faced the daunting task of elevating the status of hydropower within the socialist bureaucracy and the economic planning system. He soon discovered that, compared to thermoelectricity, hydroelectricity was neglected and significantly underfunded within the Fuel Ministry. Determined to rectify this perceived imbalance, Li raised questions about the allocation of resources and undertook to raise the profile of hydropower. He was informed that Soviet experts had advised against the large-scale development of hydroelectric projects, citing their uneconomical nature and higher cost compared to thermoelectric facilities, which could be constructed more quickly and with lower initial

---

[7] See Gail Hershatter, *The Gender of Memory: Rural Women and China's Collective Past* (Berkeley: University of California Press, 2011), 3–4.

[8] Between 1945 and 1949, Li Rui worked as a secretary for Gao Gang and Chen Yun in northeast China.

investment.⁹ Despite these obstacles, Li recognized the long-term benefits and the renewability of hydropower and sought to bring these advantages to the attention of policymakers. At that time, though, the low carbon emissions and other ecological benefits of hydroelectricity were not yet of significant concern to planners in China.

Li Rui's passion for hydropower was met with bureaucratic indifference, dashing his hopes and leaving him frustrated. Yet he held fast to the belief that only an autonomous institutional entity could bring about comprehensive progress in the field of hydroelectricity. Determined to see change occur, Li took the bold step of writing a letter to Chen Yun, bypassing his immediate supervisor at the Fuel Ministry. In the letter, he expressed his frustration with bureaucratic restrictions and called for the establishment of an independent hydroelectric institution with its own budget allocation. His persistence paid off. With the support of Chen Yun and approval from the senior Communist leader Bo Yibo, who was then the Vice Premier, the Hydroelectric Engineering Bureau was renamed the Chief Hydroelectric Construction Bureau in April 1953. As an independent body, the bureau built its own laboratory and established five regional offices across the country.¹⁰ This new independence provided a foundation for comprehensive progress in the field of hydroelectricity.

In 1955, the central government of China undertook to increase the specialization of its ministries with a view to improving administrative efficiency in industry. The Fuel Ministry was split into three separate entities: the Electric Industry Ministry, the Coal Industry Ministry, and the Petroleum Industry Ministry. The Electric Industry Ministry then created two further subordinate bureaus: the Hydroelectric Bureau and the Thermoelectric Bureau. With the establishment of these bureaus, hydroelectricity came to be seen as equal, if not superior, to thermoelectricity, at least from an institutional standpoint. Li Rui and a small group of technical consultants in the bureau played a key role in elevating the status of hydroelectricity within the administrative system of a command economy. What enabled Li Rui to achieve such institutional recognition

---

⁹ Li Rui, *Li Rui koushu wangshi* 李锐口述往事 (*Memoir of Li Rui*) (Hong Kong: Dashan wenhuachubanshe, 2016), 309–310.

¹⁰ Li, *Li Rui koushu wangshi*, 301. The six institutes were the Northeast, Southwest, and Eastern China Hydropower Engineering Bureaus, the Central South Survey Office, the North China Office, and the Northwest Hydropower Preparation Office. In 1957, the Chief Bureau set up eight survey and design institutes in Wuhan, Beijing, Shanghai, Chengdu, Changchun, Guangzhou, Kunming, and Lanzhou. These institutes were responsible for surveys and the design of hydropower projects in their respective areas.

for hydroelectricity? It is generally believed that his personal connection with Chen Yun, a prominent member of the central party elite, was a decisive factor. Li and his colleagues' tireless efforts to promote hydroelectricity among top leaders certainly also contributed to its appreciation and recognition.

Li Rui's background as a newspaper editor and propaganda cadre had equipped him with skills around effective communication and consensus-building. These skills proved invaluable in his new role as the head of the Chief Hydroelectric Bureau in Beijing. Under Li's leadership, the bureau launched a campaign to raise awareness and promote the significance of the development of hydroelectricity among both the general public and Communist cadres. With his knowledge of the power of persuasive communication, Li was able to gather support and rally key stakeholders behind his cause. This, in turn, helped lay the foundation for the eventual institutional promotion of hydroelectricity.

Starting in 1953, the Chief Hydroelectric Bureau set out to educate the public and state policymakers about the basics of hydroelectric power through the publication of the magazine *Shuili Fadian*. The magazine aimed to introduce knowledge about hydroelectricity not only to the general public but also to key members of the National Planning Commission. To make the concept of hydroelectricity more accessible, the bureau collaborated with the Beijing municipal government to build a small-scale hydropower project for exhibition purposes. Taking advantage of an existing waterfall between Beihai and Shichahai, a 10-kW hydroelectric project was built at Beihai Park and named the "Young Pioneer Hydroelectric Station." This station served as an educational tool aimed at school children, helping them understand the fundamental principles of the generation of hydroelectric power. Through this project, the Chief Hydroelectric Bureau sought to spark the interest of future generations in the hydropower industry, inspiring them to pursue careers in it.[11] By bringing hydroelectricity to life through hands-on experience, the bureau hoped to cultivate a broader understanding of and appreciation for this source of energy.

Denise Ho has argued that, during the Mao era, exhibitions in China served as powerful symbols of state authority and taught citizens how to

---

[11] Wang Yong 王永, "Shoudu Beihai gongyuan shaonian xianfengdui shuidianzhan luocheng" 首都北海公园少年先锋队水电站落成 (The Young Pioneer Hydroelectric Station at Beihai Park in the Capital Has Been Completed), *Shuili Fadian* 水力发电 (*Hydropower*) 11 (1956), 38.

participate in the revolutionary process.[12] Similarly, in the realm of economic development and infrastructure construction, the state sought to inspire enthusiasm and showcase its legitimacy through the creation of exhibitions. By 1955, China was estimated to have approximately 540 gigawatts of hydropower potential from its 1,598 rivers, with 300 gigawatts being practically exploitable. This placed the PRC second only to the Soviet Union in terms of global hydropower potential.[13] To highlight these impressive reserves and its own accomplishments, in 1957, the Chief Hydroelectric Bureau organized an exhibition of national hydropower achievements at Beihai Park in Beijing.

The exhibition was designed to make the tremendous potential of China's hydropower clear through a model of the country's major rivers. It showcased projects that were either under construction or in the planning stages, painting a picture of a promising industrial future once China was able to fully harness the power contained in its rivers. Over the course of several weeks, the exhibition attracted over 300,000 visitors and was seen by key central government leaders like Zhou Enlai and Zhu De.[14]

THE "HYDROELECTRICITY FIRST, THERMOELECTRICITY SECOND" POLICY

At the Nanning conference in January 1958, Mao Zedong summoned Li Rui to discuss Li's opposition to the Three Gorges Project proposal. Despite his passion for the development of hydroelectricity, Li believed the project was too large for China's current economic reality.[15] During this meeting, Li tried to educate Mao on the significance of hydroelectricity to China's economy and the benefits of prioritizing hydroelectricity over thermoelectricity in power industry planning. He pointed out that while many regions in China, especially in the south, lacked coal mines, they had abundant hydropower resources. Li also brought Mao's attention to the inconsistent views on water management between the Water Conservancy Ministry and the Electricity Ministry, which was hindering

---

[12] Denise Y. Ho, *Curating Revolution: Politics on Display in Mao's China* (Cambridge: Cambridge University Press, 2018).
[13] Vaclav Smil, *China's Energy: Achievements, Problems, Prospects* (New York: Praeger Publishers, 1976), 69.
[14] Li, *Li Rui koushu wangshi*, 307.
[15] Unlike Li Rui, Lin Yishan 林一山, the director of the Yangtze River Basin Planning Commission, was a proponent of the Three Gorges dam. For more detail on the debate between Li and Lin, see Yin, "Long Quest for Greatness," 334-359.

the progress of his hydro-electrification plan.[16] He emphasized the lower operating costs of hydroelectricity and its potential for long-term economic growth.[17] Mao was impressed by Li's arguments and immediately adopted the "Hydroelectricity first, thermoelectricity second" policy. He also ordered that the Water Conservancy Ministry be merged with the Electricity Ministry, thus forming the Water Conservancy and Electricity Ministry, giving it the responsibility of coordinating water conservancy and hydroelectric construction. Through this institutional change, Mao and his technical cadres aimed to implement the principle of multipurpose exploitation to drive agricultural and industrial growth.

During the first Five-Year Plan, implementing the policy of prioritizing hydroelectricity over thermoelectricity involved challenges in funding, technology, and personnel.[18] Unlike Lenin's prioritization of the electrical industry in the Soviet economy, Chinese economic planners followed the "load equilibrium theory," according to which electrical capacity would be built according to planned industry demands.[19] However, this approach failed to meet increasing demand from factories, which led to power shortages for many nonessential industries and private consumers. Energy shortages, especially that of electricity, became a hindrance to the socialist state's industrialization plans.[20]

To address this issue, the state pursued a simple solution: building more power plants, including both hydroelectric and thermal facilities. In 1958, after the merger of the Water Conservancy Ministry with the Electricity Ministry, Liu Lanbo 刘澜波, the former minister of the Electricity Ministry and now vice minister of the new Water Conservancy and Electricity Ministry, proposed a supply-oriented approach. This strategy emphasized making the electrical power industry a priority in the state's Five-Year Plans. Technical experts in general believed that electricity should be at the forefront of China's industrialization, driving the growth of other industries. However, completing large

---

[16] On the idea of "river multipurpose development," see Pietz, *Yellow River*, chapter 3.
[17] Li, *Li Rui koushu wangshi*, 331.
[18] At the Eighth Congress of Soviets in 1920, Lenin announced that "Communism is Soviet power and the electrification of the whole country." For more details, see Rassweiler, *Generation of Power*, 12–29.
[19] Liu Lanbo 刘澜波, "Wei quanguo chubu dianqihua er fendou" 为全国初步电气化而奋斗 (Fight for National Preliminary Electrification), *Renmin Ribao*, 21 June 1958.
[20] Li Daigeng 李代耕, *Xinzhongguo dianli gongye fazhan shilue* 新中国电力工业发展史略 (*Historical Outline of the Electrical Industry in New China*) (Beijing: Qiyeguanli chubanshe, 1984).

projects like the Sanmenxia Hydropower Project took a long time, and unexpected technical difficulties delayed things further and reduced their capacity electricity generation. As a result, it was very difficult for the state to solve the problem of electrical shortages in a short period of time.

On the other hand, Li Rui argued against the hasty construction of hydropower projects. After the famous Lushan Plenum in July 1959, Li was labeled a "right opportunist" and an associate of Peng Dehuai, a senior Communist military leader who was criticized by Mao at the Plenum as the head of a "counter-party clique."[21] Mao was determined to eliminate anyone who challenged his authority. During the Anti-Rightist Movement in 1957, Mao argued that 10 percent of the population – or about sixty million people – were "determined enemies of socialism." Criticisms from the intelligentsia (a category that included engineers) were perceived as a threat to the authority of the party. If members of the intelligentsia were found to have exhibited hostile activity, they faced administrative persecution. The struggle against counterrevolution was, for Mao, a long-term process and, to an extent, dependent on the success of the economic development of the PRC. Mao felt that, if China caught up with the United Kingdom and the United States in industrial production, the majority of these people would come to accept socialism. The CCP (Chinese Communist Party) aimed to convince these doubters through a combination of persuasion and repression. People labeled "rightists" were not allowed to hold management positions in offices, factories, or collective farms. They could, however, still be utilized in the construction of socialism, like how some use could be found for "scraps in factories."[22]

After he was denounced as a right opportunist, Li Rui's efforts to promote hydroelectricity were largely ignored and he was sent to the "northern wilderness" for reeducation through labor.[23] Many other hydroelectric engineers, like Luo Xibei 罗西北, faced similar criticism because of their association with Li Rui.[24] The effect was the undermining of the only-just-established professional institution for hydroelectricity

---

[21] See Li, *Li Rui koushu wangshi*.
[22] Record of conversations between the Polish delegation and PRC leader Mao Zedong, Wuhan, April 1, 1958. Wilson Center Digital Archives.
[23] Wang Ning, *Banished to the Great Northern Wilderness: Political Exile and Reeducation in Mao's China* (Ithaca: Cornell University Press, 2018).
[24] Yan Qiu 燕秋, *Wo jiale ge lieshi yigu: ji Luo Xibei de shuidian shengya* 我嫁了个烈士遗孤：记罗西北的水电生涯 (*I Married a Martyr's Orphan: Luo Xibei's Hydro Career*) (Beijing: Zhongguo dianli chubanshe, 2002).

during the period of the Great Leap Forward and Cultural Revolution. Despite this setback, hydroelectricity policy was later amended to include the development of both hydro- and thermal electricity, based on local conditions.²⁵ However, this shows that the socialist hydropower technostructure was vulnerable to political interference during the Maoist period. The endorsement of party leaders enhanced the state's capacity for hydropower development but being on the wrong side of political winds could be devastating for the functionaries of the technostructure.

LEARNING FROM THE SOVIET UNION

Despite the many geopolitical changes across the 1949 divide, transnational cooperation continued to nurture the technostructure of the Chinese hydropower nation. The animosity between the PRC and the United States made collaboration with American hydropower engineers impossible. So, China turned to the Soviet Union to enhance its technological capability to transform its rivers. In an unsent letter to Aleksei Ivanovich Rykov, the Premier of the Soviet Union from 1924 to 1930, the Russian engineer Peter Palchinsky asserted that hard science and technology were more important in shaping society than the "soft" ideology of communism. The dominant trend of the twentieth century, he wrote, "is not one of international communism, but of international technology. We need to recognize not a Komintern, but a Tekhintern."²⁶ Indeed, the Soviet Union not only supported the revolution in China but also provided crucial technical assistance to its industrialization efforts.

The outbreak of the Korean War in the early 1950s brought about significant changes to the international environment in which China operated. To secure an alignment with the Soviet Union, China implemented the "leaning to one side" policy and began referring to the Soviet Union as its "elder brother." This evoked a cultural tradition in which a younger sibling acknowledged the authority of an elder brother, who in turn was expected to nurture and safeguard the younger one.²⁷ In the

---

²⁵ Zhongguo Shuili fadianshi bianji weiyuanhui 中国水力发电史编辑委员会(ed.), Zhongguo Shuili fadianshi (*The History of Hydroelectric Power in China*), vol. 1 (Beijing: Zhongguo dianli chubanshe, 2004), 118.

²⁶ Loren R. Graham, *The Ghost of the Executed Engineer: Technology and the Fall of the Soviet Union* (Cambridge, MA: Harvard University Press, 1993), 43.

²⁷ See Jian Chen, *Mao's China and the Cold War* (Chapel Hill: University of North Carolina Press, 2001); Hua-yu Li, *Mao and the Economic Stalinization of China, 1948–1953*. (Lanham: Rowman & Littlefield, 2006); Thomas Bernstein and Hua-Yu Li (eds.), *China*

1950s, propaganda taught many Chinese citizens to admire the Soviet Union's industrial achievements and to aspire to emulate the Soviet model.[28] Lenin's idea that communism meant Soviet power plus electrification was well known in socialist China. Party media attributed the Soviet Union's success to the leadership of the Communist Party and the development of socialist economic and social structures.

Despite containing one of the world's largest hydropower reserves, China's installed power capacity lagged far behind the Soviet Union's. In 1949, by which time the Soviet Union had achieved over 10,000,000 kW hydropower capacity, the Nationalist government produced only 10,000 kW of hydropower.[29] The Chinese Communist Party attributed the Soviet Union's accomplishments to the Stalinist structures of economic and political control. It highlighted the enormous disparity between the accomplishments of the Soviet Union and those of the Nationalist government as a way of justifying the absolute leadership of the Communist Party in the field of hydropower development.

After China became involved in the Korean War, Stalin's skeptical attitude toward Mao and the Chinese Communist Party began to change, and he displayed greater willingness to assist China.[30] During a conversation in 1951, Zhou Enlai and the Soviet representative N. V. Roshchin discussed the most significant issues facing the new Chinese government. Zhou identified two problems in particular: Significant financial strain, principally due to the Korean War – with "60% of the budget going to the war" – and a lack of technically trained cadres. Zhou noted that the

---

Learns from the Soviet Union: 1949–Present (Lanham, MD: Lexington Books, 2010); Alexander V. Pantsov and Steven Levine, *Mao: The Real Story* (New York: Simon & Schuster, 2012), chapters 23–27.

[28] See Deborah A. Kaple, *Dream of a Red Factory: The Legacy of High Stalinism in China* (Oxford: Oxford University Press, 1994); Shen Zhihua 沈志华, *Sulian Zhuanjia zai Zhongguo, 1948–1960* 苏联专家在中国 (*Soviet Advisors in China, 1948–1960*) (Beijing: Zhongguo guoji guangbo chubanshe, 2003); Zhang Baichun 张柏春, Yao Fang 姚芳, Zhang Jiuchun 张久春, and Jiang Long 蒋龙, *Sulian jishu xiang Zhongguo de zhuanyi, 1949–1966* 苏联技术向中国的转移, 1949–1966 (*Technology Transfer from the Soviet Union to the People's Republic of China, 1949–1966*) (Jinan: Shandong jiaoyu chubanshe, 2004). For the electrification of the Soviet Union, see the second half of Coopersmith, *Electrification of Russia*.

[29] Shuili Fadian Jianshe Zongju Zhuanjia Gongzuoshi 水力发电建设总局专家工作室, "Sulian Zhuanjia dui woguo shuidian jianshede bangzhu" 苏联专家对我国水电建设的帮助 (Soviet Experts' Aid on Our Country's Hydroelectric Construction), *Shuili Fadian* 4 (1954), 1.

[30] Shen Zhihua, *Mao Zedong, Sidalin yu Chaoxianzhanzheng* 毛泽东,斯大林与朝鲜战争 (*Mao Zedong, Stalin, and the Korean War*) (Guangzhou: Guangdong renmin chubanshe, 2013).

Chinese revolution had developed primarily in rural areas, with senior cadres having lots of village and army experience, but that there were almost no specialists with modern engineering knowledge among party members, soldiers, officers, and state employees. The young, recently graduated students lacked experience and qualifications, and the old Nationalist technical personnel did not possess the level of knowledge required for China's industrialization. Zhou bemoaned the impact of the war, stating that the Chinese government could have addressed these problems by organizing courses, technical universities, and study opportunities for Chinese students in the USSR. However, because of the ongoing war, it would be difficult to solve this problem and it would remain a fundamental constraint on China's industrialization.[31] At Zhou's request, assistance across a comprehensive range of technical fields, including 182 Soviet hydropower advisors, was sent to China.[32]

Despite the Soviet Union's impressive hydroelectric accomplishments, its experts' authority in China was not universally acknowledged. Initially designed and directed by Japanese engineers under the Japanese puppet state of Manchukuo, the Fengman 丰满 hydropower project on the Songhua River in Jilin province was left unfinished at the time of the Japanese defeat in 1945.[33] The low-quality concrete used in its construction resulted in leakage issues after 1949, and the inexperienced Chinese technicians were unable to complete the project. Soviet engineers came to the rescue.[34] Despite the propaganda value of this activity, Chinese technicians attested to the Soviet engineers' broader assistance, which included teaching them how to tackle comparable problems on their own.

In 1953, after repairing the dam at Fengman, the Northeastern Hydropower Engineering Company began the installation of a large automatic hydro-turbine generator. However, the installation team was inexperienced and lacked the necessary expertise. Half of the team were farmers, demobilized soldiers, or common factory workers.[35] To address this knowledge gap, Soviet experts were called upon to train them. They

---

[31] Memorandum of Conversation with PRC Premier Zhou Enlai on July 24, 1951, from the Diary of N. V. Roshchin. Wilson Center Digital Archives.
[32] Zhongguo shuili fadianshi bianji weiyuanhui, *Zhongguo shuilifadianshi*, 420.
[33] See Moore, *Constructing East Asia*, 150–187.
[34] Hai Zhao, "Manchurian Atlas: Competitive Geopolitics, Planned Industrialization, and the Rise of Heavy Industrial State in Northeast China, 1918–1954" (PhD Dissertation, University of Chicago, 2015).
[35] Fan Rongkang 范荣康, "Guanche Sulianzhuanjia Jianyide Fanli" 贯彻苏联专家建议的范例 (An Example of Following Soviet Experts' Advice). *Renmin ribao*, January 23, 1954.

gave lectures and prepared instructional materials, thus forming China's first professional hydropower-turbine installation team. The Soviet experts' success in repairing the Fengman hydropower project in 1953 not only established their authority in the eyes of Chinese cadres and technicians but also reaffirmed the superiority of Soviet technology. Geopolitics certainly also played a role in shaping attitudes toward Soviet experts, not just in the hydropower industry but across other fields as well. In the 1950s, emulating Soviet technological advancements was a prevailing trend. In 1954 the official narrative claimed that the reconstruction of the Fengman hydropower station was the first victory of "proletarian hydropower technology," thus infusing political ideology into the realm of civil engineering and technology.[36] The successful completion of large public projects was seen as a measure of the legitimacy of China's political leadership.

The significant challenges encountered during the development of hydropower in the Soviet Union were rarely acknowledged by the Soviets. In the 1930s, gigantic hydroelectric projects like the Dneprostroi project were initiated before complete surveys of hydrological and geological conditions had been conducted and before their economic viability had been determined, possibly due to the "gigantomania" that characterized Stalin's bureaucracy. Although they promised high electrical output, these projects failed to deliver. There was thus some truth in the admittedly hyperbolic words of a critical Russian hydrologist: "The hay annually harvested from these lands (submerged by the reservoir), if burned as fuel, would have produced as much energy as was generated by the hydroelectric power plant."[37] Hence, in the early 1950s, some Soviet advisors cautioned China against rushing to build large-scale hydropower projects.

In this period, Zhang Tiezheng 张铁铮, the vice director of the Hydroelectric Bureau and a committed Communist revolutionary, advocated for large-scale hydropower projects. However, state leaders, including Mao and Chen Yun, were aware of the technological and financial constraints of the nascent state.[38] Despite this, other political leaders embraced Stalinist gigantomania, seeing the construction of large-scale

---

[36] Shuili Fadian Jianshe Zongju Zhuanjia gongzuo shi, "Sulian Zhuanjia dui woguo Shuidian Jianshede bangzhu," 2.
[37] Graham, *Ghost of the Executed Engineer*, 53.
[38] Li Rui 李锐, *Li Rui Riji, 1946–1979* 李锐日记 (*Diary of Li Rui*) (Fort Worth, TX: Fellow Press of America, 2008), 449.

projects as a symbol of their political ambitions and achievements. Gigantomania was certainly not solely a political issue, as the megaprojects it involved emerged from an amalgamation of growing technological capacity, favorable environmental conditions, and grandiose political ambitions. In the early years of the PRC, Soviet experts and their knowledge played a crucial role in China's technological complex, promoting the idea of multipurpose river development from 1949 to 1954, as noted by David Pietz.[39] However, we should note that this concept of multipurpose river development was not exclusively initiated by Soviet guidance in the 1950s. Decades before the arrival of Soviet advisors, the Nationalist state, its American advisors, and the Japanese had already begun to transform their river management strategies. The Soviet advisors further enhanced the country's technological capabilities. Mao and his party were certainly ambitious and, as a result, the socialist state added large-scale projects like the comprehensive development of the Yellow River to its agenda.

While Soviet experts provided assistance in China, Chinese technicians and students also traveled to the Soviet Union for training. In 1948, Li Peng (who would later become the Premier of the People's Republic) was sent to Moscow to study hydroelectric engineering.[40] In 1954, the Communist government also dispatched an electrical industry delegation to the Soviet Union, with the objective of gaining insight into the country's experience in developing its electrical industry. The delegation consisted of twenty-one members, with twelve having expertise in hydroelectricity and the remaining nine in thermal electricity. Liu Lanbo, the Vice Minister of the Fuel Industry Ministry, directed the delegation, while Li Rui, the head of the Hydroelectric Bureau, served as its vice director. The delegation spent three months in the Soviet Union, during which Li Rui documented his visit in his diary and prepared a report outlining the Soviet hydroelectric industry's accomplishments and experiences.

During their stay in the Soviet Union, the delegation attended more than thirty lectures and toured fourteen hydroelectric plants, including both completed dams and projects that were still under construction.

---

[39] Pietz, *Yellow River*, 153.
[40] Li Peng 李鹏, *Li Peng Huiyilu, 1928–1983* 李鹏回忆录 (*Memoir of Li Peng*) (Beijing: Zhongguo dianli chubanshe, 2014). Along with Li Peng, Lin Hanxiong, Cui Jun, He Yi, and Luo Xibei studied hydroelectric engineering in Moscow. For more on the career of Li Peng, see Kenneth Lieberthal and Michel Oksenberg, *Policy Making in China: Leaders, Structures, and Processes* (Princeton: Princeton University Press, 1988), 51–58.

They also visited two small-scale rural hydroelectric plants.⁴¹ Flood control was not the primary focus of river management in the Soviet Union, unlike in China. Rather, the generation of hydroelectricity was the primary function of most river projects. The visit to the Soviet Union bolstered Li Rui's belief in the need to promote multipurpose river development and to increase the share of hydroelectricity in China's energy industries. After the delegation's visit, between 1955 and 1959, China sent 68 hydroelectric technicians to the Soviet Union for training.⁴²

EXPERTS FROM THE OLD SOCIETY

In addition to the technical assistance provided by the Soviet Union in the 1950s, it is also important to acknowledge the role of a group of engineers retained from the National Resources Commission. Among them was Huang Yuxian, the director of the Hydroelectric Bureau of the Nationalist National Resources Commission, who decided to stay in mainland China after the revolution rather than flee to Taiwan. Huang was initially concerned about the Communist takeover and attempted to leave for the United States in 1949. However, through the persuasion of Chi Ch'ao-Ting, a close friend of his family and an underground Communist activist who had served as an economic adviser in the Nationalist government, Huang decided to stay.⁴³

The official historical narrative of the PRC in the 1950s emphasizes socialist reforms and Five-Year Plans, but it overlooks the tension and mistrust between the party and "experts from the old society." According to Hong Yung Lee, the party needed to restructure sociopolitical and economic systems in this period to balance the political needs of a Leninist Party with the structural requirements of economic development. The historical process of transition from revolutionary cadres to party technocrats took almost three decades and was characterized by intense struggles, torture, purges, and rehabilitations.⁴⁴ Focusing on hydropower

---

⁴¹ Li Rui, "Sulian shuili fadian jianshe de jiben qingkuang he zhuyao jingyan 苏联水力发电建设的基本情况和主要经验"(The Basic Situation and Major Experiences of Soviet Hydroelectric Construction), in *Li Rui wenji* 李锐文集 (*Collected Writings of Li Rui*) (Hong Kong: Zhongguo shehui jiaoyu chubanshe, 2009), 275–317.
⁴² Zhongguo shuili fadianshi bianji weiyuanhui, *Zhongguo shuili fadian shi*, 2, 508.
⁴³ Zhang Guangdou 张光斗, *Wode rensheng zhilu* 我的人生之路 (*The Road of My Life*) (Beijing: Qinghua daxue chubanshe, 2002), 49.
⁴⁴ Hong Yung Lee, *From Revolutionary Cadres to Party Technocrats in Socialist China* (Berkeley: University of California Press, 1991), 10 (ebook).

engineering, this section explores the role of experts from the old society in the ascent of "red engineers" and party technocrats.

During the early years of the PRC, Mao and his colleagues recognized that the Communist Party and its army lacked the scientific and technological expertise that was essential for the country's economic recovery and industrialization. To win the support of non-Communist intellectuals, Mao and the party continued the policy of the united front, which had been developed during the war. Mao was aware of the party's strengths and weaknesses and advised his cadres, "We must be humble and honest on academic problems. Pretending to know what we don't know would undermine our authority."[45]

As a new national government that inherited the functional institutions of the defeated Nationalist regime, the Communist Party placed a great deal of emphasis on technicians and engineers who had worked for the Nationalists. Mao recognized their technical expertise but doubted their political loyalty, referring to them as "capable but not politically reliable."[46] Zhang Hanying 张含英, the former director of the Yellow River Conservancy Commission during the Republican period, acknowledged that many technical personnel from the previous regime needed to adjust to the new political situation. As he explained,

> Many engineers who are from the old society still hold outdated ways of thinking that are divorced from politics and the reality of the current situation. In capitalist countries, hydraulic engineering is used to extract profits for the exploiting classes who deceive the public by concealing their true nature. They claim that technical work is apolitical and that technical personnel should distance themselves from politics. During the previous period of rule by the Nationalist reactionary clique, little water conservancy work took place, leaving most old-style technical personnel with no connection to the working people.[47]

Not all retained engineers shared Zhang's perspective, but they were focused on their work and sought ways to continue it.

---

[45] "xxx chuanda Mao Zhuxi de zhishi, 1950 xxx" 传达毛主席的指示 (xxx Delivers Chairman Mao's Instructions), in *Mao Zedong Sixiang wansui, 1949–1957* 毛泽东思想万岁 (*Long Live Mao Zedong Thought*) (Neibuziliao), 10.

[46] "Guanyu peiyang ganbu wenti de zhishi 1951.03" 关于培养干部问题的指示 (Instruction on Cadre Cultivation), in *Mao Zedong Sixiang wansui, 1949–1957* 毛泽东思想万岁 (*Long Live Mao Zedong Thought*) (Neibuziliao), 14.

[47] Zhang Hanying, "Xuexi Sulian xianjin jingyan, tigao womende sixiang yu jinshu shuiping" 学习苏联先进经验,提高我们的思想与技术水平 (Learn from the Soviet Experience, Improve Our Thought and Technology Levels), *Renmin Shuili* 人民水利 2 (1952), 6–9.

In 1950, Mao was cautious regarding the role of the party in Chinese society, but he became increasingly assertive over time. By 1954, Mao was arguing that the party must lead all aspects of society, including in professional sectors like hydropower.[48] When Li Rui joined the bureau as its director, Huang Yuxian was appointed as its chief engineer. Although Li Rui lacked experience and training in hydroelectric engineering, he was a dedicated revolutionary who vowed to lead the industry. He was also humble and recognized his lack of expertise, and so was open to input from experts from the old society.

Despite his dedication to the revolution, Li Rui faced criticism from party cadres for favoring experts from the old society over revolutionary cadres. When challenged, Li Rui defended his position by stating that "revolutionary cadres know nothing about (hydropower) technology. They had better not stand in the way like old dogs!"[49] This apparently offended some party members. This mistrust of retained Nationalist employees was common among Communist cadres, who suspected that experts like Huang Yuxian might sabotage existing infrastructure. Although Li Rui and other cadres ultimately concluded that Huang was a dedicated professional engineer rather than a counterrevolutionary, his central role in hydropower project planning and construction was marginalized in the following decades.[50]

In regional hydropower survey and design institutes, the administrative arrangement mirrored that of the Chief Bureau. A party cadre was appointed as the director and a hydropower professional served as the chief engineer, under the supervision of the director. Typically, these engineers had worked for the Nationalist government before 1949. The party considered this personnel arrangement appropriate for balancing party control with technical expertise.

Yu Kaiquan 于开泉, a skilled hydroelectric engineer, trained at Kyoto University in the 1930s. He later became the sole Chinese technician at the Japanese-controlled Fengman hydropower project in Manchukuo. After the Communist takeover, though, no government official in the region possessed knowledge of hydropower and its management. When a generator malfunctioned due to mechanical issues, the new administration

---

[48] "Dui weisheng gongzuo de zhishi, 1954.11" 对卫生工作的指示 (Instruction on Hygiene Work), in *Mao Zedong Sixiang wansui, 1949–1957* 毛泽东思想万岁 (*Long Live Mao Zedong Thought*) (Neibuziliao), 48–49.

[49] Li Rui, "Bainian huishou" 百年回首 (A Century in Retrospect), *Minjian Lishi* 民间历史, The Chinese University of Hong Kong Service Center for China Studies.

[50] Li, *Li Rui Riji, 1946–1979*, 421.

suspected sabotage by counterrevolutionaries, particularly by people who had worked for the Japanese. This prompted the deployment of armed soldiers to monitor every move of Yu and his colleagues, thereby placing the plant under military control to prevent further "wreckage." When Li Rui invited Yu Kaiquan to Beijing to report on the Fengman project, the Jilin provincial authority dispatched police to surveil him on his journey.[51]

The political distrust of the Communist authorities was disheartening, but Li Rui's pragmatism allowed many engineers to contribute their professional expertise to hydropower development in the 1950s, despite their being deprived of management positions in the field. While the retained engineers generally welcomed the support of the Soviet Union, they found relying solely on the Soviet Union for knowledge problematic. In 1953, during a conversation with the Soviet ambassador A. S. Paniushkin, the Beijing mayor Peng Zhen criticized the Chinese intelligentsia's attitudes toward the Soviet Union. He expressed the belief that the majority of the intelligentsia, who had received their education in America, England, and France, or in Chinese institutions run by people from these countries, were anti-Soviet. They openly referred to the Soviet Union as an imperialist power and asked questions like why the Chinese Changchun Railway, for example, was still owned by the Soviet Union. For Peng, this reflected their opposition to the new people's revolutionary power.[52]

Peng might have used the concerns of members of the Chinese intelligentsia to pressure the Soviet Union into returning the railway, but instead he expressed his disdain for intellectuals who had studied in capitalist countries. It is clear that Western education had a significant impact on the field of hydropower engineering in China. For example, as Table 3.1 shows, each of the "four old heads" of hydropower in China – Huang Yuxian, Zhang Changling 张昌龄, Shi Jiayang 施嘉炀, and Qin Xiudian 覃修典 – had studied in the United States during the Republican era. The Communist Party frequently assessed one's political reliability and loyalty based on family background, political history, education, and social networks. In Maoist China, these engineers' experiences of foreign study and previous employment under the Nationalist government

---

[51] Li, *Li Rui koushu wangshi*, 302.
[52] Memorandum of Conversation, Soviet Ambassador to China A. S. Paniushkin with the Chair of the City People's Government in Beijing, Peng Zhen. January 6, 1953. Wilson Center Digital Archive.

TABLE 3.1 *Education experience of selected hydropower engineers*

Name	School	Degree
Huang Yuxian	California Institute of Technology/Cornell (US)	Masters
Zhang Changling	Massachusetts Institute of Technology (US)	Masters
Shi Jiayang	Massachusetts Institute of Technology (US)	Bachelors/Masters
Qin Xiudian	Massachusetts Institute of Technology (US)/Technical University of Berlin (Germany)	Masters
Lu Qinkan	University of Colorado (US)	Masters
Yu Kaiquan	Kyoto University (Japan)	Bachelors
Zhang Guangdou	University of California at Berkeley/Harvard (US)	Masters

became a liability and were used against them in various political movements in the early decades of the PRC.[53]

Despite the Party's initial mistrust of intellectuals, they believed that their "thought reform" campaign and the policy of "unifying, educating, and transforming" 团结,教育,改造 had successfully integrated many intellectuals into the working class. As a result, the CCP granted the class status of "staff" 职员 to experts who held high-paying positions as engineers, professors, and specialists.[54] Although the party center adopted a lenient policy toward intellectuals, many lower level cadres viewed them – and especially retained Nationalist officials and experts – as members of the exploiting classes. Many party members grumbled about the generous treatment intellectuals received. In his response to the mounting complaints, Mao called for greater tactical patience in dealing with intellectuals.[55] At a conference on the "problem of intellectuals" in early 1956, Mao pointed out that China relied heavily on foreign countries for advanced machinery and equipment. He emphasized further that the country was not yet independent in industrial production and science. He criticized the disrespectful behavior of some party cadres toward intellectuals, in particular scientists and

---

[53] Yang Kuisong, *Eight Outcasts: Social and Political Marginalization in China under Mao* (Berkeley: University of California Press, 2019).

[54] Lee, *Revolutionary Cadres to Party Technocrats in Socialist China*, 31.

[55] Shen Zhihua 沈志华, *Zhonghua renmin gonghe guoshi, disanjuan, Sikao yu xuanze: cong zhishifenzi huiyi dao fanyoupai yundong (1956–1957)* 中华人民共和国史, 第三卷 思考与选择—从知识分子会议到反右派运动 (1956–1957) (The History of the People's Republic of China, vol. 3, Reflections and Choices: The Consciousness of the Chinese Intellectuals and the Anti-Rightist Campaign, 1956–1957) (Hong Kong: Research Center for Contemporary China, The Chinese University of Hong Kong, 2008), 32.

engineers, reminding them that the revolution was "now a technological revolution, a cultural revolution. Our enemies are stupidity and ignorance. It won't work without intellectuals." Zhou Enlai also argued that more intellectuals were needed and that, for the sake of socialist industrialization, they should be trusted.[56]

In 1957, after the Hundred Flowers campaign which invited criticism of the party, the political climate shifted drastically for intellectuals. Mao's rhetoric about intellectuals changed, and he started using more derogatory language. He accused intellectuals of lacking political loyalty, calling them "grass on the wall" who would follow the wind, and argued that they lacked emotional bonds with workers and peasants.[57] He also suggested that intellectuals were arrogant and needed to be taught a lesson, comparing them to dogs whose tails could be hidden with cold water, a metaphor suggesting taking a repressive approach toward the intellectuals.[58]

Mao continued to believe in reeducation as the most effective means to reform intellectuals. He expressed this in a meeting with representatives of democratic parties and various other nonaffiliated individuals. Using the analogy of hair and skin, Mao stated that the time had come for intellectuals to attach themselves to the new skin of the working class, now that the old skin of the capitalist class had been shed.[59] As criticism from rightists intensified in 1958, Mao became increasingly concerned about the role of party cadres in a variety of professional sectors. He declared the Great Leap Forward a new war against nature, stating that its goal was to "revolutionize the earth."[60] The rightists belittled Communist cadres, claiming that they were incapable of leading them. They also suggested that old cadres should be asked to retire. Mao repeated his argument about the need for new expertise, stating that it was necessary

---

[56] Shen, *Zhonghua renmin gonghe guoshi*, 3, 20.

[57] "Zai Shandong shengji jiguan dangyuan ganbuhui shangde baogao, 1957.03.18" 在山东省级机关党员干部会上的报告 (Report at the Shandong Provincial Party Cadres Meeting), *Mao Zedong sixiang wansui*, 198.

[58] "Zai Nanjing budui, Jiangsu, Anhui ersheng dangyuan ganbu huiyi shangde jianghua, 1957.03.20" 在南京部队、江苏、安徽二省党员干部会议上的讲话 (Talk at the Meeting of Nanjing Military, Jiangsu and Anhui Party Cadres), *Mao Zedong sixiang wansui*, 211.

[59] "Yueji ge minzhu dangpai fuzeren he wudangpai minzhu renshi tanhua jiyao" 约集各民主党派负责人和无党派民主人士谈话纪要 (Notes of a Conversation with Democratic Party Leaders and Independents), *Mao Zedong sixiang wansui*, 223.

[60] "Zui zuigao guowuhuiyi shang jianghua yaodian 1958.01.28/30" 在最高国务会议上讲话要点 (Summary of a talk at a State Council Meeting), *Mao Zedong sixiang wansui*, 18.

to learn in order to lead effectively.⁶¹ According to Mao, combining "red" (political commitment) and "expert" (technical knowledge) had become more urgent. Politics and professional work had to be united in every sphere of socialist construction.

How did radical Maoism affect engineers and the hydropower projects that were under construction during this period? In 1964, before the Cultural Revolution, Mao categorized Chinese intellectuals into three groups based on their relationship with the Communist Party: Engineers and technicians were placed in the highest category, followed by scientists. The third, and worst, category was comprised of social science and humanities scholars.⁶² While the mental and physical suffering of intellectuals in the last category has been widely documented, less information is available about the experiences of engineers during this period.⁶³ However, the rise of radical Maoism and the policy of "politics in command," as well as the engineering revolution, not only encouraged political struggles against individuals but also caused considerable damage to many hydropower projects that were under construction.

Like many others, Huang Yuxian was labeled a "reactionary academic authority" by radical Maoists during the "Cleansing of the Class Ranks" movement in 1968.⁶⁴ As the former director of the Hydroelectric Bureau under the Nationalist government, Huang was viewed not only as an engineer but also a staunch supporter of the Nationalist regime. His daughter's education in the United States before 1949 and his friendship with John Savage, who was seen as an American spy, raised even more questions about his political reliability.

After years of grueling struggle sessions and thought reform, Huang and his wife found themselves in exile in Qingtongxia 青铜峡, Ningxia province, in 1969. Qingtongxia was the site of one of several hydropower projects built along the upper stream of the Yellow River in the late 1950s and, as a chief engineer, Huang had played a significant role in the design and construction of the dam. Despite his expertise, though, he was no

---

⁶¹ "Gongzuo fangfa liushitiao,1958.01.31" 工作方法六十条 (Sixty Work Method Guidelines), *Mao Zedong sixiang wansui*, 24.
⁶² Chen Shaoming 陈少明, "Zuowei wentide Zhongguo zhishifenzi" 作为问题的中国知识分子 (Chinese Intellectuals as a Problem), *Kaifang shidai* 开放时代 5 (2013), 6–26.
⁶³ Judith Shapiro examines the experience of Huang Wanli, a hydraulic expert, in the Anti-Rightist Movement. See Shapiro, *Mao's War against Nature*, 48–65.
⁶⁴ On the cleansing of the class enemy movement, see Ning Wang, *Banished to the Great Northern Wilderness: Political Exile and Re-education in Mao's China* (Ithaca: Cornell University Press, 2018).

longer allowed to work as an engineer. Instead, in his late sixties by this point, Huang was relegated to cleaning restrooms at the hydropower plant. One day, as Huang worked to break up the frozen feces in the squat toilets, he and his wife, Xiong Shuchen 熊淑忱, had the following conversation: "All the squat toilets were covered with heavy ice. I had to use a hammer to break the frozen feces. It was so overwhelming," Huang said. His wife replied with a smile, "Your nose is no longer sensitive at all. Look how lucky you are. Your nose has been trained to be very tolerant after cleaning restrooms for years."[65] Despite these difficulties, Huang and his wife remained resilient and were fortunate enough to survive the political turmoil of the period. In 1978, after the end of the Cultural Revolution, they left China for the United States to be reunited with their eldest daughter. Huang never returned to China.[66]

RED AND EXPERT

In 1949, like many people employed at the Hydroelectric Bureau of the National Resources Commission, Zhang Guangdou decided to remain in Nanjing and await the arrival of the Communist forces. However, his initial interactions with the People's Liberation Army (PLA) were far from pleasant. At one early meeting, a low-ranking political instructor gathered the engineers together and bellowed, "You running dogs of the Nationalists, you have been corrupt and oppressive to the people. Now that the country is liberated and you must confess your crimes, reform yourselves and serve the people!"[67] Zhang was confused and frustrated; he considered himself to be nothing more than a hardworking engineer. He felt it was unjust that he was being accused of such crimes. Fortunately for him, these accusations were later dismissed as a misunderstanding of the Communist Party's policy toward retained engineers. The party secretary of the PLA unit that took control of Nanjing even apologized to Zhang personally and praised him as a patriotic engineer who had turned

---

[65] Huang Ailian 黄爱莲, *Wo shi Shunma* 我是顺妈 (*I am Shun's Mom*) (Shanghai: Shanghai cishu chubanshe, 2012); Li Xinchun 李欣春 and Zhou Baiyuan 周百源, "Woguo shuili fadian shiyede xianquzhe Huang Yuxian xiansheng" 我国水力发电事业的先驱者黄育贤先生 (Mr. Huang Yuxian, Pioneer of China's Hydropower Enterprise), *Chongren wenshi ziliao* 崇仁文史资料 3 (1991), 67–73.

[66] One of Huang's daughters was sent to study in the United States during the Republican period, with the help of John Savage. She has lived there ever since.

[67] Zhang, *Wode rensheng zhilu*, 50.

over a large number of documents related to hydropower to the new authority (while he absentmindedly picked at his toenails).[68]

With an eye toward advancing his career, Zhang requested a transfer from the Hydroelectric Bureau to a faculty position at Tsinghua University in Beijing. As one of the top universities in China, Tsinghua was being reorganized in the early years of the PRC to cultivate engineers for the country. Joel Andreas has examined the rise of "red engineers" in the Communist bureaucracy through a comprehensive study of the university during the Maoist period.[69] Along with many other intellectuals, Zhang was required to undergo "thought reform" while contributing his professional knowledge to the training of "red" engineers for the fledgling state.[70] During a struggle session in 1952, he was criticized by his Tsinghua students for advocating the adoption of American science and technology and opposing the Soviet experience of development. Although Zhang believed that questions concerning technology should not be limited by national borders, he was pressured to confess and to reflect on his beliefs. Despite the Communist Party's claims of success in reforming intellectuals, Zhang remembered being emotionally hurt and ideologically confused by the thought reform process he experienced in the 1950s.[71] Nonetheless, like many others, he followed the prevailing "culture of accommodation" and submitted himself to the party's authority for both political and material reward.[72]

With the aim of strengthening the party, the leaders of the CCP began to train their own technical cadres and to recruit well-educated, advanced intellectuals. Zhou Enlai declared that 80 percent of intellectuals were supportive of the CCP and that they formed a "formidable force" in the program of socialist construction He encouraged the party to improve the living and working conditions and the political status of these loyal intellectuals. Following Zhou's guidance, An Ziwen 安子文, the director of the party's Organization Department, instructed lower-level party committees to "first accept well-known specialists and authorities and then investigate their qualifications."[73] In 1956, despite Zhang's family's

---

[68] Zhang, *Wode rensheng zhilu*, 50. [69] Andreas, *Rise of the Red Engineers*.
[70] Aminda Smith, *Thought Reform and China's Dangerous Classes* (New York: Rowman & Littlefield, 2012).
[71] Zhang, *Wode rensheng zhilu*, 60–61.
[72] On the "thought reform" and "culture of accommodation," see Jeremy Brown and Paul Pickowicz (eds.), *Dilemmas of Victory: The Early Years of the People's Republic of China* (Cambridge, MA: Harvard University Press, 2010).
[73] Lee, *Revolutionary Cadres to Party Technocrats in Socialist China*, 33–34.

landlord-class background, the party secretary of the Hydraulics Department approached Zhang and convinced him to be among the first group of faculty members at Tsinghua to join the Communist Party.[74] This was likely due to his flexible personality and political sensitivity. From that point on, Zhang appeared to be a model red engineer, possessing both professional expertise and the required political attributes.[75]

However, being a party member did not exempt Zhang from political struggles in the coming decades. As an engineer who had received formal training in the United States, Zhang was skeptical of the "leaning to one side" policy in scientific and technological affairs in the 1950s. He complained that there was no room for "a hundred flowers" in the field of science and technology.[76] He quickly realized that the ultimate goal of the "hundred flowers" campaign was to settle on a single authority: Marxism.[77] It is unclear whether Zhang became a "true believer" overnight. Given his educational background and professional experience, though, it would be unfair to label him a simple opportunist. During the "pull the white flag, raise the red flag" campaign in universities and other institutions in the spring of 1958, Zhang was praised in the Tsinghua newspaper for his political progress as a Communist professor.[78]

Despite this, when the Cultural Revolution began, Zhang quickly became a target of the Red Guards. He was accused of being a counterrevolutionary and a capitalist and his prior study experiences in the United States before 1949 and his work with Soviet experts in the 1950s were used as evidence of his support for both American imperialism and Soviet revisionism.[79] In order to avoid being beaten by the Red Guards, Zhang complied with them and was forced to undergo labor reeducation by cleaning restrooms. Then, in 1968, as the party began its nationwide purge of class enemies, Zhang was subjected to brutal interrogations by the worker-controlled Mao Zedong Thought Propaganda Team. The team questioned his motivation for participating in the

[74] Donald J. Treiman and Andrew Walder, "The Impact of Class Labels on Life Chances in China," *American Journal of Sociology* 124, no. 4 (2019), 1125–1163.
[75] In 1954, before joining the Communist Party, through the introduction of his colleague Meng Zhaoying, Zhang Guangdou joined the Jiusan Society, a democratic party which consisted mainly of intellectuals, especially scientists and engineers.
[76] Shen, *Zhonghua renmin gonghe guoshi*, 226.
[77] Shen, *Zhonghua renmin gonghe guoshi*, 230.   [78] Andreas, *Rise of the Red Engineers*.
[79] On the Sino-Soviet split and Sino-US hostility, see Chen, *Mao's China and the Cold War*; Shen Zhihua and Douglas Stiffer (eds.), *Cuiruo de lianmeng: Lengzhan yu Zhongsu guanxi* 脆弱的联盟: 冷战与中苏关系 (Frail Alliance: The Cold War and Sino-Soviet Relations) (Beijing: Shehui kexue wenxian chubanshe, 2010).

construction of hydropower plants in southwest China during the 1940s, which he claimed were built to support the War of Resistance against Japan. The team accused him of building these power plants to provide energy for manufacturing ammunition to be used against the Communists. Zhang's work with American engineers like John Cotton and John Savage in the 1940s and with Soviet experts in surveying the Heilongjiang River for hydropower planning in the 1950s was also viewed with suspicion. The propaganda team accused him of being both an American spy and a Soviet spy.[80] These accusations left Zhang speechless and with no chance to defend himself.

In addition to questioning Zhang about his past actions, the propaganda team sought out testimonies concerning the "counterrevolutionary" activities of other engineers. In his memoirs, Zhang documents in detail Huang Yuxian's criticism of the Communist Party and his attempt to leave the mainland in 1949.[81] He also reveals the immense pressure he faced during the Cultural Revolution to betray his former colleagues. Although he never admitted to exposing Huang's disdain for the Communist Party before 1949, it is reasonable to think that he did so under duress in order to redeem himself.[82] Despite the challenges he faced during the Cultural Revolution, Zhang's political savvy proved to be essential to his long-term professional success in the post-Mao years.

At the height of the Cultural Revolution, Mao called for a revolution in higher education. He recognized that "science and engineering universities were still needed." But he felt that "the program length should be shortened."[83] He demanded a fundamental change in college admissions policies, with merit-based examinations being replaced by class identity and family background as the most important criteria of admission.[84] According to Mao's directives on the "proletarianization of higher education," students from peasant, worker, and PLA-soldier backgrounds should form the core of every student cohort. The revolution in education also required the combination of study and production.

---

[80] Zhang, *Wode rensheng zhilu*, 103.   [81] Zhang, *Wode rensheng zhilu*, 47.
[82] Zhang, *Wode rensheng zhilu*, 104.
[83] Mao Zedong, "Guanyu daxue jiaoyu gaigede yiduan tanhua" 关于大学教育改革的一段谈话 (A Talk on Higher Education Reform), in Zhonggong zhongyang wenxian yanjiushi 中共中央文献研究室 (ed.), *Jianguo yilai Mao Zedong wengao* 12 建国以来毛泽东文稿 (*Manuscripts of Mao Zedong since 1949*) (Beijing: Zhongyang wenxian chubanshe, 1998), 505.
[84] "Guanyu peiyang ganbu wentide zhishi, 1951.03" 关于培养干部问题的指示 (Instructions on Cadre Cultivation), in *Mao Zedong sixiang wansui, 1949–1957*, 14.

In the late 1960s, then, the Hydraulics Department of Tsinghua University, including Zhang Guangdou, was sent to the Sanmenxia hydropower project on the Yellow River to provide technical support.[85] In the so-called experimental class, all of the 230 enrolled students were either workers at the Sanmenxia project or peasants from the surrounding areas, many of whom had only graduated from elementary school and some of whom were totally illiterate.[86] At first, Zhang was prohibited from teaching and required to engage in manual labor like an ordinary worker. Because of his professional training, however, he was eventually allowed to teach classes. Despite the shortened study period and the varying intellectual abilities of the students, many of them respected Zhang as their teacher and formed good relationships with him. Still, it would be difficult to claim that this experimental class successfully cultivated effective red engineers for China's hydropower development.

PRACTICING MAOISM IN HYDRO INFRASTRUCTURE CONSTRUCTION

Following the Nanning meeting, the central government made a decisive move to drive the development of hydropower forward. In February 1958, the Water Conservancy and the Electric Industry Ministries were merged to bring about a more unified and efficient approach to what was felt to be an important sector. Mao saw the need for swift action and, at the Eighth National Plenum, called for the Great Leap Forward to begin. He believed that speed was the key to success in this ambitious endeavor.[87] Mao felt that, in order to catch up with industrialized nations like the United Kingdom, China needed to increase its electricity generation capacity rapidly. According to him, forty million kilowatts of electricity were needed, which was ten times the current capacity. To achieve this goal, he emphasized the need for the simultaneous development of both hydroelectricity and fossil fuel electricity.[88] The party's Central Committee echoed Mao's vision and announced in *People's Daily*

---

[85] William Hinton, *Hundred Days War: The Cultural Revolution at Tsinghua University* (New York: Monthly Review Press, 1972), 266.

[86] Shi Yun 石耘, "Qinghua daxue shuilixi zai Sanmenxia" 清华大学水利系在三门峡 (The Hydraulics Department of Tsinghua University at Sanmenxia), *Henan Wenshiziliao* 河南文史资料 1 (2011), 4–20.

[87] "Lizheng gaosudu" 力争高速度 (Work for High Speed), *Renmin ribao*, June 21, 1958.

[88] "Zai zuigao guowuhuiyishangde jianghua, 1958.01.28" 在最高国务会议上的讲话(Talk at the Highest State Council), *Mao Zedong sixiang wansui, 1958–1960*, 12.

that all water conservancy projects must include plans to generate electricity.[89]

Between 1958 and 1960, the height of the Great Leap Forward, the state undertook a massive effort to build 53 medium- and large-scale hydropower projects. Each of these projects required a significant initial investment of over one million *yuan*. By the mid 1960s, though, only twenty of these projects were complete and capable of generating electricity.[90] The intense eagerness to meet the goals of the Great Leap Forward led to the implementation of the so-called *sanbian* 三边 policy, which aimed to speed up the process by conducting surveys, designs, and construction simultaneously.[91] However, this haste resulted in many mistakes and setbacks. Fifteen of the original set of projects were suspended in the early 1960s because of poor quality or a lack of materials, and eighteen were eventually abandoned altogether. This represented a significant waste of resources, as the mobilization of thousands of workers, the consumption of materials, and the spending of all that money ultimately produced zero kilowatt hours of electricity. One particularly notable example of this was in Sichuan, where the government began a dozen hydropower projects in 1958, mobilized over 100,000 laborers, and used vast amounts of resources. Due to the lack of proper hydrological and geological surveys, poor planning, and unrealistic goals, however, most of these projects were suspended. Only one was completed during the Great Leap Forward.[92]

The rapid growth of hydropower in China faced many challenges, one of which was the shortage of essential construction materials such as cement, steel, and timber. In response, the state and technical cadres issued a call for a material revolution in hydropower construction, encouraging the use of innovative methods and alternative materials. Christopher Sneddon has called the building of large dams in the twentieth century a global concrete revolution.[93] In China, there was not only a

---

[89] "Neng fadian de shuili gongcheng douyao fadian" 能发电的水利工程都要发电 (Every Qualified Hydraulic Project Shall Generate Electricity), *Renmin ribao*, December 4, 1958.
[90] Li, *Xinzhongguo dianli gongye fazhan shilue*, 291.
[91] Liu Yanwen 刘彦文, *Gongsi shehui: Yintao shangshan shuili gongchengde geming, jitizhuyi yu xiandaihua* 工地社会：引洮上山水利工程的革命集体主义与现代化 (Revolution, Collectivism and Modernization in China: A Case Study of the Yintao Water Conservancy Project in Gansu Province) (Beijing: Shehui kexue wenxian chubanshe, 2018).
[92] Zhongguo shuili fadianshi bianji weiyuanhui, *Zhongguo Shuili fadianshi* 1, 141.
[93] Sneddon, *Concrete Revolution*.

concrete revolution but also, in the late 1960s, an effort to revolutionize concrete. For example, clinker-free cement was promoted for use in the interior of dams, and bamboo was proposed as a replacement for steel and timber. In addition to using modern concrete dam technology, the state also emphasized the construction of more affordable and easily built earthen, rock, and other types of dams that could be constructed using local materials and human and animal muscle power.[94] By 1972, 12,517 dams over 15 meters high had been completed, with the majority being smaller dams between 15 and 30 meters in height. Of these, the majority were earth-filled structures, many were stone masonry and rock-filled dams, and only seventy-three were built with concrete.[95]

The Communist Party relied heavily on mass labor mobilization as a tool for development, especially in infrastructure projects like hydropower.[96] Despite receiving aid from its socialist allies in the 1950s, many hydropower construction sites in China still lacked advanced machinery. Instead, tens of thousands of peasant laborers worked tirelessly using simple tools like chisels, picks, shovels, weigh-beams, and locally made scaffolding. To conserve scarce resources like steel, laborers used innovative methods to build cofferdams. At the Danjiangkou dam, for example, over 100,000 people from Henan and Hubei provinces used a mixture of wheat stalks, straw, and earth to build a 1.3 kilometer long and 13 meter-high clay, sand, and stone cofferdam in just six months, without the aid of any machinery.[97]

The construction of the large concrete dams at Xin'anjiang 新安江 and Sanmenxia saw mechanization levels of 48 percent and 64 percent, respectively. However, in the building of earthen and rock dams, human labor remained the predominant force.[98] Unfortunately, in many cases, the so-called revolution in technique simply meant the employment of labor-intensive methods, which often neglected workers' health and safety. During the 1960s, Mao believed that China faced immanent military threats from both the Soviet Union and the United States. Covell Meyskens' study of the Third Front project demonstrates that political calculations often overruled technical considerations. Any concerns regarding the feasibility of exchanging human labor for machinery

---

[94] "Geshuini gangcaihe mucai deming" 革水泥, 钢材和木材的命 (Revolutionize Cement, Steel, and Timber), *Shuili fadian* 水力发电 18 (1958), 3–5.
[95] Smil, *China's Energy*, 81.   [96] Pietz, *Yellow River*; Liu, *Gongsi shehui*.
[97] Smil, *China's Energy*, 79.
[98] Zhongguo shuili fadianshi bianji weiyuanhui, *Zhongguo Shuilifadianshi* 1: 135–136.

and technical expertise were met with binary Cold War politics. Party officials dismissed critics by accusing them of betraying the developmental path of socialist China and of supporting the bourgeois economic methods of the capitalist world and the revisionism of the Soviet Union.[99] The photo of workers at the Liujiaxia 刘家峡 dam construction site (Figure 3.1) serve as a reminder that human labor played a significant role in the building of mega dams. The power and ingenuity of human labor was instrumental in creating such gigantic industrial structures.[100] In pursuit of these ambitious projects, muscle power and Maoist politics were highly regarded in state propaganda.

In engineering and construction circles, much debate occurs about which methods and materials are best for a particular job. In China during the Maoist period, that debate took on a whole new level of intensity. Those who were hesitant to use the local methods and materials were not just criticized – they were labeled "right-leaning conservatives." The most radical Maoists believed that the technological revolution should not rely on well-trained and well-paid engineers and technicians, but rather on ordinary workers.[101]

This Maoist approach to construction certainly had its downsides, as illustrated by the experiences of engineers like Huang Yuxian and Zhang Guangdou. The Maoist practices of social classification and class struggle deprived many well-trained engineers of the chance to apply their technical skills to large-scale projects. Consequently, many of those projects had abnormally high rates of quality problems and accidents. Take, for example, the Liujiaxia dam construction site, a massive hydropower project on the upper stream of the Yellow River. Because of a shortage of tools, party cadres ordered workers to tamp the poured concrete with their feet.[102]

---

[99] Covell Meyskens, *Mao's Third Front: The Militarization of Cold War China* (Cambridge: Cambridge University Press, 2020), 122–164; Covell Meyskens, "Rethinking the Political Economy in Mao's China," *Positions: Asia Critique* 29, no. 4 (2021), 809–834, 826.

[100] Covell Meyskens, "Labour," in *Afterlives of Chinese Communism*, ed. Christian Sorace, Ivan Franceschini, and Nicholas Loubere (Canberra: Australian National University Press, 2019), 107.

[101] "Jishu geming yidingyao fadong qunzhong" 技术革命一定要发动群众 (We Must Mobilize the Masses in Technological Revolution), *Renmin ribao*, June 24, 1968. Mao sought to eliminate the gap between mental and manual labor and build an egalitarian society in the process of industrialization. One famous example would be the Daqing oil field, see Hou, *Building for Oil*.

[102] Zhu Sulian 朱淑莲, "Liujiaxia shuidianzhan guangyao jiuju Huanghe sishichun" 刘家峡水电站光耀九曲黄河四十春 (Liujiaxia Hydropower Project Shines over the Yellow River for Four Decades), *Guojia dianwangbao* 国家电网报, September 3, 2009.

FIGURE 3.1 Struggle session at the Liujiaxia Dam site (Photo by API/Gamma-Rapho via Getty Images)

If that was not bad enough, there was also a shortage of cement, so party technical cadres reduced the amount of cement in the building mix and replaced it with local materials. That section of the dam had to be blown up and rebuilt a few years later.[103]

CONCLUSION

Unlike during wartime, when hydropower was directly associated with national survival and strengthening, hydropower was incorporated into the socialist state building effort under Communist rule. With the unfolding of political movements and the ideological shifts they entailed, the engineers experienced extreme duress in their professional practice.

---

[103] *Liujiaxia shuidian changzhi* 刘家峡水电厂志 (*Gazetteer of Liujiaxia Hydropower Plant*) (Lanzhou: Gansu renmin chubanshe, 1999), 38; Zhang Zhihui 张志会, "Liujiaxia shuidianzhan gongchengjianshe de ruogan lishifansi" 刘家峡水电站工程建设的若干历史反思 (Historical Reflections on the Construction of the Liujiaxia Hydropower Project). Gongcheng yanjiu 工程研究 5, no. 1 (2013), 58–70.

During the era of great acceleration, hydropower technology spread rapidly across the world, in many ways transcending social and political differences.[104] But the impact of politics on societies' approaches to concrete dam construction cannot be overlooked. Politics, technology, and the environment have all played crucial roles in shaping and reshaping people's daily lives as well as the waterscapes in which they live.

After the victory of the Bolshevik Revolution, Lenin had hoped that politics would take a backseat so that engineers and agronomists could take center stage. However, neither Russia nor China saw this vision realized.[105] Through the analysis of the civil engineering community, particularly the National Resources Commission, William Kirby argues that there was no technocracy during the Republican era. Ying Jia Tan continues this line of inquiry into the 1950s, arguing that it was not an "engineers' dictatorship" but rather a cooperative effort between Communist cadres and electrical engineers.[106] This chapter finds that, while there was tension, there was also collaboration between the ruling party and hydropower engineers. Party leaders were pragmatic enough to establish productive working relationships with these experts in developing the technostructure of the fledgling hydropower nation, at least before the rise of radical Maoism during the Great Leap Forward and after. Mao and his followers, critical of the division between expertise and politics, used both punitive and affirmative measures like thought reform, struggle sessions, hard labor, and educational revolution to control the hydropower engineers for the benefit of the party, regardless of the achievements and setbacks in hydropower development. Thus, the party had effectively tamed these experts – if not the rivers – for their use and abuse. Entanglement with Maoist politics was a troubling characteristic of the hydropower nation in the early decades of the socialist state.

While some scholars have attempted to quantify the negative impact of Maoist political movements on China's long-term economy, this chapter has not attempted anything like a systematic quantitative analysis.[107] It is noteworthy, though, that both the number of dams being built and

---

[104] Sneddon, *Concrete Revolution*.   [105] Kirby, "Engineering China," 153.
[106] Tan, *Recharging China in War and Revolution*, 163.
[107] Zhaojin Zeng and Joshua Eisenman, "The Price of Persecution: The Long-Term Effects of the Anti-Rightist Campaign on Economic Performance in Post-Mao China," *World Development* 109 (2018), 249–260.

installed electrical capacity grew steadily during the Maoist period.[108] The question of what this progress cost remains unanswered. The radicalism of Maoist ideology during the Great Leap Forward and the Cultural Revolution played a contradictory role in the party-state's pursuit of hydropower. While the Maoist regime prioritized hydropower in the state's agenda, the political repression of engineers undermined that very enterprise. This tension was only resolved with the fading of Maoist ideology and the emergence of a new generation of red engineers. The formula of "red and expert" would only bear fruit after Mao's reign. Top-level Chinese leaders like Li Peng, who studied hydropower in the Soviet Union in the 1950s, and Hu Jintao, who majored in hydropower at Tsinghua University in the early 1960s, symbolize the rise of socialist technocrats in the Chinese party-state system.[109] This rise is essential to understanding the expansion of the hydropower nation's capacity.

---

[108] Zhongguo shuili fadianshi bianji weiyuanhui, *Zhongguo shuili fadianshi*, 1, 127–129, 172–174.

[109] Li Cheng and Lynn White, "Elite Transformation and Modern Change in Mainland China and Taiwan: Empirical Data and the Theory of Technocracy," *China Quarterly*, no. 121 (1990), 1–35.

4

The Great Leap of Small Hydro

In 1959, at the height of the Great Leap Forward, the Changchun Film Studio released "Young People in Our Village" 我们村里的年轻人, a film that chronicles the incredible achievements of a group of young, determined villagers. Despite facing enormous obstacles, the young people persevere in their efforts to build a canal that will solve the water shortage crisis in Kongjiazhuang, a village in Shanxi Province. The film was an instant sensation, inspiring and galvanizing viewers across the country to engage in their own water conservation efforts. Despite the catastrophic effects of the policies of the Great Leap Forward on the country's food supply, it was purported that the film was received with broad enthusiasm. As a result, Chen Huangmei 陈荒煤, the director of the Film Bureau, approached Su Li 苏里, the film's director, and commissioned a sequel. Mao was calling for *sihua* 四化 (four transformations) in rural China, which included technological advancement in water conservation, fertilization, mechanization, and electrification.

The sequel centers on the construction of a small hydroelectric station on the canal whose construction was covered in the first film, which would help to achieve rural electrification. The young people of the village work tirelessly to build the hydroelectric station, relying on their ingenuity and on the Maoist principle of self-reliance to teach themselves the basics of hydroelectricity. Maolin, a talented youth who had only attended primary school, even constructs a wooden turbine. The film culminates in the electrification of the people's commune and the celebration of the young people's achievements, including the happy marriages that result from their joint efforts.

As a piece of state propaganda, the film was intended to highlight the transformative changes that were occurring in rural China under the guidance of the Communist Party. By interweaving romantic subplots with stories of rural collectivization, water conservation, and small hydropower campaigns, the film aimed to inspire enthusiasm for socialist small hydropower initiatives. Despite its overtly ideological agenda, the film nonetheless provides insight into the reality of rural electrification efforts during the Maoist period. Alongside large-scale projects like the Sanmenxia and Xin'anjiang schemes, small hydropower initiatives flourished in China's countryside.[1]

The benefits of small hydropower were tangible for rural residents, who used the technology to process grain and to light their homes at night. But the construction of small hydropower stations has not been fully explored by historians of modern China. In her study of scientific farming during the Mao era, Sigrid Schmalzer identifies the *tu-yang* binary in Maoist discourse, with *tu* denoting local, rustic, and practical knowledge, and *yang* referring to foreign, Western, and professional knowledge.[2] During the Great Leap Forward, Mao contrasted the *tu* approach to technology with the *yang* approach, with small hydropower representing *tu* and large concrete dams embodying *yang*. In Maoist propaganda, small hydropower was portrayed as an accessible and easy-to-implement solution that could be built and operated by peasants with limited education using local materials. This approach was in line with Maoist ideology, which prioritized self-reliance and mass mobilization, as well as Mao's political objectives. In contrast, large-scale hydropower projects were associated with professionalism and foreign influence, which made them ideologically and politically inferior to small

---

[1] According to the International Center on Small Hydro Power, there is no agreed upon definition of "small hydropower" worldwide. However, small hydropower generally refers to hydropower stations with an installed power capacity below 10 MW; "mini hydro" is less than 500 kW; and "micro hydro" is less than 100 kW. Available at www.inshp.org/detail.asp?RID=8&BID=80, last accessed May 25, 2017. (One MW (megawatt) = 1,000 kW.) The definition of small hydropower has also changed over time. In the 1950s, a "small" hydropower project meant one that had less than 500 kW installed capacity. By the 1960s, that had risen to 3,000 kW. By the end of the 1960s, the capacity cap had gone up to 12,000 kW. Here I define all projects with an installed capacity below 12,000 kW as small hydropower, and I will not make fine distinctions between small, mini, and micro-sized projects.

[2] Sigrid Schmalzer, *Red Revolution, Green Revolution: Scientific Farming in Socialist China* (Chicago: The University of Chicago Press, 2016), 34–36.

hydropower projects. During the Cultural Revolution, the Maoist approach to socialist industrialization was epitomized by the "small, indigenous, mass" 小土群 campaign, which extended to various industries, including the energy industry.[3]

"Smallness" during the Maoist period was characterized not just by the physical size of hydropower projects, but also by the approach taken in building them. The Maoist principle of "small, indigenous, mass" was applied to the energy industry, including hydropower, as part of the drive for socialist industrialization. This approach was most famously taken in the "backyard furnace" campaign, although, in the context of the steel-making industry, it resulted in failure. In the development of small hydropower projects, however, it was more successful and it continued to shape the industry until the end of the 1970s.

This chapter explores the role of small hydropower in China's rise as a hydropower nation. It analyzes how Maoist mass campaigns and a Maoist politics of technology expanded and altered the conventional technocratic approach.[4] The "model county" of Yongchun in Fujian Province will serve as a case study to illustrate how the rise of the hydropower nation in China was driven not just by the worldwide "concrete revolution" but also by Maoist mass campaigns.

---

[3] See Shapiro, *Mao's War against Nature*; Smil, *China's Energy*.

[4] The most widely read book on small hydropower in China is by Tong Jiandong 童建栋, former director of the International Center of Small Hydro Power. Tong briefly discusses the history of small hydropower in China from the early twentieth century, but his book focuses primarily on the management and engineering of small hydropower in contemporary China. See Tong Jiandong, *Zhongguo xiaoshuidian* 中国小水电 (*Small Hydropower in China*) (Beijing: Zhongguo shuili shuidian chubanshe, 2006). Other sources for this chapter include Gao Jun 高峻, "Ershi shiji wushi niandai Fujian xiaoshuidian jianshe de xingqi" 二十世纪五十年代福建小水电建设的兴起 (The Rise of Small Hydropower in Fujian in the 1950s), *Dangshi yanjiu yu jiaoxue* 党史研究与教学 6 (2009), 67–75. Sources in English include the following: Based on official reports, Robert Carin summarizes the general progress that the Communists had made in rural electrification by the early 1960s, see Robert Carin, "Rural Electrification," in *Contemporary China, VI 1962–1964* ed. Stuart Kirby (Hong Kong: Hong Kong University Press, 1968), 11–21; Vaclav Smil mentions small hydropower briefly in his comprehensive study of China's energy industry; see Smil, *China's Energy*. Leslie T. C. Kuo and Robert C. Hsu also briefly introduce the construction of small hydropower in rural China in their studies of Chinese agriculture in the PRC; see Leslie T. C. Kuo, *Agriculture in the People's Republic of China: Structural Changes and Technical Transformation* (New York: Praeger Publishers, 1976), 236–243; Robert C. Hsu, *Food for One Billion: China's Agriculture since 1949* (Boulder: Westview Press, 1982), 82–85.

MAKING HYDROPOWER A MASS CAMPAIGN

In January 1956, Mao threw his support behind the "National Program for the Development of Agriculture from 1956 to 1967." This plan prioritized the expansion of agricultural production over a more capital-intensive approach. Mao believed that the state should leverage the large labor force present in rural China to achieve growth. In summarizing this approach, Jürgen Domes states that combining political campaigns with production campaigns would, for Mao, allow for quantitative and qualitative "leaps" rather than gradual and measured development.[5] At the same time, following the Soviet model, electrification was seen as the most effective way to boost productivity. Before the Water Conservancy Ministry and the Electricity Ministry were merged in 1958, the former had already taken on the task of developing small hydropower projects in rural areas.

Small hydropower was seen as a key aspect of agricultural mechanization efforts in China, as it would provide electrical power that could improve irrigation and the processing of agricultural products.[6] By June 1956, 361 rural hydroelectric stations had been built in the country, primarily in Sichuan, Fujian, Zhejiang, and Guizhou provinces.[7] The Water Conservancy Ministry aimed to support rural communities in building an additional 1,000 small hydroelectric stations, with a combined capacity of 30,000 kilowatts, to supply electricity to over half a million rural households. But the Ministry faced significant challenges in implementing this ambitious plan, as its resources were limited. The principle behind the plan was that local communities would build their own small hydroelectric stations with state assistance of various kinds.

The greatest challenge China faced in completing the task of rural electrification was the acute shortage of technical personnel. To address this problem, the Water Conservancy Ministry provided training to 400 technical cadres in Shanxi, Fujian, and Sichuan provinces, where

---

[5] Jürgen Domes, *Socialism in the Chinese Countryside: Rural Societal Policies in the People's Republic of China, 1949–1979*, translated by Margitta Wending (London: C. Hurst & Co., Ltd., 1980), 21. Deng Zihui, the most prominent expert in agriculture in the government, was criticized by Mao as a rightist for his conservative attitude toward rural collectivization and this proposal.

[6] On agricultural mechanization, see Benedict Stavis, *The Politics of Agricultural Mechanization in China* (Ithaca: Cornell University Press, 1978).

[7] Luo Xuan 罗漩, "Kan shuidian fadian jianshe zhanlan" 看水力发电建设展览 (Visit the Hydropower Construction Exhibit), *Renmin ribao*, October 13, 1957.

small hydropower was relatively developed. These cadres later became a driving force behind the nationwide rural hydro-electrification campaign.[8] Before the Great Leap Forward, the technocratic approach was still generally believed to be the most efficient way to achieve rural electrification. However, because of technical and financial constraints, mass participation and cost sharing were also necessary for infrastructure construction. Mao Zedong encouraged both class struggle and the "mass line," which resulted, in rural areas, in a combination of the technocratic approach and mass participation in the development of small hydroelectric stations. Hydroelectricity was still a new concept for most of the rural population in the 1950s, but the use of hydropower, as in water mills and water trip hammers, had a long history in China. Local people often leveraged existing waterpower facilities, such as local water mills, in building small hydroelectric stations, drawing on an indigenous technological legacy.

Gordon Bennett's study of mass movements in Communist China suggests that these movements contributed more to overall economic growth than they took away.[9] While Bennett's conclusions about campaign-style economic management may be true for the 1960s and 1970s, the 1950s tell a different story. During this period, tens of millions of peasants died due to widespread famine, highlighting the shortcomings of radical Maoist ideology and policies.[10] The 1950s was also a period of rivalry on the world stage between the Soviet-led socialist camp and the United States-led capitalist camp. In 1957, Soviet Premier Nikita Khrushchev proposed that the Soviet Union would surpass the United States in gross domestic production within fifteen years. Mao followed suit, proposing that China could surpass Britain within the same timeframe, thereby setting an ambitious target for China's economic growth.

The rapid growth of electrical power output was crucial to facilitating industrial and agricultural progress in other areas. During the height of the water conservation movement, from 1957 to 1958, over 1,592 reservoirs with a total capacity of 25.9 billion cubic meters were built. These

---

[8] "Jinnian quanguo jiang xingjian yiqianduo zuo xiaoxing shuidianzhan" 今年全国将兴建一千多座小型水电站 (One Thousand Small Hydropower Stations Will Be Built This Year Nationwide), *Renmin ribao*, February 10, 1956.

[9] Gordon Bennett, *Yundong: Mass Campaigns in Chinese Communist Leadership* (Berkeley: University of California Press, 1976).

[10] Yang Jisheng, *Tombstone: The Great Chinese Famine, 1958–1962* (New York: Farrar, Strauss and Giroux, 2013); Felix Wemheuer, *Famine Politics in Maoist China and the Soviet Union* (New Haven: Yale University Press, 2014).

reservoirs were estimated to have the potential to produce a minimum of one million kilowatts of electricity. In reality, though, only 200 of these projects were equipped with hydroelectric systems, and these had a total of 100,000 kilowatts of capacity – just one tenth of the expected amount.[11] Champions of hydroelectricity lamented the building of dams without the ability to generate electricity as a waste. In line with the principle of multipurpose exploitation, the party encouraged the inclusion of hydroelectricity production in water conservancy efforts, stating as a matter of policy that "all qualified hydraulic projects shall generate electricity." To increase the capacity of already completed reservoirs, the party also asked localities to increase the height of dams or to build additional upstream reservoirs if the outflow was insufficient to drive turbines. In many areas, particularly in northern China, agricultural irrigation consumed most of the available water, which made it difficult to establish stable reservoir reserves for the production of electricity. Some proponents of hydropower even suggested adjusting cropping systems in irrigated areas if the impact on grain yields could be limited, to ensure a stable source of water for electricity production.[12]

In the 1950s, most power plants – which were a legacy of pre-1949 efforts to build industrial infrastructure – were located in or near major urban centers and provided electricity exclusively to cities and towns.[13] Most rural areas in China still relied on kerosene lanterns for lighting. Many peasants had never seen an electric light bulb, let alone learned how to generate electricity. In the popular imagination of rural residents, electricity was associated with foreign, urban, elite, and professional elements.

Despite the efforts of local elites and the National Resources Commission to promote hydroelectricity, it remained beyond the reach of most people. Nevertheless, the Communist central government urged people to build hydroelectric projects of various sizes, ranging in capacity from a few kilowatts to tens of thousands, which were dubbed "electrical flowers." During the Great Leap Forward from 1958 to 1961, officials transformed small hydropower into a top-down initiative, yet they still relied heavily on local resources for its implementation. Despite the "poor and blank" state of rural areas, as they were described, the state expected peasants to achieve rural electrification largely on their own.

---

[11] On the water conservancy campaign, see Pietz, *Yellow River*.
[12] "Neng fadian de shuili gongcheng douyao fadian."
[13] See Wright, "Electric Power Production in Pre-1937 China."

Lack of technical knowledge was a major early obstacle to making the construction of small hydroelectric stations a movement that involved the masses. Despite the aforementioned training of hundreds of technical cadres in 1956, there were still not enough of them to keep pace with the ambitions of the Great Leap Forward. To address this problem, at a national conference of mid-size water conservancy and hydroelectric projects in Zhengzhou in 1958, the Party Central Committee asked every prefecture in the country to build a model hydroelectric station with a capacity of 500–1000 kilowatts. These models were meant to serve as examples for the masses to follow.[14]

In the city of Lüda in Liaoning province, for instance, people were mobilized to participate actively in the campaign to generate electricity. Factories in light industries, such as textiles, were encouraged to generate their own electricity. As a coastal city, it was also possible that Lüda could harness tidal energy for electrical power, although electrical engineers in China had not yet found an efficient way to generate stable electricity from tides. During the mass campaign, it was claimed that textile workers, middle school students, and teachers successfully built two tidal power electrical stations in the city. According to Maoist ideology, the "masses" were considered to be smarter than "experts" in what was referred to as a "revolution in electricity." People from many professions, including doctors, nurses, firefighters, security guards, and others, joined in the electrification campaign. Various sources of energy – industrial wastewater, drainage water, drinking water, heating steam, wind, coal, and methane – were used to drive electricity generators.[15] Due to the pervasive shortage of steel and other metals, people started to reuse materials like iron waste and wood to manufacture turbines and generators. This was endorsed by state media as a "technological revolution."

To mobilize the rural population, state media and officials promoted the idea that the task of building hydroelectric projects was not one that was reserved for specialists and technical cadres, and that it was not a mysterious process. Maoists argued that relying on experts to accomplish things was a superstition and encouraged localities to mobilize indigenous resources to build small, middle-sized, and large hydroelectric projects.[16] They suggested that every commune should make use of any reservoirs,

---

[14] "Neng fadian de shuili gongcheng douyao fadian."
[15] Hu Ming 胡明, "Qunzhong bandian" 群众办电 (The Masses Produce Electricity), *Renmin ribao*, November 15, 1958.
[16] "Neng fadian de shuili gongcheng douyao fadian."

rivers, and creeks that were available, and that "every waterhead should be used to generate electricity."[17]

On August 16, 1958, the Agriculture Ministry held the first national conference on rural electrification in Tianjin. It proposed a preliminary "five counties, one hundred communes" rural electrification mass campaign, in which each province would choose five counties and one hundred communes to be electrified first, which would then serve as models for the rest to follow. Following the conference, a mnemonic formula was circulated to convey the major principles for rural electrification:

1. Give priority to small-sized projects and agricultural production
2. Center on each commune
3. Plan comprehensively
4. Combine various sources of energy
5. Give priority to hydroelectricity
6. Rely on the masses
7. Build stations with frugality
8. Emphasize both construction and management
9. Follow the spirit of "more, faster, better, and more economical."[18]

The national policy for rural electrification was not created out of nothing. Like key policies in other fields, it was based on successful local practices in the countryside. Following Gordon Bennett's analysis, in a mass movement, new policy ideas were first conceived, then debated and negotiated, and finally decided upon, although we do not in this case have access to the exact details of the policymaking process. Based on the local experience of electrification in Yongchun, which I will examine later in this chapter, it seems clear that the key points of the policy were distilled from practical experience.[19]

However, bureaucratic quotas like the "five counties and one hundred communes" target, which was imposed across the country, disregarded local differences. The implementation of state policies followed the hierarchical structure of the Communist Party, and existing scholarships showed that career incentives, loyalty to Mao, and inner conviction were all essential to local cadres' compliance with central directives during the

---

[17] Liu, "Wei quanguo chubu dianqihua er fendou."
[18] "Guangfan yongdianli daiti laoli, quanguo jiangxianqi nongcun dianqihua gaochao" 广泛用电力代替劳力，全国将掀起农村电气化高潮 (Widely Using Electrical Power to Replace Labor Force, a Nationwide High Tide of Rural Electrification Is Coming), Renmin ribao, August 18, 1958.
[19] Bennett, Yundong, 39.

Great Leap Forward.[20] The CCP Organization Department's tactics of cadre deployment and strict bureaucratic discipline effectively placed local cadres under the control of the party center.[21]

The central government's goal for rural electrification during the Great Leap Forward was perceived by local cadres as the bare minimum target. Fearing criticism as "rightists," they very often set out even more ambitious targets for their localities, regardless of any challenges posed by hydrological conditions, technical difficulties, or equipment shortages. In 1957, according to official figures, rural China only had 20,000 kilowatts of installed electricity capacity; the goal of the Great Leap was to increase it to 900,000 kilowatts.[22] Because of pressure from above, many projects were started before proper preparation and surveys had been done, which resulted in many projects being abandoned before they were complete.

The drive to meet artificially high targets and widespread exaggeration of achievements was evident in the small hydropower campaign. In Suining 遂宁 in Sichuan province, for example, the prefectural party committee initially planned to build 600 small hydroelectric stations in a single county. However, the cadre responsible for the county expressed concern about meeting this goal because of a shortage of labor and materials. In response, the prefectural party secretary challenged the cadre to build 1,000 stations, and then a further 1,000 if they were needed. In the end, the county was instructed to build 2,000 small hydroelectric stations. To spur enthusiasm, the prefectural committee dangled the promise of building factories in the county that fulfilled its hydroelectric station quota first. They also promised that a chemical fertilizer factory would be built once hydroelectricity was available, and that this would greatly benefit the local peasants.[23]

---

[20] James Kai-Sing Kung and Shuo Chen, "The Tragedy of the Nomenklatura: Career Incentives and Political Radicalism during China's Great Leap Famine," *American Political Science Review* 105, no. 1 (2011), 27–45; Dali Yang, Huayu Xu, and Ran Tao, "The Tragedy of the Nomenklatura? Career Incentives, Political Loyalty and Political Radicalism during China's Great Leap Forward," *Journal of Contemporary China* 23, no. 89 (2014), 864–883.

[21] Daniel Koss, *Where the Party Rules: The Rank and File of China's Communist State* (Cambridge: Cambridge University Press, 2018), 244–249.

[22] "Woguo kaishi xiang dianqihua maijin" 我国开始向电气化迈进 (Our Country Is Marching toward Electrification), *Renmin ribao*, May 15, 1958.

[23] *Dagao qunzhong yundong quanmin ban shuidian* 大搞群众运动全民办水电 (*The Mass Campaign of Hydropower Construction*) (Beijing: Shuili shuidian chubanshe, 1959), 5.

Despite the state's promise to provide assistance for local hydropower construction, self-sufficiency was the only option for many communities. With limited state support, local materials and native knowledge were crucial to the success of the small hydro campaign. Lacking steel and iron, communities turned to wood and bamboo to build turbines. And without sufficient state loans, local communes often turned to rural households for fundraising. In Sichuan, again because of technical and financial limitations, communities built conventional watermills first to process grains and other agricultural products, thereby accumulating funds to purchase and install electrical generators and motors later.[24] In Shuanglong Commune in Zhejiang Province, people converted five existing watermills into small hydroelectric stations in just fifty days.[25] Even when funds were available, it was often difficult to find electrical generators and motor manufacturers to meet the sudden demand for electrical equipment during the hydro-electrification campaign. With a lack of trained electrical technicians, many communes relied on traditional craftsmen, such as carpenters, blacksmiths, and masons, who thereby transformed themselves into "proletarian engineers." With their traditional skills, this group of local experts began to build small hydroelectric stations and to manufacture turbines, generators, and motors in their own unique ways.[26] In the following section I will explore the mass campaign in Yongchun, a model county for small hydropower during the Maoist period.

MODEL COUNTY OF SMALL HYDROPOWER: YONGCHUN

In Maoist China, when a national policy was announced as a part of a mass movement, the state would encourage other administrative units to adopt it by promoting a representative unit at the national level. This approach was known as the "expanding the point to the plane" strategy (*yidian daimian* 以点带面).[27] In the case of the rural small hydropower

---

[24] "Woguo kaishi xiang dianqihua maijin."
[25] "Gaosuduo xiang dianqihua maijin, Jin Xuecheng weiyuan de fayan" 高速度向电气化进军, 金学成委员的发言 (Marching toward Electrification with High Speed: Speech by Jin Xuecheng), *Renmin ribao*, April 10, 1960.
[26] Jeremy Brown mentions the contribution of a downsized worker in the electrification of his home village near Tianjin. However, more research is needed to estimate the role of workers in rural electrification nationwide. See Jeremy Brown, *City Versus Countryside in Mao's China: Negotiating the Divide* (Cambridge: Cambridge University Press, 2012), 103–104.
[27] Bennett, *Yundong*, 39.

campaign, Yongchun county in Fujian province was selected as a model county for other places to follow. What factors led to Yongchun being chosen as a model for small hydropower?

Yongchun is located in the middle-southern region of Fujian province and falls under the administration of Quanzhou city. Its terrain generally slopes downward from west to east, and over 70 percent of the county's land is mountainous. Due to the presence of Pacific Ocean monsoons, Yongchun receives an average annual precipitation of 1,681.6 mm. The majority of the rainfall occurs between May and August, with only 11 percent of the annual precipitation occurring between October and January.[28] This heavy precipitation combined with the mountainous topography means that several significant streams run through the county. Four of the largest are Taoxi, Huyangxi, Yiduxi, and Kengzaikouxi, which have an estimated combined total annual runoff of between 1.5 and 1.8 billion cubic meters.[29] These factors together provide the county with the natural conditions necessary to develop small hydropower.

Small hydropower had also already been introduced to Yongchun in the Republican period, which provided a technical foundation for its larger-scale development in the post-1949 period. In 1926, Lin Miaoqing 林妙庆, a resident of Guiyang 桂洋 village, purchased a 5-kW electrical generator from Shanghai to light the local school. Initially powered by gasoline, the generator was connected to a water wheel on the riverbank in 1930 when gasoline became scarce, thus creating the first hydroelectric station in Yongchun. By 1949, the county had already constructed five small hydroelectric stations to provide residential illumination, giving the local society a certain degree of experience with hydroelectricity prior to the Great Leap Forward.[30] Yongchun was not the only county with hydroelectricity before 1949, though, as previous chapters have described the earliest hydroelectricity efforts in Yunnan and Sichuan in the 1910s and 1920s. While the county's natural environment and historical legacy were important factors, it was the proactive response of local cadres to the central government's call that gave Yongchun the chance to stand out and gain national recognition as a model county for small hydropower.

[28] Yongchun xianzhi bainzhuan weiyuanhui 永春县志编撰委员会, *Yongchun xianzhi* 永春县志 (*Yongchun gazetteer*) (Beijing: Yuwen chubanshe, 1990), 123.
[29] Yongchun xianzhi bainzhuan weiyuanhui, 301.
[30] Yongchun xianzhi bainzhuan weiyuanhui, 303.

As I mentioned in the previous section, the National Water Conservancy Conference of 1955 promoted the development of small hydropower as a practical way to increase grain yields and address energy shortages in rural China. The "Outline of National Agricultural Development for 1956 to 1967" also emphasized the generation of electricity from all hydraulic projects in which that would be possible. In 1955, Yongchun county leader Fan Jinfu 范进甫 conducted a survey of hydropower in the county and developed a preliminary plan for rural electrification to implement the central government's new policy. The plan was submitted to Liu Shaoqi, whose response was positive. The Ministry of Water Conservancy then dispatched officials to Yongchun to assess its small hydropower development. Impressed by its already existing small hydropower projects, the Ministry of Water Conservancy selected Yongchun as a model county for the national campaign on rural electrification, with a particular emphasis on small hydropower development.

For the mass campaign style of small hydropower construction to succeed, the state needed to train more hydropower technicians. In February 1956, a training session for representatives from eight southern provinces on small hydropower technology was conducted in Yongchun. Each province was asked to send several young people who had graduated from middle school for basic training. During the program, technical instructors and trainees worked together to construct five small hydroelectric stations. In addition to designing small hydropower plants, the state also expected trainees to learn how to manufacture essential equipment on their own. To facilitate this, in March 1956, the Water Conservancy Bureau of Fujian province organized a training session in Yongchun on how to manufacture wooden turbines, using wood propeller turbines and wood pressure pipes as models for trainees from other counties to emulate.[31]

In the public narrative of the small hydropower campaign, the wood propeller turbine was frequently praised as one of the most notable achievements of local technology. However, it may not have been a completely indigenous technology: A Soviet design for a wooden turbine had been introduced to China in 1956.[32] Xie Yusheng 谢玉生, a technical

[31] Fujiansheng difangzhi bianzuan weiyuanhui 福建省地方志编纂委员会, *Fujian shengzhi: shuilizhi* 福建省志：水利志 (*Fujian Provincial gazetteer: Water Conservancy*) (Beijing: Zhongguo shehui kexue chubanshe, 1999), 156.
[32] On the introduction of the Soviet design for a wooden propeller, see *Nongcun xiaoxing shuidianzhan cankao ziliao* 农村小型水电站参考资料 (*Materials on Rural Small Hydropower Stations*) (Beijing: Shuili chubanshe, 1956), 33–55.

worker at Yongchun's agricultural tool factory, successfully made a wooden turbine with local materials and thus became known as "the *tu* expert." It is unknown whether he was inspired by the Soviet design or he developed the technology entirely on his own. Another "*tu* expert," Zheng Shenglin 郑声廪, attended the small hydropower training session as an electrical student and later designed five small hydroelectric stations. Zheng became the "chief engineer" of the mass hydro-electrification campaign in Yongchun during the Great Leap Forward, organizing short-term training sessions that provided skills to over 300 rural technicians. This training enabled indigenous technicians to become skilled workers capable of implementing the electrification campaign by bringing it to their own communes and production brigades.[33] Not everyone had the opportunity to receive training, though; most trainees were already relatively skilled carpenters, blacksmiths, or masons in their own communities. For example, Huang Ziduan 黄自端, originally a carpenter, transformed himself into a specialist on hydroelectric equipment installation on the basis of the short-term training he had received. To address equipment shortages at a local agricultural tool factory, workers used a 200,000 *yuan* donation from a returned overseas Chinese merchant, You Yangzu 尤扬祖, who had been born in Yongchun, to begin producing electrical motors and generators.[34] Although they lacked any experience in designing electrical equipment, the workers at the factory, primarily carpenters and blacksmiths, disassembled an electrical motor they had borrowed in order to replicate its construction and create new ones.

Alongside the state's promotion of self-reliance and the use of local materials and knowledge for small hydropower projects, in 1958 the Yongchun county government nonetheless opted to import four sets of hydropower equipment from Austria. These machines were used to provide electricity to the county seat, where the local government was located.[35] This decision, which prioritized the needs of the industrial

---

[33] "Yongchun shuidian shiye dafazhan" 永春水电事业大发展 (The Great Achievement of the Yongchun Hydropower Enterprise), *Fujian ribao* 福建日报 (*Fujian Daily*), January 21, 1959.

[34] Zhang Jiaqin 张加芹, "Yongchun xiaoshuidian fazhan jianjie" 永春小水电发展简介 (Brief Introduction to Small Hydropower Development in Yongchun), *Yongchun wenshi ziliao* 永春文史资料 24 (1994), 60.

[35] Quanzhoushi shuilishuidianju 泉州市水利水电局, *Quanzhoushi shuilizhi* 泉州市水利志 (*Water Conservancy in Quanzhou*) (Beijing: Zhongguo shuilishuidian chubanshe, 1998), 277.

sector and urban residents over the rural population, was at odds with the state's purported goal of creating a more egalitarian society.[36]

Yongchun county, following the central government's Five-Year Plan for rural electrification, developed its own county-level plan in 1958. The plan called for the construction of 260 small hydropower stations, which would provide the electricity required to process over 100 billion *jin* of wheat, sweet potatoes, and beans, as well as 30 million *jin* of animal feed, 3,000 cubic meters of timber, and 1,000 *dan* of tea. The use of hydroelectric-powered threshing machines was also envisioned.[37] In 1958 and 1959, thirty-seven new small hydropower stations were built, with nineteen of them fully financed by members of agricultural brigades and communes and the remainder subsidized by bank loans or government aid.

Initially, commune cadres were successful in rallying support for the projects from the masses. However, the limited financial means of the local population in Yongchun made them hesitant to contribute in monetary ways. At the same time, while the central government supported rural electrification in principle, little financial assistance was available to local communities.

Local communes in Fujian province utilized a technological complex that allowed them to manufacture turbines and conduits using local timber and pebbles without needing cement. But they lacked the expertise and resources to manufacture generators and other electrical equipment on their own. With funds limited, many communes built a waterpower plant – similar to a conventional watermill – first and charged members of the community to use it to process grain. After they had earned enough funds in this way, communes were then able to purchase electrical equipment and upgrade the project into a hydroelectric plant. According to a 1958 survey of small hydropower plants in Fujian, the cost of building such a plant was between 400 and 1,200 *yuan* per kilowatt. The generator and transmission equipment represented 40–60 percent of the total cost. By following the two-step approach, the total cost of building a 10–20 kilowatt hydropower plant would be between 2,000 and 8,000 *yuan*, which was seen as affordable to most agricultural communes.[38]

---

[36] Social inequality is often intertwined with energy extraction, consumption, and its environmental and health ramifications. See Andrew Needham, *Power Lines: Phoenix and the Making of Modern Southwest* (Princeton: Princeton University Press, 2014).

[37] "Xiang nongcun dianqihua de daolu maijin" 向农村电气化的道路迈进 (Marching toward Rural Electrification), *Fujian ribao*, April 5, 1958.

[38] Shanghai shuili fadian shejiyuan 上海水力发电设计院, "Minzhe liangsheng xiaoxing shuidianzhan diaocha baogao" 闽浙两省小型水电站调查报告 (Report on Small Hydropower in Min and Zhe Provinces), *Shuili fadian* 8 (1958), 41–43.

As more small hydropower plants were completed, many communes adjusted their approach to electrification. Yongchun county was a pioneer in the use of the principle of "using electricity to produce electricity" (*yidian yangdian* 以电养电) by using the charges earned by the stations to expand their electrical capacity or to build additional stations. This approach became known as "like a hen delivers eggs, one station produces another" (母鸡生鸡蛋，一站生一站), vividly illustrating the self-sustaining nature of the system.[39] Unlike the "making steel campaign," which resulted in a significant waste of resources, the small hydropower campaign initially followed a pragmatic approach that maximized the use of local resources and expertise while remaining mindful of financial limitations.

Although small hydropower stations were built in Yongchun county to boost grain output and labor efficiency, they failed to save farmers from the devastation of famine between 1958 and 1960. The state's grain acquisition policies deprived farmers of their most basic necessities, causing widespread suffering and illness. Women were particularly vulnerable, experiencing edema, amenorrhea, and other serious health problems that often led to their deaths. In some cases, bodies remained unburied for days as surviving villagers lacked the strength to dig graves or carry coffins. Despite the urgent need to prioritize agriculture and alleviate the suffering of the local population, many cadres were preoccupied with small hydropower and "making steel" campaigns, which often involved unscientific and counterproductive instructions that undermined agricultural production. Peasants were forced to increase the density of crops to boost production, for example, which often resulted in there being no harvest at all. The pressure to achieve production targets was so intense that one brigade chief at the Chengguan Commune in Yongchun committed suicide.[40] Local cadres persisted with their misguided policies, knowing the potentially devastating consequences of sending peasants' rations away to the central state.

But the relentless drive for production growth came from the top of the political hierarchy, and those who failed to meet their assigned targets risked being labeled "rightists." In Fujian province, provincial cadres

---

[39] "Xiaoshuidian zhixiang: Fujian Yongchuxian" 小水电之乡—福建省永春县 (Hometown of Small Hydropower: Yongchun, Fujian Province), *Fujian shuili shizhi ziliao* 福建水利史志资料 2 (1984), 13-17, 14.

[40] Li Zili 李自力, "Sannian zanshi kunnan zai Yongchun Chengguan gongshe" 三年暂时困难在永春城关公社 (Three Difficult Years in Chengguan Commune, Yongchun), *Yongchun wenshi ziliao* 21 (1994), 35-43.

threatened to denounce cadres who were unable to complete their assigned tasks, using tactics similar to those employed during the violent land reform campaigns of the early 1950s. Under such political pressure, many local cadres made the tragic decision to sacrifice the lives of peasants. The consequences of these policies were disastrous, resulting in a famine that claimed countless lives and leaving a lasting legacy of suffering in the region.

During the Great Leap Forward, the socialist state mobilized the masses to join campaigns like the steel campaign and enforced high rates of grain acquisition, taking limited resources away from peasants or wasting them in failed efforts to produce more resources for the state. Although small hydropower stations were not as widespread as the backyard furnaces so powerfully associated with the steel campaign, both campaigns resulted in the waste of human labor and local resources, as well as environmental destruction, albeit to different degrees.[41] Unlike the steel campaign, which was at odds with agricultural production, the preliminary rural electrification plan aimed to increase agricultural output. Specific targets for the construction of small hydropower plants and the installation of electrical capacity were assigned to counties and communes by higher authorities. As a top-down campaign, though, electrification did not prioritize peasants' daily needs. In addition to contributing labor service, commune members were asked to donate construction materials. Women, in addition, were often expected to give away their jewelry to raise funds for public projects.

Although small hydropower stations permitted an expansion in the amount of land that could be irrigated and brought electric lighting to many rural households, the campaign's successes were meaningless without an adequate supply of grain and a living population. The small hydropower campaign did not cause the famine in Yongchun, but its model of construction and form of organization exacerbated the situation. Top leaders and the bureaucratic system bear ultimate responsibility for the devastating consequences of the Great Leap Forward, but the campaign's approach to building small hydropower did not help.

By 1960, more than 9,000 small hydroelectric stations had been built in rural areas across China. The Great Leap Forward, however, ended in catastrophe for rural communities, prompting a period of economic adjustment and of reflection on the campaign style of mass mobilization

---

[41] Shapiro, *Mao's War against Nature*, 80–86.

under the leadership of moderates in the party. The radical Maoist line of building small hydroelectricity with mass participation was brought to a halt. With Mao's return to the national political stage in the lead up to the Cultural Revolution in 1966, the campaign for small hydropower was revived. Despite the failures of the Great Leap Forward, the state again promoted small hydropower as a way of promoting rural electrification and development. In the late 1950s, Yongchun county had served as a model for small hydropower in the south; in the 1970s, Qinglong 青龙 county in Hebei province was promoted as another model county in the north.

In 1975, the Water Conservancy and Electricity Ministry released an educational film called "Small Hydroelectricity in the Mountainous Countryside," which described the experiences of Qinglong county as a model for building hydroelectric stations despite the presence of natural and geological disadvantages. Although the county had many small creeks and springs, they were not large enough to drive turbines. Undeterred, the people of Qinglong dug canals in the mountains to channel the creeks and springs into a single larger flow.

Being in northern China, Qinglong experiences unevenly distributed precipitation throughout the year. To ensure a steady supply of electricity in the dry season, residents built reservoirs to conserve water during the summer. Unlike Yongchun, which had fast-flowing rivers with significant hydropower capacity, the rivers in Qinglong flowed slowly, limiting their potential for generating power. To overcome this disadvantage, people straightened them to increase their velocities. By sharing the experiences of Qinglong, the film encouraged people to overcome obstacles posed by nature and to build hydroelectric stations with innovation and creativity. The key to success, the film suggested, was for people to "liberate their minds" and to find new ways to harness the power of water to drive turbines. Despite the challenges of building hydroelectric stations in mountainous countryside, the film argued, the experiences of Qinglong county demonstrated that, with determination and ingenuity, anything was possible.[42]

The Wenzhangzi brigade built the first small hydroelectric station in Qinglong in the winter of 1957 by containing spring water in the mountains. With an installed capacity of 20 kW, it was a modest beginning.

---

[42] Guo Yi 郭仪, "Keke mingzhu fangguangcai: caise kejiaopian 'Shancun xiaoshuidian' guanhou" 颗颗明珠放光彩—彩色科教片 '山村小水电' 观后 (Every Pearl Shines: Review of the Color Scientific Education Film "Small Hydroelectricity in the Mountainous Countryside"), *Renmin ribao*, May 9, 1975.

In 1966, though, the Donghaocun brigade launched a new offensive aimed at taming the Qinglong River. Brigade members dug a 2,300 meter long conduit and a 1,900 meter long dam to supply an 84-kW hydroelectric station. By using electricity-powered pumps, they were now able to irrigate more than 60 percent of their land, leading to a significant increase in crop yields. Inspired by the success of Donghaocun, the Toudaohe brigade in Qinglong reclaimed 10 *mu* of rice paddy from wasteland during the "learning from Dazhai" movement in the 1970s. They dug a 1,500 meter long ditch to divert river water for irrigation and installed a self-made wooden turbine to power two mills for grain processing. They then used the funds accumulated by the mills to purchase a steel turbine and an electrical generator, which equipped a 2.8-kW hydroelectric station. With the electricity thus generated, a water pump was installed to deliver water up to the hills, which significantly increased the amount of irrigated land, from 20 *mu* to 150 *mu*. This also raised the average yield from 139 kg per *mu* to 201 kg per *mu*.

Encouraged by the successes of Donghaocun and Toudaohe, the county's Cultural Revolution Commission urged other communes and brigades to follow their models, promoting the principles of "managing water and generating electricity, promoting mechanization with electricity, and obtaining electricity and grain from water" (治水办电, 以电促机, 向水要电, 向水要粮). By 1977, eighty-four small hydroelectric stations had been built in Qinglong with a total capacity of 5,632 kW. More than 17,300 *mu* of land was irrigated with electric pumps, and hydroelectricity was used to power the county's agricultural tool factories.[43] The lessons learned from Qinglong demonstrated the potential of small hydroelectricity to promote rural development and improve people's lives.

In the 1970s, China was in the throes of another revolutionary campaign known as "learning from Dazhai." The goal of this campaign was to inspire people to work hard, learn from successful models of socialist development, and conquer nature through collective effort. One of the most ambitious projects to emerge from this campaign was the Hongqi (Red Flag) Canal in Linxian 林县, Henan province. The Hongqi Canal had been built to address local drinking and irrigation needs, but the leaders of Chengguan Commune saw its potential as a source of energy

---

[43] Sun Zhaoming 孙肇明 and Han Hai 韩海 (eds.), *Qinglong Manzhu zizhixian shuilizhi* 青龙满族自治县水利志 (*Water Conservancy in Qinglong Manchu Autonomous County*) (Tianjin: Tianjin daxue chubanshe, 1993), 69.

and a symbol of socialist progress. They set out to build not just one, or two, but twenty-seven hydroelectric stations along the canal, which was 6 kilometers long, using the power generated to drive improvements in industry and agriculture.[44] By 1973, ten stations were up and running, and more were under construction. The people of Chengguan worked tirelessly to dig canals, build dams, and install turbines, driven by a shared commitment to the revolutionary ideals of the time. As the years passed, the Hongqi Canal grew into a sprawling network of eighty-three hydropower stations with a total capacity of 17,000 kW. From the largest unit, with an output of 1,600 kW, to the smallest, which churned out a modest 5 kW, these stations represented a triumph of collective labor according to a report produced by the United Nations Industrial Development Organization in 1981.[45]

In the context of the often chaotic Cultural Revolution, the introduction of small hydropower into rural communities represented a significant technological innovation for agricultural growth, despite limited material and capital investments. As illustrated in Figures 4.1 and 4.2, the number and capacity of hydropower stations in rural areas tripled across the 1970s, highlighting the basic success of this initiative in promoting rural electrification and improving agricultural efficiency.[46]

SMALL IS NOT ALWAYS BEAUTIFUL

Small hydroelectric stations brought significant benefits to rural areas, particularly in terms of agricultural production. The rural electrification plan was designed to support and improve farming by increasing irrigation, grain processing, and other sideline economic activities. According to official data, many places saw remarkable increases in labor efficiency and agricultural yields. In the 1970s, as David Pietz has

---

[44] Lin Feng 林风, "Hongqi qupan" 红旗渠畔 (On the Banks of the Hongqi Canal), *Renmin ribao*, December 9, 1973.
[45] United Nations Industrial Development Organization, Group study tour in the field of medium and small-scale hydropower plants to the People's Republic of China, final report, UNIDO/IO.467 (September 16, 1981), 6.
[46] Various scholars have pointed out the improvement of people's living standards and agricultural growth under the commune system during the Maoist period. See, for example, Dwight Perkins and Shahid Yusuf, *Rural Development in China* (Baltimore, MD: Johns Hopkins University Press, 1984); Joshua Eisenman, *Red China's Green Revolution: Technological Innovation, Institutional Change, and Economic Development under the Commune* (New York: Columbia University Press, 2018).

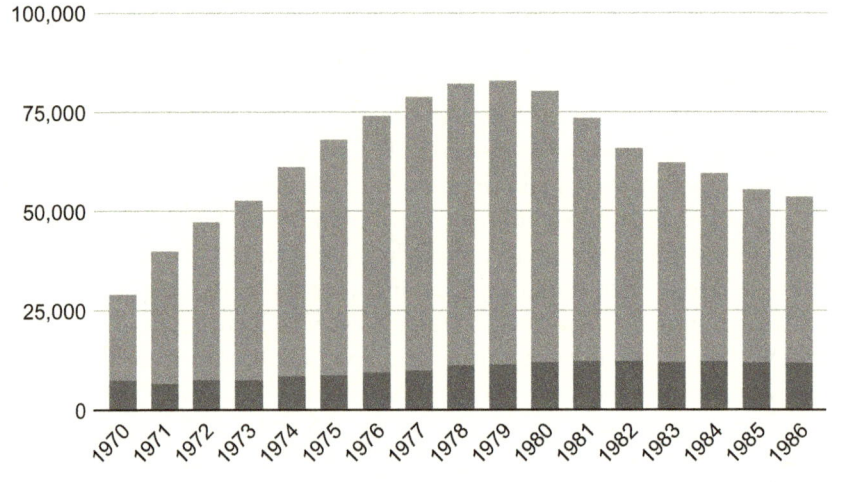

FIGURE 4.1 Estimated annual total number of hydropower stations in rural China, 1970–1986 (bottom: Built and owned by Communes/Xiang; top: Built and owned by Brigades/Villages)[47]

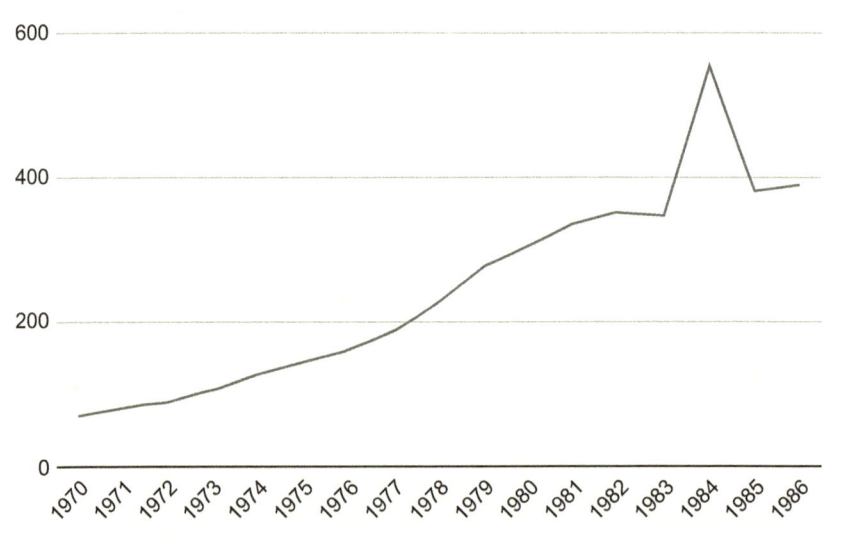

FIGURE 4.2 Electrical generation capacity of rural hydropower (unit: 10,000 kilowatts)[48]

shown, ground water was widely exploited and diesel-fueled electric pumps were generally used for agricultural irrigation in northern

---

[47] *Zhongguo nongcun jingji tongji daquan (1949–1986)* 中国农村经济统计大全 (*Economic Statistics for Rural China*) (Beijing: Nongye chubanshe, 1989), 321.
[48] *Zhongguo nongcun jingji tongji daquan (1949–1986)*, 321.

China.⁴⁹ However, in places like Yongchun and Qinglong, where hydropower was available, hydroelectricity powered pumps that could transmit river water to land at higher altitudes. Yongchun county, for example, increased its irrigated land by more than 1,000 *mu*, and this extra land produced more than 100,000 *jin* of additional grain. Moreover, electric lights could now be used on farmland at night to lure and exterminate moths and bollworms, which decreased the damage caused to crops by pests. Aquaculture also benefited from hydroelectricity, which generated heat to speed up the production of fish roe.⁵⁰

Electricity created by small hydroelectric stations built during the Cultural Revolution provided rural communities with a new energy source to power agricultural mechanization, which significantly eased onerous tasks like rice and flour milling. This freed up time for peasants to receive education and to engage in sideline production. It also facilitated political literacy.⁵¹ Electricity also enabled access to radio broadcasts and for many rural households it was the first time they had electric bulbs to illuminate their homes at night. In addition to the introduction of public education, improved access to medical services, and agricultural innovation, the improvement of rural residents' living standards during the Cultural Revolution can be attributed in part to the rural electrification facilitated by small hydro. This is remarkable, considering that the period is mostly remembered for political turmoil.⁵²

Small hydropower also had a significant impact on the roles of women in the rural economy, as it relieved them of the burdensome task of processing flour and rice. Women had been responsible for these tasks, but electric machines took them over, allowing women to spend more time on other forms of collective agricultural production. This prompted the state to proclaim that "thanks to collectivization, women can now get out of the kitchen, and thanks to electrification, women can now get out of the mill house."⁵³ As Gail Hershatter has pointed out, though, rural

---

⁴⁹ Pietz, *Yellow River*; Huaiyin Li, *Village China under Socialism and Reform: A Micro-History, 1948–2008* (Stanford: Stanford University Press, 2009), 229–258.
⁵⁰ Yongchunxian shuiliju dianlike 永春县水利局电力科 (Yongchun County Water Control Bureau Electric Power Office), *Yongchunxian nongcun shuili shuidian ziliao* 永春县农村水利水电站资料 (*Materials on Rural Water Conservancy and Hydropower in Yongchun County* (1964).
⁵¹ "Woguo kaishi xiang dianqihua maijin."
⁵² In an ethnographic study of one village, Mobo Gao describes many improvements in the rural population's standard of living during the Cultural Revolution. See Mobo Gao, *Gao Village: Rural Life in Modern China* (Honolulu: University of Hawai'i Press, 2007).
⁵³ Quanguo Nongye Zhanlanguan Shuiliguan 全国农业展览馆水利馆 (ed.), *Nongcun Shuidian* 农村水电 (*Rural Hydro*) (Beijing: Nongye chubanshe, 1960), 8.

collectivization actually increased women's workloads, and electric light bulbs extended their work hours.[54] With this new technology, women could perform household chores, such as needlework, at night. Thus, small hydropower, as a new presence in the rural economy, did indeed alter the spatial and temporal dimensions of gendered work, but the effects were more complex than the official narrative suggested.[55]

Although small hydropower construction brought about positive changes in rural life – increases in agricultural productivity and improvements in living standards, for example – the campaign form of the movement also created many problems. One major issue was the feverish rush to build hydroelectric stations without careful planning and preparation. In Suining county, Sichuan province, for example, proposed projects were approved summarily by the county committee, and infrastructure work began just days later. This left little time for proper studies and preparations. The result was low-quality construction and neglect of safety at construction sites. As a result, many electrical generators and transformers failed because of shoddy construction and mismanagement. Some hydropower stations functioned for a time, but many were eventually abandoned because of their poor quality.[56]

In some cases, a lack of gasoline and diesel oil made it impractical to use internal combustion engines to drive generators. In Chongqing, for example, many small hydropower projects were abandoned because of technical difficulties or equipment shortages. By 1962, only five of the fifteen stations that had been built since the beginning of the Great Leap Forward were still operational.[57] Similarly, in Yongchun, a model county for small hydropower, the Qingyuan project was abandoned halfway through because of a lack of funds to complete it. Local people also complained that too much labor and money was devoted to hydroelectric projects, taking these resources away from agricultural production.[58]

---

[54] Hershatter, *Gender of Memory*; Micah Muscolino, "Water Has Aroused the Girls' Hearts: Gendering Water and Soil Conservation in 1950s China," *Past and Present* 255 (2021), 351–387.

[55] Francesca Bray, *Technology and Gender: Fabrics of Power in Late Imperial China* (Berkeley: University of California Press, 1997).

[56] Shuili dianli jianshe zongju xinan gongzuozu 水利电力建设总局西南工作组, *Dagao qunzhong yundong quanmin banshuidian* 大搞群众运动全民办水电 (*Mobilizing the Masses to Build Hydroeletricity*) (Beijing: Shuili dianli chubanshe, 1959).

[57] Chongqingshi nongji shuidianju 重庆市农机水电局, *Chongqingshi shuilizhi* 重庆市水利志 (*Water Conservancy Gazetteer of Chongqing*) (Chongqing: Chongqing chubanshe, 1996), 187.

[58] *Fujian shengzhi: shuilizhi*, 145.

In an investigation carried out in Fujian Province in 1962 and 1963, it was found that less than 40 percent of the installed capacity was operating normally, and that more than half of the power lines failed to meet safety requirements. By the 1980s, in Fujian province alone, more than 5,000 of the small hydroelectric stations built during this rural electrification campaign had been abandoned.[59] Even worse, the lack of professional planning and management sometimes led to devastating tragedies. For instance, in the Tanbu district of Guangning 广宁 county in Guangdong province, one hydroelectric station was built on a hillside without a proper geological study having been done. Its poor construction resulted in a landslide that destroyed the station and killed twelve people.[60]

In addition to the significant economic and human losses it entailed, the small hydro campaign also had a negative impact on the environment. Generating hydroelectricity requires a specific natural setting: An adequate river flow, adequate levels of precipitation, and an appropriate topography and geology. Not all regions are suitable for these types of projects, particularly in the north of China and on the plains where the environment is not conducive to the success of small hydroelectric initiatives. Despite the significant investment of time and resources, many projects failed to produce the desired results because they were built in a campaign-style manner without proper consideration of the local hydrology. By the late 1970s, drought conditions in parts of China led to a reduction in river flows, causing many hydroelectric stations to fail to operate effectively. By 1985, then, only twenty-five of the original eighty-four stations in Qinglong county were still in operation, with a combined capacity of just 649.8 kW.[61]

The small hydro campaign also resulted in the fragmentation of rivers and the destruction of river ecologies. The construction of small hydro infrastructure combined with demand for timber led to deforestation and the loss of vegetation along these waterways. The campaign also exacerbated the existing problem of soil erosion, which had been a major environmental challenge in China for much of the twentieth century. Without proper environmental impact assessments and without the

---

[59] *Fujian shengzhi: shuilizhi*, 157.
[60] *Guangdongsheng shuilizhi* 广东省水利志 (*Water Conservancy Gazetteer of Guangdong Province*) (Guangzhou: Guangdongsheng shuili dianliting, 1994), 511.
[61] *Qinglong Manzhu zizixian shuilizhi*, 76.

implementation of rehabilitation measures, small-scale hydropower projects made the soil erosion issue significantly worse in various river basins.

The electricity generated by these small hydro facilities was also used to pump water for irrigation, which led to an expansion of agricultural land and, accordingly, increased deforestation. While areas at higher altitudes benefited temporarily, small hydro facilities created new environmental problems for communities at lower elevations. In Yongchun, the construction of small hydropower plants on the upper stream of the Jin River resulted in deforestation and soil erosion, which caused sediment to accumulate in the lower stream. This then led to riverbed silting and an increased risk of flooding.[62] Further, seasonal variation in precipitation necessitated the building of reservoirs to retain water, which then resulted in the drying up of streams at lower points in the river and the loss of water resources for local communities and wildlife. In some cases, multiple hydropower plants were built along the same river, which caused fragmentation and also turned once thriving waterways into lifeless machines for the generation of electricity. The cascading effects of the exploitation of small hydropower had a devastating impact on river ecologies, putting multiple species at risk.

Despite these challenges, small hydropower nonetheless played an important role in providing electricity to rural communities before the broad expansion of the national grid in the 1980s. In the 1990s, hydroelectricity was adopted as a means of combating deforestation by replacing the use of wood and other organic fuels in rural China, thereby helping to mitigate soil erosion in mountainous areas. In the long-term, the small hydro-electrification campaign of the Maoist period was a catalyst for an energy transition in rural China. It also laid the foundation for China's emergence as a leading force in the global small hydropower industry in the 1990s and for the formation of an all-encompassing hydropower nation.

In the 1970s, E. F. Schumacher, a British economist, advocated for the use of what he called intermediate technology, later known as appropriate technology, as a means of achieving sustainable development.[63] Among the set of technologies described as appropriate, small hydro has traditionally been seen as environmentally friendly. Recent studies by

---

[62] Quanzhoushi shuilishuidianju, 58–59. The silt percentage of the Jin River increased 48 percent from the 1950s to the 1960s and 146 percent between the 1950s and 1970s.

[63] Schumacher, *Small Is Beautiful*.

hydrologists and geographers, as well as the history of the small hydropower campaign, however, pose a challenge to this perception. Despite small hydro being promoted as a sustainable source of energy, it brings many of the negative environmental impacts of large dams. These include changes to the flows of rivers, increased loss of usable water through evaporation and seepage, the creation of impediments to the movements of aquatic organisms, thermal stratification, changes to sediment loading and nutrient levels, and the loss of terrestrial habitat with the creation of artificial lake habitats. The ecological benefits of small hydropower may not be as clear as once thought.

The environmental impact of small hydropower is even more concerning when measured per kilowatt of energy produced.[64] The geographer Thomas Ptak has developed the concept of the "furtive hydroscape" to help us understand the implications of small hydropower development.[65] Unfortunately, the adverse social and environmental effects of small hydro campaigns have long been obscured by official narratives rooted in the social and political contexts of the Great Leap Forward and the Cultural Revolution. The building of large dams and their significant environmental impacts have also overshadowed the impact of the growing number of small hydropower projects. As a result, the social and environmental consequences of small hydropower have remained hidden from public view. While this study cannot provide a comprehensive ethnographic examination of small hydropower in China, given the biased nature of state records and the limitations placed by that state on access to relevant materials, it can offer insight into the role of small hydropower projects in China's energy transformation and in changes to its river systems. By illuminating the impact of small hydropower projects as well as large ones, this exploration hopes to contribute to a more

---

[64] Kelly Kibler and Desiree Tullos, "Cumulative Biophysical Impact of Small and Large Hydropower Development in Nu River, China," *Water Resources Research* 49 (2013), 3104–3118; M. Premalatha, Tabassum-Abbasi, Tasneem Abbasi, and S. A. Abbasi, "A Critical View on the Eco-friendliness of Small Hydroelectric Installations," *Science of the Total Environment* 481 (2014), 638–643; Roland Jansson, Christer Nilsson, and Brigitta Renofalt, "Fragmentation of Riparian Floras in Rivers with Multiple Dams," *Ecology* 81, no. 4 (2000), 899–903; Mingyue Pang, Lixiao Zhang, Sergio Ulgiati, and Changbo Wang, "Ecological Impacts of Small Hydropower in China: Insights from an Energy Analysis of a Case Plant," *Energy Policy* 76 (2015), 112–122.

[65] Thomas Ptak, "Towards an Ethnography of Small Hydropower in China: Rural Electrification, Socioeconomic Development and Furtive Hydroscapes," *Energy Research & Social Science* 48 (2019), 116–130, 119.

complete understanding of the complex issues involved in China's rise as a hydropower nation.

THE POLITICIZATION OF SMALL HYDROPOWER

Even the seemingly mundane subject of hydropower could become a contentious issue in the Maoist period. In 1963, the Soviet-educated energy expert Xu Shoubo 徐寿波 published an article in *People's Daily* in which he argued that the efficiency of hydroelectricity and thermoelectricity generation was impacted by the social and political systems underlying it. According to Xu, the full development of hydropower, and in particular hydroelectric projects, was impeded in capitalist countries because of the existence of private land ownership. Socialist countries, on the other hand, having abolished private property rights, could not face opposition from landowners. This made hydroelectricity a more viable option. Xu argued that only a socialist state had the ability to fully utilize natural rivers to support national economic growth.[66]

The global development of hydroelectricity in the twentieth century casts doubt on the validity of Xu's argument. Despite being capitalist, countries like the United States, Japan, and various European nations saw substantial development in hydroelectricity. Thus, the potential for energy production is not determined by political and economic systems. In the second half of the twentieth century, the increase in demand for energy due to the "great acceleration" resulted in a convergence of state planning and market forces, both aiming to increase the efficiency of energy production.[67] The methods used to achieve this goal varied according to the political system. As the small hydropower campaign demonstrates, the formation and implementation of energy production plans were greatly influenced by social and political factors.[68]

Local party officials clearly recognized the political significance of rural electrification. The provision of electric light was seen as a tangible

---

[66] Xu Shoubo 徐寿波, "Shuili fadian yu huoli fadian" 水力发电与火力发电 (Hydroelectricity and Thermoelectricity), *Renmin ribao*, March 6, 1963.
[67] On the definition of the "great acceleration," see John McNeill and Peter Engelke, *The Great Acceleration: An Environmental History of the Anthropocene since 1945* (Cambridge, MA: Belknap Press, 2014), 4.
[68] Covell Meyskens re-emphasizes the centrality of muscle power and mass mobilization in socialist China's industrial policy. See Meyskens, "Rethinking the Political Economy of Development in Mao's China."

manifestation of the Communist Party's leadership and the success of socialist policies in rural communities. They stated:

> The political impact of small hydro is immeasurable. Many peasants view the installation of electric light as a landmark moment, believing that they are witnessing the manifestation of socialism and are inspired to work harder. Electrification has illuminated the minds of the peasants, instilling a belief that agricultural labor will not remain stagnant forever.[69]

In other words, socialism is not only local, but also electrical.[70]

Francesca Bray argues that agricultural science and technology are powerful ways for a state to maintain social order and to imbue the subjectivities proper to that order.[71] Similarly, beyond the meeting of practical energy needs, the Maoist regime used small hydropower as a tool to further its ideology. Many widely distributed pamphlets from the 1950s to the 1970s featured Mao's words, "Socialism not only liberates workers and production materials from the old society, but also liberates the natural world that the old society was incapable of utilizing."[72] Mao's assertion that the previous society was unable to harness the natural world was not entirely true: China had a long history of waterpower use, as discussed in Chapter 1. But he was at least partially correct in his assessment of the greater efficiency of socialism. It is important to note that socialism in this context should not be considered simply as a political ideology, but rather as a systematic state-led mobilization of human labor to extract resources from the natural world. In other words, socialism is defined not only in terms of a specific set of social and political relationships but also as a form of human–nature interaction.

During the campaign to build rural hydroelectric stations, the official narrative de-emphasized professional or foreign engineering knowledge. Instead, amateur and local methods were celebrated. In the small hydropower campaign, communities joined forces with local technicians and party officials to produce both electricity and political culture, whether consciously or unconsciously. Despite Mao's proposal that rural development should "walk on two legs," the absence of specialized tools and

---

[69] My translation of a quotation found in Quanguo Nongye Zhanlanguan Shuiliguan, *Nongcun shuidian*, 2.
[70] Hershatter, *Gender of Memory*, 13–15.
[71] Francesca Bray, "Science, Technique, Technology: Passages between Matter and Knowledge in Imperial Chinese Agriculture." *British Journal for the History of Science* 41, no. 3 (2008), 319–344, 327.
[72] *Jijifazhan nongcun xiaoxing shuidianzhan* 积极发展农村小型水电站 (*Actively Develop Rural Small Hydropower*) (Beijing: Renmin chubanshe, 1972), 1.

knowledge presented many challenges to the mass campaign. To give just one example, the ventilation, lighting, and pneumatic drills needed to excavate tunnels were largely absent. Rather than addressing these problems, the official narrative emphasized voluntarism and revolutionary ideology, prioritizing the masses' morale and enthusiasm over their safety and the quality of projects. Approval from engineers and experts was deemed unnecessary. In Suining, for example, every project was decided upon by the county committee, with local officials calling this approach "authentic Marxism-Leninism."[73] On the other hand, the professional way of building power plants, which did not actively involve the masses except in construction labor, was referred to as the "scholarly way" (*xiucaishi* 秀才式). During the Cultural Revolution, this "scholarly way" was criticized as an aspect of the "capitalist line," as opposed to the egalitarian and proletarian line represented by the mass campaign. These differing approaches to building hydroelectricity came to be key elements in political struggles among top leaders.

In Yongchun county, in an attempt to realize what they felt was the revolutionary politics of the proletariat, officials encouraged every town, commune, brigade, and school to construct their own hydroelectric stations. This resulted in the creation of a variety of different kinds of stations: "militia hydroelectric stations," "youth hydroelectric stations," "young pioneer stations," and "women's hydroelectric stations." The local government claimed that by 1958 the county had built over 400 such stations. During the Great Leap Forward, peasants worked in their fields during the day and built hydroelectric stations at night, driven on by the state's call to harness every flow of water as a source of electricity. On Tianma Mountain, for example, despite a flow of a mere 0.03 cubic meters/second flow and a 15-meter high waterfall, local peasants built a three-horsepower hydropower station to process sweet potatoes.[74]

In a broader context of the American invasion of Vietnam in the 1960s and the Sino-Soviet conflict over Zhenbao Island in 1969, Mao called on the Chinese people to "prepare for war, prepare for famine, and work for the people."[75] As an aspect of the "small third front" project – a state policy aimed at shifting industries to rural and mountainous areas for the

---

[73] *Dagao qunzhong yundong quanmin banshuidian*, 9.
[74] *Dagao qunzhong yundong quanmin banshuidian*, 43.
[75] On the Third Front project, see Judd C. Kinzley, "Crisis and the Development of China's Southwestern Periphery: The Transformation of Panzhihua, 1936–1969," *Modern China* 38, no. 5 (2012), 559–584; Meyskens, *Mao's Third Front*.

purpose national defense in the event of an invasion – small hydropower was seen as a critical source of energy for national security if war should break out. Because of its geographic dispersion, the small hydropower campaign, while improving the efficiency of agricultural production, was also propelled by the state's bureaucratic system in support of this goal, with the support of local technicians and the masses. The campaign transformed not just large rivers but also local streams and creeks into energy sources that were legible to and manageable by state authorities.

Although small hydropower was primarily facilitated by local resources, technicians, and laborers, it was initiated and largely controlled by the party-state during the Great Leap Forward and the Cultural Revolution. As the state expanded its reach into rural China, the party reorganized human relationships, both social and economic, through collectivization. Like the intertwining of a "red revolution" and a "green revolution," discussed by Sigrid Schmalzer, rural electrification in the form of small hydropower campaigns expanded the reach of the socialist state by transforming human–river relationships, albeit, until the 1980s, on a limited scale.[76] Under the state's guidance, local communities worked to build a hydropower infrastructure, converting running waters into electricity through turbines and generators, and delivering electricity to many rural households through newly installed powerlines. Although the Maoist period ended with de-collectivization in rural China in the late 1970s and early 1980s, the human–river relationship defined by the generation of electricity persisted and continued to grow beyond the state's ideological shift in that period.

CONCLUSION

During the early years of the Cultural Revolution, while cities and towns were plagued by political turmoil and the activities of the Red Guards, the Communist Party prioritized agricultural production for all rural brigades and production teams in order to prevent a repetition of the disaster that occurred during the Great Leap Forward.[77] Most rural residents embraced the technological innovations promoted by party officials in line with Maoist ideology, as local cadres and peasants shared the goal of improving agricultural production. The ease with which small hydropower could be built and its low cost made it a viable option for rural

---

[76] Schmalzer, *Red Revolution, Green Revolution*.
[77] Li, *Village China under Socialism and Reform*, 141.

electrification both in China and elsewhere, in contrast to the mega dams that required high levels of professional expertise and significant capital investment. Although Maoist politics largely came to an end with the death of the chairman, small hydropower, despite its many challenges, persisted and continued to grow in the post-Mao period.

In 1979, the United Nations Industrial Development Organization (UNIDO), with its mission to alleviate poverty through sustainable and inclusive industrial development, organized a study tour on medium and small-scale hydropower plants in China in collaboration with the PRC's Ministry of Water Conservancy. The tour consisted of fifteen engineers and officials from fifteen developing countries, including India, Nigeria, and Suriname. As trained engineers, the visitors noted several problems, including a general lack of knowledge about the water regime, spillage, and excessively high voltage levels. They also noticed that, during their visit at least, none of the small hydropower plants were operating at full capacity. Ferruh Anik, the head of the Design Department of the Electrical Power Resources Surveying and Development Administration of Turkey, noted that hydropower data in China was generally presented in kilowatts, which indicate the total energy quantity and installed capacity, but not in kilowatt-hours, which is a more practical measure of dependable and usable energy. He argued that, without this data, "it will not only be impossible to ascertain whether or not the energy requirements are met, but there will also be no means of working out a daily and annual operational plan for the power plants in keeping with the curves of energy demands." Anik thus concluded that China's small hydropower plants were "unfortunately operated in a haphazard way."[78] This observation highlights a significant problem with the small hydropower campaign, which prioritized the growth of output quantity over the reliability and efficiency of the technologies involved. In the post-Mao period, this approach would need to change.

Despite these challenges, the group was impressed by China's small hydropower development, which they described as remarkable. They were amazed by the artificial creation of preconditions for energy production where they had not been provided by nature.[79] Mansour Aydin, a director of power plant construction from Syria, was inspired by the

---

[78] United Nations Industrial Development Organization (UNIDO), Group study tour in the field of medium and small-scale hydropower plants to the People's Republic of China, final report UNIDO/IO.467 (September 16, 1981), 16.

[79] UNIDO, 5.

tremendous effort made by the Chinese people to maintain their independence and to control their own initiatives, relying on their own hard work and frugality to build their country.[80] Raymond Niekoop, an engineer from Suriname, believed that small hydropower could lead to good levels of development. He noted that China's small hydropower infrastructure and equipment were simple and affordable and that, while their levels of efficiency might be lower than that found in Western countries, the meaningful application of this appropriate technology held great promise.[81] Such technology was particularly important for developing countries with limited funds available for capital investment and technological innovation, but with many remote areas in need of development. Overall, the group saw China's achievements in small hydropower as a positive example for their own countries.[82]

As this chapter shows, the technological complex most commonly found in the small hydropower campaign was a mixture of indigenous and foreign elements. On one hand, it was comprised of local technical personnel like carpenters, blacksmiths, and masons, local materials like timber and pebbles, and existing waterpower facilities. On the other hand, the widespread use of wooden propellers was widely seen as a success story for indigenous technology. And in Yongchun, a returned overseas Chinese businessman named You Yangzu helped a local agricultural tool factory to import machines from abroad, which allowed the factory to upgrade itself to a level where it could produce electrical motors and generators.

Based on Mao's slogan of "walking on two legs," the binary of *tu* and *yang* is often used to distinguish the two basic technological complexes in the study of hydropower construction. As David Pietz has shown, the *yang* technological complex is typically associated with large dam projects, while the *tu* technological complex is linked to the mass water conservancy movement and the development of small hydropower.[83] However, a closer examination of the small hydropower campaign reveals that the *tu* complex was not purely indigenous but rather a

---

[80] UNIDO, 18.   [81] UNIDO, 34.
[82] In 1992, the United Nations Industrial Development Organization proposed to share the Chinese experience in small hydro with the United Nations General Assembly. Two years later, the United Nations and the Chinese government co-founded the International Center on Small Hydro Power, establishing its headquarters at West Lake in Hangzhou. The Center was the first international organization to set up its headquarters in China.
[83] Pietz, *Yellow River*.

combination of both *tu* and *yang* technological factors. This can be seen as echoing the state's call to "raise *tu* and *yang* together" 土洋并举.[84]

In the 1950s, the Communist state in China employed a campaign style of small hydropower construction, employing a mixed technological complex. Despite the significant role that small hydropower played in rural and peripheral areas, it has often been overlooked in historical discussions of China's hydropower energy. The choice to pursue small hydropower was both pragmatic – considering available technological capacity and the natural environment – and political – driven by Mao's vision for economic growth and an egalitarian society. Political and environmental factors, as well as human technology, shape technology's potentials and limitations. The small hydropower campaign, with its emphasis on mass participation, the mobilization of labor, and the use of indigenous materials and technologies, made the socialist hydropower nation accessible to rural residents and grounded in their reality. While it had mixed consequences during the Maoist period, as I've shown, it paved the way for far more extensive rural electrification in the 1980s, ultimately establishing China as a leading nation in small hydropower.

---

[84] In Sigrid Schmalzer's study of China's agriculture, the blurred boundary between *tu* and *yang* is personalized in relation to two agriculturalists: Pu Zhiqiang and Yuan Longping. See Schmalzer, *Red Revolution, Green Revolution*.

PART III

A HUGE SETBACK
The Sanmenxia Dam

5

Silt and Hydroelectricity

On the Yellow River, not far from the ancient city of Shaanzhou 陕州, two rocky islets jutted out of the water, standing as a testament to the legendary feats of Yu the Great, who had supposedly cut them to divert the great flood recorded in early Chinese historical texts. Locals had christened the three narrow gates formed by these islets as the Gate of Men, the Gate of the Spirits, and the Gate of the Gods, together creating the Three Gates Gorge, or Sanmenxia. The constraint placed on the Yellow River by the rocky gorge had turned it into a turbulent torrent, which made navigation through the passage on wooden boats or sheepskin rafts extremely dangerous. On the north bank, near Sanmenxia, one could regularly see smoke from burning incense and hear the crackle of firecrackers emanating from the Temple of Yu the Great, perched on a hill. On a busy day, hundreds of small wooden boats would wait to pass through the "gates" but, before they did so, the boatmen would stop to burn incense at the temple. At the front of the hall where Yu was worshipped stood a pair of iron ducks, nearly as tall as an adult. One of the boatmen would be chosen to throw a copper coin into a hole on the back of one of the iron ducks. If the iron duck quacked, it was a bad omen, and passing through Sanmenxia that day would be extremely dangerous. If the duck remained silent, it was considered auspicious, and the journey eastward was believed to be safe.[1] For centuries, the temple stood as a symbol of reverence for the mighty river, until 1957, when the incense

---

[1] Ba Mu 巴牧, *Sanmenxia de chuanshuo* 三门峡的传说 (*Folktales of Sanmenxia*) (Beijing: Zuojia chubanshe, 1958), 8–14.

and firecrackers were replaced by the clatter of steel cranes and the roar of bulldozers and trucks laden with rocks and concrete. The Temple of Yu the Great would soon be overshadowed by a "temple of modernity," a massive concrete dam that sliced the Yellow River in two.[2]

In the 1950s, the Sanmenxia hydropower project stood out as a testament to the might of the Chinese hydropower nation, set to become, once it was completed, the largest hydropower plant in China. Planning and building the project involved the collective labors of a multitude of state leaders, domestic and foreign engineers, local cadres, reservoir migrants, journalists, artists, and other citizens of the People's Republic. The scope of the social and political involvement in a single river engineering project was unprecedented. Despite technical, social, and environmental setbacks, the grand concrete dam had been regarded as a totem of the socialist nation. In Chapters 3 and 4, we explored the evolution of the hydropower nation largely in chronological sequence. In this section, I focus on a single project – Sanmenxia – as a way of exploring the social and environmental aspects of the hydropower nation. At the same time, I will also examine the domestic and international political dimensions of the Sanmenxia project. Through an analysis of the political, social, and environmental factors that both drove and were caused by the Sanmenxia hydropower project, I illuminate the unparalleled magnitude of the hydropower nation.

David Pietz's research has explored the Chinese government's commitment to transforming the Yellow River from a source of devastation, "China's Sorrow," into a source of agricultural and industrial growth. Unlike conventional studies of water conservancy, Pietz emphasizes internationalization and the idea of "multipurpose exploitation" in managing the Yellow River.[3] With a particular focus on the Sanmenxia hydropower project, I seek to highlight the significance of hydroelectricity in twentieth-century river engineering in China.

Between the late 1920s and early 1940s, Japanese engineers attempted to develop hydroelectric power on the Yellow River by constructing dams. After the Communists took power in 1949, they argued that the main purpose of the Sanmenxia project was flood control. However, with

---

[2] Jawaharlal Nehru, the first prime minister of independent India, used the term "temples of modern India" to describe the Bhakra Nangal Dam and other large industrial infrastructure projects. See Shripad Dharmadhikar, *Unraveling Bhakra: Assessing the Temple of Resurgent India: Report of a Study of the Bhakra Nangal Project* (Badwani: Manthan Kendra, 2005).

[3] Pietz, *Yellow River*.

TABLE 5.1 *Main stages of the Sanmenxia project*

Year/ Context	1940/ Japanese Invasion	1949/ Communist China led by Mao	1956/Sino– Soviet Alliance	1969/Project Reconstruction
Engineers	Japanese	Chinese Communists	Soviets and Chinese	Chinese
Claimed Major Function	Electricity Generation	Flood Control	Multi-Purpose	Multi-Purpose
Electrical Capacity (kW)	1,123,000 (design)	N/A	1,100,000 (design)	250,000

influences coming from the Soviet Union and the Electric Power Ministry, they began to emphasize the generation of electrical power as well, with the goal of installing infrastructure with a capacity to generate 1,000,000 kW. Due to the nature of the river and other constraints, though, the leaders' and engineers' ambitions were frustrated. By 1969, only 250,000 kW capacity had been installed and the power output was unstable (see Table 5.1). In contrast to previous water conservancy efforts, the design and construction of the Sanmenxia project was dominated by an obsession with hydropower, which resulted in serious reservoir silting and profound social disruption.

SANMENXIA BEFORE THE DAM

In separate research, Ling Zhang and Ruth Mostern have examined the natural (and unnatural) history of the Yellow River and its basin in its middle and lower reaches before the twentieth century.[4] It appears that, before the construction of the dam, Sanmenxia was of little significance in imperial Chinese history, being noted primarily for its natural uniqueness. The main focus of the late imperial state was the maintenance of dikes and the whole hydraulic complex associated with the Grand Canal in its lower reaches. In the 1950s, when the new state laid out its comprehensive plans for the Yellow River and the Sanmenxia project in particular, the primary concern was the safety and agricultural productivity of the North China Plain. While it is important to see the Yellow River as a whole if we are to

---

[4] Zhang, *River, Plain, and State*; Mostern, *Yellow River*.

understand its history and challenges, examining specific nodes and sections, and in particular the areas that would be submerged by the project, can provide a local context with which to examine the convergence of environment and technology in the late 1950s.

In fact, the rocks and the section of the river at Sanmenxia have been marked by imperial Chinese states and their people for centuries, leaving behind a rich history. As early as the Han dynasty, the state transported grain from the North China Plain to its capital, Chang'an, along the Yellow River. But the rapids and submerged rocks at Sanmenxia were a serious threat, leading to many boats being wrecked and boatmen drowned. To ensure the supply of grain, the imperial government organized the construction of a plank road along the face of the northern cliff at Sanmenxia. Even then, long lines of barge haulers risked their lives to help boats loaded with grain pass through the gorge safely. In the Tang dynasty, a short canal was dug on the north bank, which significantly reduced the risk of passing through Sanmenxia.[5] With the historical shift of China's political center to the east after the Tang dynasty, though, the political and economic value of the Sanmenxia passage declined and it was gradually forgotten.

In the twentieth century, a dam at Sanmenxia was first proposed as a means of flood control. Advocates claimed that it could control more than 90 percent of the river's flow. The unique geological structure of the rocks under and along the course of the river at Sanmenxia made it an ideal site for a large concrete dam. Investigations by Chinese geologists in the 1950s showed that the riverbed consisted of diorite porphrite, created in geological time by the tectonic movements of the earth. Diorite porphrite is hard, water-resistant, and appears at Sanmenxia in the shape of an intrusive sheet between 90 and 130 meters thick. Moreover, unlike most parts of the middle and lower streams of the Yellow River, which are miles wide, the narrowest section of the Yellow River at Sanmenxia was only 120 meters, which from an engineering perspective made it an ideal location for a concrete dam.[6] Despite the river's historical significance and natural characteristics, the proposed dam at Sanmenxia would soon transform it and the landscape around it beyond recognition.

---

[5] Zhongguo Kexueyuan Kaogu Yanjiusuo 中国科学院考古研究所, *Sanmenxia caoyu yiji* 三门峡漕运遗迹 (*Sanmenxia Canal Transportation Relic*) (Beijing: Kexue chubanshe, 1959).

[6] Jiang Daquan 姜达权, "Sanmenxia baduande xuanding" 三门峡坝段的选定 (Selecting the Sanmenxia Dam Site), *Shuiwen dizhi gongcheng dizhi* 水文地质工程地质 no. 12 (1957), 1–3.

Silt and Hydroelectricity

JAPANESE PLANNING

Although the Sanmenxia Dam was not built until the late 1950s, the idea was proposed years before. In 1929, in his book *The Yellow River Conservancy*, the Japanese explorer Ogashi Heiriku 小越平陸 proposed the building of hydroelectric projects at Hukou and Sanmenxia. Although he wasn't a hydraulic specialist, Heiriku traveled along the Yellow River to explore why it brought so many disasters to the Chinese people. Under the influence of Pan-Asianism and as a proponent of the idea of the "unification of the yellow race against the white race," he called for Japan to assist the Chinese people in remaking the Yellow River basin, which was viewed in Pan-Asianist thought as the cradle of East Asian civilization.[7]

A few years later, in 1933, Li Yizhi 李仪祉, the director of the Yellow River Conservancy Commission of the Nationalist government, who had studied hydraulic engineering in Germany, proposed the construction of a dam between Hukou and Mengjin in Henan province to manage the flow of the river. In 1935, S. Elisson, a Norwegian engineer employed as a consultant for the Nationalist government, proposed after conducting field investigations that a dam be built at Sanmenxia rather than at two possible alternative sites: Baili Hutong and Xiaolangdi, located at the lower stream of Sanmenxia.[8] With the 1933 flood of the Yellow River firmly in mind, as well as its frequent historical dike breaches, all of these proposals primarily aimed to protect the central plain from devastating floods.

In October 1940, when most of North China was under military occupation by Japan and its Chinese collaborators, a group of Japanese engineers launched an investigation of the Yellow River. They exclaimed,

The ancestors of the Chinese people have been credited with the miraculous accomplishment of building up the Great Wall in the North and the Grand Canal running across its lowlands in the East. But, despite the most strenuous efforts made by Emperor Da Yu and other national leaders in past history, they had not succeeded in finding an effective means of control of the Yellow River, which had been a constant threat to the people in North China from time immemorial.[9]

---

[7] Ogashi Heriku, *Koga chisui* (Tokyo: Seikyosha, 1929), 160.
[8] Zhao Zhilin, "Jianguoqian Sanmenxia gongcheng yanjiu" (Studies of the Sanmenxia Project before 1949), *Zhongguo shuili fadian shiliao* 3 (1991), 13.
[9] *A Brief for the Control of the Yellow River and Construction of San Meng Shar Dam*, Percy Othus Collection on Chinese Hydrology, Special Collection Library, Pennsylvania State University, 3.

This report argued that it was time for the Japanese to regulate the river and to make it serve the interests of both the Chinese people and the Japanese empire. Despite the differing motives behind the proposals, they all viewed the construction of the Sanmenxia Dam as a way of providing flood control and generating hydroelectric power.

The Japanese survey team consisted of 26 specialists, including ten hydroelectric experts, indicating that their primary interest lay in exploiting hydroelectric power to serve the Japanese empire.[10] Although the team expressed an interest in the welfare of the Chinese people, it was clear that the investigation was aimed at making preparations for the production of hydroelectric power. The team created a cascade development plan for the river, stretching from its upper stream in Inner Asia to its lower stream in the North China Plain. They estimated that building dams on the Yellow River could eventually result in the creation of 8,000,000 kW of hydroelectric capacity. In this period, Japanese engineers believed that a nation's strength was directly linked to the quantity of energy it could command, and they had almost fully developed hydroelectric power along Japan's major rivers. They expected that hydroelectric power would also soon be fully developed in Korea and Manchuria. Given the distribution of Japanese military forces, these engineers felt that their most urgent tasks were building dams at Fengman on the Songhua River and at Sup'ung on the Yalu River.[11] But they believed that the next step should be the development of the Yellow River, which could provide valuable resources for building the Greater East Asia Co-Prosperity Sphere if managed properly.[12]

As part of the Japanese plan for developing the Yellow River, eleven possible dam sites were identified along the middle part of the river, with Sanmenxia selected as the first priority.[13] The Japanese engineers referred

---

[10] Toa Kenkyujo dai 2 chosa linkai hokushi linkai dai 4 bukai 東亞研究所第二調查委員會內地委員會第四部會, Koga suiryoku hatsuden keikaku hokokusho 黃河水力發電現地出張報告 (The Yellow River Hydropower Survey) (January 1941), 2.

[11] See Moore, *Constructing East Asia*, chapter 4. The mobilization of rivers for hydroelectric development during wartime could be seen in other parts of the world as well. One example is Canada's rapid development of hydroelectricity during the Second World War. See Evenden, *Allied Power*.

[12] Toa Kenkyujo dai 2 chosa linkai hokushi linkai dai 4 bukai 東亞研究所第二調查委員會內地委員會第四部會, *Koga suiryoku hatsuden keikaku hokokusho* 黃河水力發電計劃報告書 (Report on the Yellow River Hydropower Development Plan) (Tokyo: Toakenkyujo, May 1941), report 5, 2.

[13] Toa Kenkyujo dai 2 chosa linkai hokushi linkai dai 4 bukai, report 3, 9.

to it as "one stone killing five birds":[14] A dam at Sanmenxia would not only generate hydroelectricity but also provide flood control and irrigation for agriculture, improve navigation, and provide water for industries in the region. In addition, hydroelectricity from the Yellow River would permit more efficient excavation of mineral resources in the North China region, especially coal in Shanxi. The Japanese engineers believed that the production of sufficient energy would enable them to stabilize the political and economic situation in North China.

Because of the limited capacity of reservoirs in Japan, hydroelectric projects there typically required the supplementary operation of fossil fuel power plants. The planned capacity of the Sanmenxia reservoir was three times greater than that of Biwa Lake, the largest reservoir in Japan, which excited the Japanese engineers. However, large reservoirs involved tremendous inundation of residential areas. The areas designated for the Sanmenxia project's reservoirs were heavily populated. Still, Japanese engineers, believing that the benefits of hydroelectricity and flood control outweighed the costs of inundation, suggested sticking to the standard height of the dam. The dam's height, then, from dead water level to full water level, would be no less than 70 meters. Anything less would seriously diminish the dam's hydroelectric production capacity.[15]

The cost of constructing the Sanmenxia Dam was estimated to be around US$174,000,000, with an additional $176,000,000 allocated for land acquisition and property compensation for residents of the areas to be submerged. The Japanese engineers were so optimistic about the potential benefits of the dam that they suggested that the entire expense could be written off in less than a year. They wrote, "When hydroelectric power plants are erected, we can start productive enterprises on a large scale, and turn the quiet countryside into booming industrial districts."[16]

According to the Japanese designs, the dam would be constructed to 70 meters above water level, with the assumed effective water head at

---

[14] Koga chisui- hatsuden 黃河治水-發電 (Yellow River Management: Electricity Generation), file no. 5-1645, Second Historical Archives of China, Nanjing.
[15] Koga suiryoku hatsuden notokucho 黃河水力發電之特色 (Characteristics of Yellow River Hydropower), file no. 5-685, Second Historical Archives of China, Nanjing.
[16] *A Brief for the Control of the Yellow River and Construction of San Meng Shar Dam*, Percy Othus Collection on Chinese Hydrology, Special Collection Library, Pennsylvania State University, p. 6. After the end of World War II, the Yellow River Conservancy Commission of the Nationalist government translated part of the Japanese materials into English for their American consultants. This is why this document uses US dollars as its monetary unit.

maximum load at roughly 64 meters. Thus, the maximum installed capacity was estimated to be 1,123,000 kW, and maximum power output could reach 5,000,000,000 kW per year. The cost of the power generated, including compensation paid to the owners of submerged areas, would be 0.73 cents per kW, which was much cheaper than fossil fuel electricity. Building such a massive concrete dam and hydropower plant obviously required a reliable supply of materials. It was seen as an important project for the goal of building the Greater East Asian Co-Prosperity Sphere and according to the plan all construction materials would, whenever available, be obtained from ready stocks in Japan, Korea, Manchuria, and North China.

Although many Japanese were enthusiastic about the prospects for developing the Yellow River, the eruption of the Pacific War and defeats on the battlefield eventually led to the collapse of the Japanese empire. In 1945, the Japanese engineers had no choice but to leave China, leaving their investigation and planning materials in the hands of the Chinese and their allies.

The Yellow River, much like rivers in Europe and North America, was not immune to the effects of World War II.[17] While the Nationalist government broke the Yellow River dike in 1938 in an attempt to stop the advancing Japanese army, the Japanese themselves were interested in using the Sanmenxia project as a way to further their own wartime agenda of building the Japanese empire in East Asia.[18] Although they were ultimately unable to put their plans into action, their investigations and planning were taken over by the Nationalists and their American consultants. The Yellow River consultancy board for the Nationalist government acknowledged that the Japanese plan "contains many excellent proposals and shows evidence of careful, intelligent study," but ultimately they deemed the cost of the reservoir to be too high.[19] Further, the Japanese had failed to appreciate fully the issue of silting in the reservoir, which would have considerably shortened its lifespan. In light of this, John Cotton, a member of the American consulting board

---

[17] See Blackbourn, *Conquest of Nature*; Evenden, *Allied Power*; Giacomo Parrinello, "Systems of Power: A Spatial Envirotechnical Approach to Water Power and Industrialization in the Po Valley of Italy, ca. 1880–1970," *Technology and Culture* 59 (2018), 652–688.

[18] Muscolino, *Ecology of War in China*.

[19] The Yellow River Consultant Board, *Preliminary Report on Yellow River Project* (January 17, 1947), Percy Othus Collection on Chinese Hydrology, Special Collection Library, Pennsylvania State University, 19–20.

for the Nationalist government, recommended that no dam be constructed at Sanmenxia until effective soil conservation measures could be designed and implemented across the entire affected region.[20]

EARLY SOIL AND WATER CONSERVATION EFFORTS

Many of the Yellow River's problems stem from severe soil erosion on the Loess Plateau.[21] Across the last centuries of China's imperial period, the environmental deterioration of the plateau grew progressively worse, which caused an increase in summer rainfall that in turn washed away more loess. As a result, the Yellow River carried this sediment to its lower reaches as runoff. Because of the reduced flow of the river on the North China Plain, these sediments were then deposited in the riverbed, leading to the formation of what is known as a "suspended river." The lower stream of the river is commonly described as especially prone to silting, overflows, and diversions. As a result, most historical accounts of the Yellow River have focused on flood control in its lower reaches. Scholars studying medieval China from the Tang to the Song have argued that the imperial state's approach to managing the Yellow River issue was not comprehensive. Although certain imperial officials identified erosion on the Loess Plateau as the root cause of the problem, China's political economy at the time impeded effective action.[22]

In 1923, the American soil conservationist Walter C. Lowdermilk visited northwest China for the first time. He meditated on the causes of the struggles of Chinese peasants along the Yellow River and concluded that "silt was the great enemy causing this endless, hopeless struggle! Silt had defeated the courageous toiling farmers, valiant as they were!"[23]

---

[20] In Wang Huayun's memoir, he refers to Japanese planning during one of his meetings with Mao Zedong. See Wang Huayun, *Wode Zhihe Shijian* (*My Practice of River Management*) (Zhengzhou: Henan kexue jishu chubanshe, 1989), 145.

[21] On loess erosion, see Vaclav Smil, *The Bad Earth: Environmental Degradation in China* (Armonk: M. E. Sharpe, 1984), 38–57.

[22] See Cen Zhongmian 岑仲勉, *Huanghe bianqianshi* 黄河变迁史 (*History of the Yellow River*) (Beijing: Renmin chubanshe, 1957); Randall A. Dodgen, *Controlling the Dragon: Confucian Engineers and the Yellow River in Late Imperial China* (Honolulu: University of Hawai'i Press, 2001); Yao Hanyuan 姚汉源, *Huanghe shuilishi yanjiu* 黄河水利史研究 (*Study of the Yellow River Conservancy*) (Zhengzhou: Huanghe shuili chubanshe, 2003); Zhang, *River, Plain, and State*; Mostern, *Yellow River*.

[23] Walter C. Lowdermilk, *Soil, Forest, and Water Conservation and Reclamation in China, Israel, Africa, and the United States*, vol. 1 (Berkeley: University of California Berkeley, Regional Oral History Office, 1969), 62.

In an attempt to find an effective way to control the silt and thereby the river, in 1925 Lowdermilk and his colleagues set up three runoff plots in Shanxi to measure erosion on the Loess Plateau. Although this experiment yielding little in the way of quantitative results, Lowdermilk observed:

> When the runoff from the rain of an inch an hour rushed down off the slopes, there was a roar because the runoff was full of soil and debris and boulders. One could hear the boulders striking each other, sort of a muffled sound like cannonading. The smaller gravel hitting against the boulders sounded like machine-gun fire. In other words, these were torrential flows – we called the mud flows, they were so powerful.[24]

It was necessary that erosion be reduced where it took place:

> But the situation which makes it so difficult is that demands of the people to grow food on the land are now so high they can't permit trees to grow. If they need fuel, they'll go out and pull up the trees they have planted for fuel, and to cook their food. So fuel has become a part of their food supply, and as I have said many times, we have to be in possession of a certain amount of abundance to act in an intelligent way in the conservation of our resources, for a starving farmer will eat his seed grain.

Considering the economic and environmental conditions of this region, then, Lowdermilk concluded that "it is a tremendous problem for which there isn't any ready and rapid solution. It would require a consistent and continuing program based on measures that will work."[25]

Meanwhile, Li Yizhi, a hydraulic specialist trained in Germany, also understood that silting was the most challenging problem of Yellow River conservancy. He proposed to build check dams and to dig ditches along the Loess Plateau to prevent soil from flowing into the river.[26] In 1942, as a consultant for the Nationalist government, Lowdermilk and his Chinese colleagues established erosion prevention demonstration projects near Tianshui, in the upper Wei River valley in southern Gansu province. They planted experimental strip crops of alfalfa, rye grass, and sweet clover mixture to absorb rain on sloping land. This prevented runoff on slopes by up to 24 percent.[27] It did not work well on the steep walls of

---

[24] Lowdermilk, *Soil, Forest, and Water Conservation and Reclamation*, vol. 1, 86.
[25] Lowdermilk, *Soil, Forest, and Water Conservation and Reclamation*, vol. 1, 95.
[26] Li Yizhi, *Huanghe gaikuang ji zhiben tantao* 黄河概况及治本探讨 (*Overview of the Yellow River and Exploration of Fundamental Solutions*) (Kaifeng: Huanghe shuili weiyuanhui, 1935).
[27] Lowdermilk, *Soil, Forest, and Water Conservation and Reclamation*, vol. 2, 395.

gullies, though. They proposed the use of airplanes to drop clay pellets containing the seeds of grasses, shrubs, and trees, together with fertilizer. According to their research, it would also be necessary to build thousands of soil-saving dams to collect silt in flat stretches of great gullies. This would build up alluvial areas for farming and reduce the silt load in streams.[28]

After he had established demonstration projects in the northwest, Lowdermilk attempted to obtain aid from the American government to establish nurseries and to conduct even more thorough research on the Loess Plateau. Unfortunately, his proposal was denied by the US government. The American agricultural attaché responded: "The United States government is not interested in doing this sort of thing for Chinese farmers. All we are interested in is getting from China tung oil, raw silk, tea and hog bristles."[29] Disappointed by this attitude, Lowdermilk was never given the opportunity to expand his soil erosion prevention project in northwest China to a larger scale.

Before Lowdermilk's arrival, local peasants had long developed the practice of bench terracing to safeguard the land from erosion. In Lowdermilk's view, however, this measure lacked "scientific exactness [and] has not been sufficient to prevent serious erosion damage in many places."[30] When they learned about the soil conservation effects produced by Lowdermilk's demonstration projects, some farmers asked Lowdermilk and his colleagues to do the same thing on their sloping lands. After 1949, despite his return to the United States, Lowdermilk was delighted to learn that his method of soil erosion prevention was still practiced by Chinese peasants under the new Communist government.[31] The construction of the Sanmenxia hydropower project, though, would bring unprecedented challenges to water and soil conservation efforts on the Loess Plateau.

SOVIET ASSISTANCE

In February 1950, the signing of the Sino-Soviet Treaty of Alliance, Friendship, and Mutual Assistance marked the Chinese Communist Party's formal adoption of the Soviet model for industrial development.

---

[28] Lowdermilk, *Soil, Forest, and Water Conservation and Reclamation*, vol. 2, 396.
[29] Lowdermilk, *Soil, Forest, and Water Conservation and Reclamation*, vol. 2, 404.
[30] Lowdermilk, *Soil, Forest, and Water Conservation and Reclamation*, vol. 2, 398.
[31] Lowdermilk, *Soil, Forest, and Water Conservation and Reclamation*, vol. 2, 410.

Two years later, in 1952, the Water Conservancy and Fuel Industry Ministries approached the central government with a proposal to enlist Soviet experts to help plan the management of the Yellow River. Given the Soviet Union's impressive track record in the development of hydroelectric power, the State Council decided to include the control and exploitation of the Yellow River in their request for assistance. For the Soviet government, this represented an opportunity to showcase their "Tekhintern" capabilities and, in a bid to outshine the United States in the context of the Cold War, to demonstrate their ability to transform one of the world's most challenging rivers.

Before the Soviet experts arrived, the State Planning Commission created a Yellow River research group, comprised mainly of people from the Water Conservancy and Fuel Industry Ministries. The group's main task was to gather and translate data and other materials on the Yellow River for the use of the Soviet experts. On January 2, 1954, these experts arrived in Beijing and, after studying the basic materials on the Yellow River, suggested that comprehensive planning be commenced while field surveys were still being conducted.[32] The State Council established the Yellow River Planning Commission in April 1954, drawing technocrats from various departments including the Ministries of Geology, Forestry, Transportation, Railways, and Agriculture. With the establishment of this commission, the Yellow River was no longer the sole concern of the Water Conservancy Ministry. Now, multiple state departments were involved in comprehensive river valley development efforts.

Managing the Yellow River was a complex challenge that could only be met through the collective effort of multiple functional institutions. Different departments, though, had differing goals and orientations. For the Water Conservancy Ministry, flood control was the top priority, while the Fuel Industry Ministry saw its primary responsibility as the development of hydroelectric power. The high silt content of the Yellow River made it difficult to exploit the river's full potential. Although the shared goal was to develop the Sanmenxia Dam as a multipurpose project, achieving a reasonable balance between flood control, the costs of reservoir inundation, and hydroelectricity output was necessary. Achieving

---

[32] The leader of the group was the vice chief engineer of the Leningrad Hydropower Design Institute of the Soviet Union's Electricity Ministry. Group members included hydrotechnical specialists, hydrology and hydropower calculation specialists, construction specialists, project geologists, irrigation specialists, and navigation specialists. The group had no silting, water, and soil conservation expert or reservoir inundation expert.

this balance would require a thorough understanding of the characteristics of the entire river and the drawing of clear distinctions between primary and secondary goals. The Soviet experts' focus on hydroelectric achievement inadvertently upset this balance, although not permanently. Alongside the development of political movements in China, this led to the greatest setback in the country's long efforts to control the river in the twentieth century, resulting in the displacement of hundreds of thousands of people who had been living along the upper reaches of the river above the dam.[33]

The Yellow River planning group included experts from the Soviet Electrification Ministry, specifically the Hydroelectric Design Academy. After doing field surveys along the river, the Soviet experts recommended Sanmenxia as the most suitable site for a dam, despite the fact that it would involve submerging more land than other possible locations. "None of the other sites could bring comprehensive benefits to the river valley like Sanmenxia," remarked one expert. "In order to eliminate the threat of flood, we must have reservoirs. It is impossible to build a reservoir without population displacement."[34] Still, the central state chose to divide the project into multiple stages, in order to give local governments ample time to prepare for the significant displacement that would result and to attempt to win over local bureaucrats who opposed the project.

In October 1954, the Planning Commission published the *Comprehensive Development Planning of the Yellow River Report*, which proposed the construction of forty-six dams across the main stream and the tributaries of the Yellow River. Sanmenxia was identified as the top priority for the first phase of the project, given its ability to improve flood control, sediment detainment, irrigation, the generation of electricity, and navigation simultaneously. In the middle reaches of the river, only Sanmenxia was deemed suitable for multipurpose development.[35] The Soviet Union agreed to design the dam and to provide much of the necessary equipment. The report was then submitted to the State Council and the Central Political Bureau for review and discussion. Most state and party leaders felt that with Soviet assistance the plan was feasible and thus

---

[33] See Leng Meng, "The Battle for Sanmenxia," *Chinese Sociology and Anthropology* 31, no. 3 (1999), 4–98.

[34] *Huanghe Sanmenxia shuili shuniuzhi* (*Annals of the Yellow River Sanmenxia Hydro Station*) (Beijing: Zhongguo dabai kequanshu chubanshe, 1993), 30.

[35] Huanghe zonghe liyong guihua jishu jingji baogao (Technical and Economic Report of the Multipurpose Development Planning of the Yellow River).

submitted it to the National People's Congress for discussion. With no public opposition, the Congress passed the plan.

Despite the challenges noted by Japanese and American engineers prior to 1949, the Communist government ultimately decided to undertake the highly ambitious Sanmenxia hydropower project. The task of developing a specific design was assigned to the Leningrad Design Academy of the Soviet Union. In 1956, Wang Huayun proposed that the Soviet design team be invited to work in China so that it could train Chinese engineers in design. The Soviets rejected the proposal. Instead, the Leningrad Design Academy invited Chinese delegates to visit the Soviet Union.[36]

The Yellow River report recommended a regular water level for the Sanmenxia reservoir of 350 meters above sea level, but the engineers from the Leningrad Design Academy argued that, to ensure the dam's continued functionality for at least 100 years, the regular water level should be no lower than 360 meters. The initial Soviet design aimed for multipurpose development by creating a reservoir with a capacity of 65 billion cubic meters to prevent floods, balance the river's flow, and provide water for the irrigation of 40,000,000 *mu* of farmland on the North China Plain throughout the seasons. The project also included a large-scale hydroelectric plant with a capacity of more than 1,100,000 kW and annual output of 6 billion kWh with coverage of Shaanxi, Shanxi, Hebei, and Henan provinces. This energy grid would power heavy industries and contribute to China's industrialization.[37]

Propaganda in China claimed that the Soviet experts were sincerely devoting themselves to managing the Yellow River. The former Nationalist commander and at the time the secretary of the Water Conservancy Ministry Fu Zuoyi praised them by saying that "they love China as much as they love their native country."[38] They worked hard, not only contributing to the survey work and design of hydraulic projects but also sacrificing their leisure time to deliver lectures and write textbooks for their Chinese colleagues. Their intention was to share their knowledge and they strove to help Chinese technicians learn more

---

[36] Wang Huyun zhuren zai Sanemnxia gongchengju qicaide bufen wenjian digao (Parts of Manuscripts of Director Wang Huayun at the Sanmenxia Project Bureau) (1956) file no. 1-3-1956-0052Y, the Yellow River Conservancy Commission Archives, Zhengzhou.

[37] A. A. Kololev, "Huanghe Sanmenxia shuili shuniu" (Sanmenxia Hydropower Project on the Yellow River), *Zhongguo shuili* 7 (1957), 5.

[38] Fu Zuoyi 傅作义, "Sulian dui Zhongguo shuili shiye de bangzhu" 苏联对中国水利事业的帮助 (Soviet Assistance to Chinese Hydraulic Enterprise), *Renmin ribao*, October 22, 1957.

quickly. One Soviet expert expressed delight that their work would contribute to the socialist enterprise of their "Chinese brothers."[39]

As Zhang Hanying, the vice minister of the Water Conservancy Ministry, put it, the Soviet experts were also "people sent by Stalin."[40] Their activities were thus deeply connected to international geopolitics. Once the proclaimed fraternal relationship between the Soviet Union and the People's Republic broke down, then, their scientific internationalism came to be overshadowed by the ideological dispute between the two Communist parties.[41] In July 1960, when the Soviet experts were suddenly recalled by their home country, most of them expressed puzzlement, being unclear about what had happened between the two countries.[42] For their part, despite their disappointment at the departure of the Soviet experts, Chinese technicians expressed gratitude for their assistance. Many projects, though, including the Sanmenxia Dam, remained incomplete. The departure of the Soviets meant that Chinese engineers were left to complete the mega-project on their own. The fact that the Soviets had delivered the hydro turbine assemblage to China without any instructions or data was especially troublesome. But this sparked national pride among the Chinese engineers, who vowed to bring the project to completion.[43]

THE CHALLENGE OF SILTING AND DEBATES OVER THE SOVIET DESIGN

In the early 1950s, effective methods of water and soil conservation on the Loess Plateau were widely recognized by hydraulic specialists as preconditions for building a durable reservoir on the Yellow River. To find such methods, the Yellow River Conservancy Commission of the new

---

[39] Shuili Fadian Jianshe Zongju zhuanjia Gongzuoshi, "Sulian zhuanjiadui woguo shuidian jianshede bangzhu," 7.
[40] Zhang Hanying, "Sidalin pailaideren zenyang zai Zhongguode heliushang gongzuozhe" (How People Sent by Stalin Work on Chinese Rivers), *Renmin ribao*, November 25, 1952.
[41] On the Sino–Soviet split in general, see Odd Arne Westad (ed.), *Brothers in Arms: The Rise and Fall of the Sino-Soviet Alliance, 1945–1963* (Washington, DC: Woodrow Wilson Center Press, 1998).
[42] Zhang Peiji, "Sanmenxia jianshe zhongde Sulian zhuanjia" (Soviet Experts in the Sanmenxia Project), in *Wanli Huanghe diyiba* (Zhengzhou: Henan renmin chubanshe, 1992), 309.
[43] Qu Zhide and Bai Wenying, "Hanhao wolunji, weiguo zhengguang (Welding the water turbine well, honoring the Motherland)," *Wanli Huanghe diyiba* (Zhengzhou: Henan renmin chubanshe, 1992), 462.

government established three experimental stations in Tianshui, Xifeng, and Suide, thus inheriting Lowdermilk's legacy. In 1955, Wang Huayun toured Shanxi and Shaanxi to study local water and soil conservation practices. In his report, Wang described the experiences of Daquanshan in Yanggao county and Jiajiayuan in Liulin county of Shanxi province, and Jiuyuangou in Suide county of Shaanxi province. According to him, these locations provided both biological and engineering methods to prevent water and soil erosion. These included paving terraces, building soil-saving dams, digging wells, building small and medium-sized reservoirs, and cultivating trees and grasses. If these specific methods could be combined with the strong leadership of the Communist party, Wang was confident that the goals of the water and soil conservation project on the Loess Plateau could be achieved by 1967.[44]

In another experiment, the commission built a silt trap reservoir at Jiuyuangou, estimating that it would take ten years for it to become fully silted. The silting process was much faster than anticipated, though: the reservoir was fully silted after only three years. This experiment was a warning for the prospective Sanmenxia reservoir. During a State Council meeting on May 27, 1957, Zhou Enlai expressed his concern: "According to the Jiuyuangou experience, it is possible that the Sanmenxia reservoir could quickly become clogged as well. Although the project has already begun, I feel uneasy about it."[45]

In 1954, the Yellow River conservancy plan was created. It required the completion of 165 million *mu* of field engineering on the Loess Plateau, the improvement of 220 million *mu*, the cultivation of 20 million *mu* of forest, the construction of 720,000 check dams and silt trap dams, and the development of 4.9 million *mu* of irrigated land. This was especially crucial to the provinces of Shaanxi, Shanxi, and Gansu, as it would reduce soil erosion and the flow of silt into the Yellow River by at least one third.[46] With rural collectivization and the launch of the Great

---

[44] Wang Huayun 王化云, "Huangtu qiuling gouhequ shuitu baochi kaocha baogao 黄土丘陵沟壑区水土保持考察报告 (Investigation Report of Water and Soil Conservation on the Loess Plateau)," *Xin Huanghe* 新黄河 12 (1955), 26–31.

[45] Zhonggong Zhongyang wenxian yanjiushi中共中央文献研究室(ed.), *Zhou Enlai nianpu, erjuan, 1949–1976* 周恩来年谱1949–1976 (*A Chronicle of Zhou Enlai, vol. 2 (1949–1976)*) (Beijing: Zhongyang wenxian chubanshe, 1997), 45.

[46] "Huanghe shuili weiyuanhui 1954 nian shuitu baochi gongzuo huiyi jielun" 黄河水利委员会1954年水土保持工作会议结论 (Conclusion of the 1954 Yellow River Conservancy Commission Water and Soil Conservation Working Meeting)," *Kexue tongbao* 科学通报 3 (1955), 19–21.

Leap Forward in the late 1950s, thousands of rural construction teams were organized to take charge of water and soil conservation efforts. These teams were comprised of laborers from local communes. In a very short period, hundreds of thousands of acres of terraced land were paved, and thousands of soil-saving dams were built. Because of the lack of professional instruction and comprehensive planning, however, many of the silt trap dams collapsed in the following years as a result of rainstorms.[47]

The Soviet engineers who worked on the Sanmenxia project were also aware of the silting problem. They estimated that around 80 percent of the silt would remain in the reservoir, while the remaining 20 percent would flow downstream. They felt that water and soil conservation efforts on the Loess Plateau could reduce the amount of sediment flowing to the reservoir by up to 50 percent. If this could be accomplished, they predicted that a little over half of the total capacity of the reservoir, or about 34 billion cubic meters, would become silted in the next fifty years. Placing this against the many benefits promised by the project, Soviet engineers were confident that this slow pace of silting was acceptable.[48]

Then, in 1957, a debate occurred about the Soviet design of the Sanmenxia Dam, with some parties voicing opposition to the project. In his contribution to the debate, Huang Wanli 黄万里, a hydraulic professor at Tsinghua University, acknowledged the importance of water and soil conservation in the region. He also argued, though, that it should not be seen as a complete solution to the problem of the Yellow River. Huang argued that the high amount of silt in the river was a natural principle, and that containing sediment in the reservoir would violate this principle and potentially cause ecological problems in the dam's upper stream region.[49] Despite Huang's concerns, the project proceeded. To mitigate the potential negative effects of silting, Huang suggested that the reservoir maintain a lower regular water level and that the diverting conduit at the base of the dam should be preserved in case of severe silting

---

[47] Hao Ping, "A Study of the Construction of Terraced Fields in Liulin County, Shaanxi Province in the Era of Collectivization," in *Agricultural Reform and Rural Transformation in China since 1949* ed. Thomas DuBois and Huaiyin Li (Leiden: Brill, 2016), 101–114.

[48] Kololev, "Huanghe Sanmenxia shuili shuniu," 6.

[49] Huang, "Duiyu Huanghe Sanmenxia shuiku xianxing guihua fangfa de yijian" (Comments on the Current Principles of Operation of the Sanmenxia Reservoir on the Yellow River)," *Zhongguo shuili* 8 (1957), 26–29.

in the future.⁵⁰ Huang was later criticized as a rightist by Mao Zedong and his followers, though, which meant that his contributions were downplayed.⁵¹ Along the same lines, according to his memoir, Zhang Guangdou also expressed concerns about the project but eventually chose to conform to avoid political risk.⁵²

Mao had a deep-seated distrust of Chinese intellectuals, describing many of them at one point as "little Chinese Nagys."⁵³ During the Great Leap Forward, he called the Chinese expert class and intelligentsia in industrial plants and universities "an intelligentsia of bourgeois heritage" who used the freedom of speech they had been granted to viciously attack the Party and to promote the restoration of capitalism.⁵⁴ Based on his experiences of this struggle, Huang wrote an allegorical story, "The Road Turns Over." This is what prompted Mao to denounce Huang as a rightist on June 8, 1957. However, it wasn't Huang's opposition to the Soviet high dam design that led to his identification as a rightist, but rather his criticism of the Communist Party.⁵⁵ Even so, after his denunciation, no other hydraulic engineer publicly supported his technological concerns.

In Shaanxi province, raising the reservoir's water level from 350 to 360 meters would entail the submersion of much of the most fertile land, affecting up to 3.25 million *mu* of land and 870,000 people (see Map 5.1). Provincial officials, including the provincial party secretary Zhang Desheng 张德生, expressed concern about the long-term impact of a fully silted reservoir in fifty or a hundred years, if not sooner. The rising riverbed of the Yellow River on the central plain had, after all, been a concern for centuries and, according to the Soviet design, containing silt in the reservoir merely moved the problem to the Guanzhong Plain and

---

⁵⁰ Huang, "Duiyu Huanghe Sanmenxia shuiku xianxing guihua fangfa de yijian"; Also see Zhao Cheng 赵诚, *Huang Wanli de Changhe gulu* 黄万里的长河孤旅 (*Huang Wanli's Lonely Journey down the Long River*) (Xi'an: Shaanxi renmin chubanshe, 2013), 68–82.

⁵¹ Zhao, *Huang Wanli de Changhe gulu*, 94–101. Also see Shapiro, *Mao's War against Nature*.

⁵² Zhang, *Wode rensheng zhilu*, 74.

⁵³ Imre Nagy was a Hungarian politician who became the leader of the 1956 revolution against the Soviet Union and the communist political system in Hungary. He was executed following the Soviet invasion of Hungary.

⁵⁴ Record of Conversation between Polish Delegation and PRC Leader Mao Zedong, Beijing, October 14, 1959. Wilson Center Digital Archives.

⁵⁵ See Shapiro, *Mao's War against Nature*, 53–54; Huang Wanli, "Huacong xiaoyu" 花丛小语 (Murmuring among Flowers), in *Huang Wanli wenji* 黄万里文集 (*Anthology of Huang Wanli*) (Beijing: Huang Wanli wenji bianjichuban xiaozu, 2001).

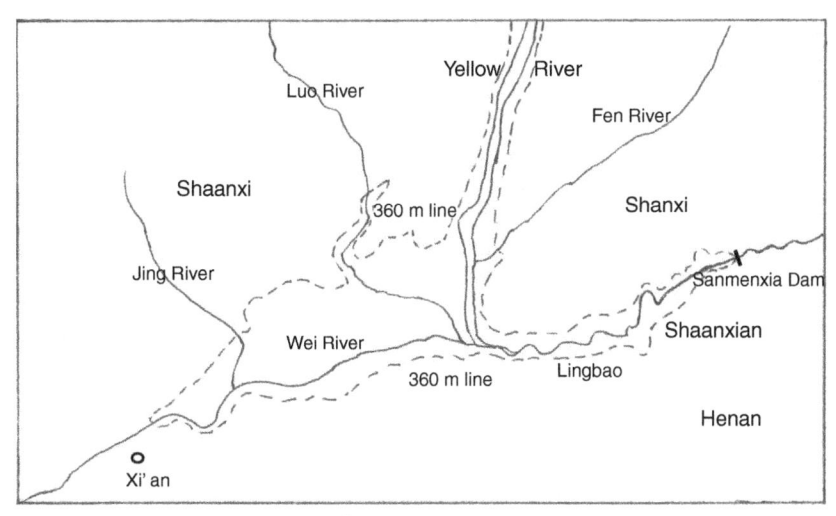

MAP 5.1 Sanmenxia Reservoir area

the Wei River Valley. Officials in Shaanxi thus preferred fixing the regular water level of the reservoir as low as possible.

After receiving a number of complaints from Shaanxi officials and residents, Zhou Enlai promptly summoned experts to Beijing to discuss the Soviet design for the Sanmenxia project. During this meeting, Wen Shanzhang 温善章, a technician in the Hydroelectric Development Bureau, suggested flushing all sediment down to the lower stream instead of detaining it in the reservoir. He also proposed a regular water level of 336–337 meters. This, he suggested, would minimize the cost of submerging land and moving the population in the valley, even if it meant lower hydroelectric output.[56] Despite Wen's suggestions, most experts at the meeting were captivated by the promised benefits of the high dam design. They supported the Soviet design with a regular water level of 360 meters and adhered to the principle of keeping sediment in the reservoir. They argued that flushing sediment out of the reservoir was misguided, as doing so would not alleviate the problem of the rising riverbed of the lower stream.[57] Furthermore, Wen's proposals would significantly reduce the

---

[56] *Huanghe Sanmenxia shuili shuniuzhi*, 48.
[57] "Sanmenxia shuili shuniu taolunhui" (Discussion Meeting on the Sanmenxia Hydropower Project), *Zhongguo shuili* (July 1957). Judith Shapiro suggests that the idea of "Shengrenchu, Huangheqing" (when a Sage emerges, the Yellow River will run clear) was influential in the design of the project (to please Mao). However, there is no direct evidence that this idea determined the design of the dam. From the perspective of

comprehensive benefits promised by the project and lessen its political and economic significance. These experts argued that once a large project like Sanmenxia had been decided upon, it was imperative that it contribute to the agricultural and industrial development of the nation.

By early 1958, much of the Sanmenxia project had been constructed, but disagreements about the anticipated water level were causing delays in the design process. In an effort to reconcile the different opinions, Zhou Enlai and other state leaders visited the project site in April of that year and called a meeting of all concerned parties. During this meeting, the delegates from Shaanxi argued for a reduction of the height of the dam and a corresponding lowering of the anticipated water level, with a view to preserving valuable arable land. Meanwhile, the Water Conservancy and Electricity Ministries, the Yellow River Conservancy Commission, and the project's own Bureau advocated for the initial design, in order to maximize hydroelectric power and create the largest possible reservoir for the absorption of silt. Peng Dehuai and Xi Zhongxun, both senior party leaders, emphasized the importance of the comprehensive nature of the initial design. For them, transforming the Yellow River was vital to promoting China's international status, achieving industrialization, and the goal of catching up to Great Britain within fifteen years.[58] Zhou, however, took a more pragmatic approach. He acknowledged that while the Communist government was committed to controlling the river, technology and human knowledge were not yet advanced enough for a single project, however massive, to turn the harmful river into a beneficial one. Zhou thus believed that a successful design process required an appropriate combination of politics and technology. While politically committed to conquering the river, Zhou was cautious about the uncertainties around soil conservation in the steppe and along the river. This led him to advocate a relatively conservative approach, leaving some leeway for future adjustments in the design.[59]

---

engineering and flood control in North China, the rising riverbed had been a concern of hydraulic engineers and officials long before the Mao era.

[58] "Peng Dehuai fuzongli zai 1958 nian Sanmenxia shuilishuniu shuiku huiyi shangde jianghua" 彭德怀副主席在1958年三门峡水利枢纽水库会议上的讲话 (Vice Premier Peng Dehuai's Talk at the Sanmenxia Project Meeting in 1958); Xi Zhongxiong mishuzhang fayan 习仲勋秘书长发言 (Secretary Xi Zhongxiong's talk at the meeting), April 24, 1958. Sanmenxia Municipal Gazetteer Office. Although Peng was later seen as a major opponent of Mao's Great Leap Forward, he was optimistic about the Sanmenxia project in 1958.

[59] Zhou Enlai zongli zai Huanghe Sanmenxia shuili shuniu shuiku huiyi shangde zongjie fayan 周恩来总理在黄河三门峡水利枢纽水库会议上的总结发言 (Premier Zhou Enlai's

Zhou Enlai played a major role in finding a compromise among stakeholders over the design of the Sanmenxia Dam. He recognized that the primary goal of the project was flood control and the prevention of dike breakage in the river's lower stream. Zhou was also sympathetic to Shaanxi province's concerns about the loss of arable land and the displacement of its population. After discussions between all the delegates, Zhou led the effort to reach a compromise that satisfied all parties. The revised design sought to maintain the anticipated water level of 360 meters, but it should not exceed 340 meters until after 1967. To address Shaanxi province's concerns, the dead water level was reduced to 325 meters, and the "most desirable" water level was set at 335 meters. The goal of containing silt in the reservoir remained. To achieve sustainable hydropower generation on the Yellow River, all experts agreed that it required massive water and soil conservation efforts on the Loess Plateau, where most of the river's sediment originated. The American consultancy group hired by the Nationalist government in 1946 had predicted that soil conservation could reduce erosion by 0.3 percent annually, which meant that it would take 300 years to eliminate sediment buildup in the river completely.[60] Through mass mobilization, officials in Shaanxi aimed to reduce soil erosion by 20 percent before 1967 and 50 percent by 2010.[61] While some engineers believed that conservation efforts would reach a virtuous tipping point where silt would no longer pose a threat, the rapid silt buildup of the Sanmenxia reservoir after it was built in the early 1960s suggested that such efforts had fallen far short of expectations.[62]

Talk at the Sanmenxia Project Meeting), April 24, 1958, Sanmenxia Municipal Gazetteer Office.
[60] Zhao, "Jianguoqian Sanmenxia gongcheng yanjiu," 18.
[61] *Huanghe Sanmenxia shuili shuniuzhi*, 49.
[62] Martin Reuss, "Seeing Like an Engineer: Water Projects and the Mediation of the Incommensurable," *Technology and Culture* 3 (2008), 531–546. On the role of negotiation in public works, in this case, negotiation over the dam design involved Soviet engineers, Chinese engineers, the Chinese central government, and local governments. But the people who were impacted on the most, reservoir residents, were not directly involved. This negotiation was not simply a top-down process steered by the central government. Instead, it was initiated by engineers and local officials, who had their own practical concerns. Despite the fact that modifications were made in the construction process, the original Soviet high dam design persisted. Thus, we can claim that this negotiation process ended with the triumph of high modernists. Years later, with the emergence of serious silting problems and the reverse tide of reservoir population displacement, the hydropower project had to reconcile itself with the river and the indigenous population. Bao Heping 包和平. Gongchengde shehuiyanjiu: Sanmenxia gongchengzhongde zhenglunyu jiejue 工程的社会研究：三门峡工程中的争论与解决

CONSTRUCTING SANMENXIA WITH PROPAGANDA

At the opening ceremony of the Sanmenxia project on April 13, 1957, Fu Zuoyi, the Minister of Water Conservancy, gave a speech that emphasized the indispensable role of the leadership of the Chinese Communist Party in making the project possible. The decorations that accompanied the ceremony also reflected changing Chinese attitudes toward nature. Since the Ming dynasty, one cliff in the gorge had been inscribed with the characters *qiaobi xiongliu, guifu shengong* (steep gorge and formidable torrent, the work of the spirits and the gods), which described the river's magnificent natural power. At the Sanmenxia project's opening ceremony, however, this inscription was replaced with a new one, *genzhi shuihai youri, Huanghe bianqing youqi*, which promised the imminent elimination of floods and the flow of a clear river. Although the ideal of a clear-running Yellow River had a long history and persisted in the People's Republic, this revision reflected the growth of technological hubris in subordinating nature to human purposes that characterized much of twentieth-century world history (see Figure 5.1).

As in the Soviet Union and the United States, the leaders of Communist China held a utilitarian view of the natural world. This extended to politics as well as the economy.[63] The benefits promised by the hydropower project under the indispensable leadership of the Communist Party mesmerized the people. Unlike river conservancy in previous polities, mega hydropower projects in twentieth-century China were promoted in propaganda media and affected the daily lives of millions of people who lived far from rivers. Through official media, the people of the People's Republic were made aware of this unprecedented plan. Thus, in addition to engineers and state cadres, millions of ordinary people and hundreds of thousands of people who lived along the river inevitably got involved in the construction of the dam. State propaganda institutions used the Sanmenxia hydropower project to project the strength of the new regime and win the people's support.

Following the State Council's announcement of the Yellow River conservation plan, *People's Daily* published an editorial entitled "A Great Plan to Change the Yellow River has been Declared." The editorial

---

(*Social Study of Engineering: Debates and Solution of the Sanmenxia Project*) (Hohhot: Neimenggu jiaoyu chubanshe, 2007).

[63] Dorothy Zeisler-Vralsted, *Rivers, Memory and Nation-Building: A History of the Volga and Mississippi Rivers* (New York: Berghahn Books, 2015); Pietz, *Yellow River*.

FIGURE 5.1 Sanmenxia Dam under construction (*Renmin Huabao* 3.1962)

explained the historical and contemporary context of the Sanmenxia project and highlighted its numerous expected benefits. It would create a reservoir larger than Lake Tai, which would be utilized for flood control, irrigation, the generation of hydroelectric power, the improvement of navigation, and reducing sediment in the river. The media portrayed a bright future for the river basin; there was no mention of the social or technological challenges that might arise from the project.[64]

Another report in *People's Daily* also painted a bright picture of the future of the Sanmenxia project. One engineer was quoted to say that the project was just the first step in exploiting the country's massive 540,000,000 kW hydropower reserves.[65] Wang Huayun, the vice director of the bureau in charge of the project, emphasized the project's impressive production of energy: It would generate 6,000,000,000 kWh of hydroelectricity annually and reduce the country's annual use of coal by over 3,600,000 tons.[66] These claims highlighted the significant contribution

---

[64] Hua Shan 华山, "Sanmenxia: yige gaibian Huanghe ziran mianmao de weida jihua xuanbule!" 三门峡：一个改变黄河自然面貌的伟大计划宣布了 (Sanmenxia: A Great Plan to Change the Yellow River has been Announced), *Renmin ribao*, July 21, 1955.

[65] Ye Jianyun 叶剑韵, "Sanmenxia de mingtian" 三门峡的明天 (The Future of Sanmenxia), *Renmin ribao*, November 26, 1956.

[66] Ye, "Sanmenxia de mingtian."

the project would make to the country's energy production. But the media portrayal of the Sanmenxia project was idealistic, with no acknowledgment of potential negative consequences.

The day after the opening ceremony of the Sanmenxia project, an editorial titled "Everybody come to support Sanmenxia!" appeared in *People's Daily*. Unlike the earlier excessively optimistic reports, this editorial acknowledged the enormous amount of work required to make the project a success. As a megaproject in a planned economy, the article argued, the Sanmenxia project required the cooperation of a number of different state departments and local governments. In a socialist system, the state was required to organize all of the project's support facilities, including transportation, housing, food, and even barber services for workers.[67] The editorial called on all state institutions and ordinary people to support the project. Henan province, where the project was located, assumed most of the support duties. Under instruction from the local government, nearby agricultural cooperatives expanded their vegetable cultivation area from 1,500 to 4,000 *mu*. The project became the focus of the area's social and economic activities.[68]

Journalists from *Xinhua Press*, *People's Daily*, and *Henan Daily* observed and reported on the project's progress on-site. The National Writers Association organized writers' visits to the project, encouraging them to write poetry and prose about it in order to build a supportive social atmosphere. Central and local radio and television stations also highlighted the project's significance. As part of the propaganda campaign associated with the Great Leap Forward, the Central News Documentary Film Factory produced color documentaries called *Sanmenxia* and *Great Changes on the Yellow River* in 1958 and 1961 and screened them nationwide.

Due to the overwhelming state propaganda that successfully imprinted its significance in the minds of the majority of the people, Sanmenxia became an important element of popular consciousness. The reports and works of literature that were produced about it were not objective introductions. Rather, they aimed to build up confidence in the socialist state and in the Communist Party. From this perspective, the Sanmenxia project was not only a landmark of professional hydraulic engineering, but also of the party's social engineering ambitions.

[67] Editorial, "Dajia lai zhiyuan Sanmenxia a!" 大家来支援三门峡啊 (Everybody Come to Support Sanmenxia!), *Renmin ribao*, April 14, 1957.
[68] "Zhiyuan Sanmenxia" 支援三门峡 (Support Sanmenxia), *Renmin ribao*, April 18, 1957.

In the several months before and after the project began, the project bureau office and the project union received more than 3,000 letters from workers, farmers, cadres, soldiers, and even Young Pioneers from all over the country. The state had made the megaproject a part of elementary education, and the entire society was broadly mobilized to participate in the conquering of the Yellow River. The letters wished for the dam to be completed soon, so that the Yellow River could be transformed into a boon for the nation. One letter from the students of Shiyan Elementary School of Chongming county, Shanghai, for example, reported that they picked up ears of wheat on their way to and from school, sold them for 5.33 *yuan*, and donated the proceeds to the project.[69] State propaganda portrayed the project as promising a prosperous future to the industrial workers and peasants of the region, and the nation evidently recognized its sublimity. Through the Sanmenxia project, the state successfully integrated its legitimacy and authority with the all-encompassing hydropower nation.

RECONSTRUCTIONS

After the dam was completed, between September 1960 and March 1962 the reservoir began to store water, but the silting process was much more severe than expected.[70] In just three years, 1.53 billion tons of sediment had been deposited in the reservoir. This blocked the estuary of the Wei River, the Yellow River's largest tributary. This resulted in unexpected flooding and the salinization of 670,860 *mu* of farmland on the Guanzhong Plain.[71]

To address this problem, the project authority had to change the operating principles and the structure of the dam. At a meeting held in Zhengzhou in February 1962, it was decided that the policy of "detaining water and sediment" would be changed to "detaining water and flushing sediments." Unless a clear danger of flooding in the lower stream existed,

---

[69] Ji Wenxuan 季文选, "Quanguo zhiyuan Sanmenxia" 全国支援三门峡 (The Whole Country Supports Sanmenxia), in Zhongguo renmin zhengzhixieshang huiyi Sanmenxia shi weiyuanhui中国人民政治协商会议三门峡市委员会 and Zhongguo shuilishuidian dishiyi gongchengju中国水利水电第十一工程局 (eds.), *Wanli Huanghe diyiba* (Zhengzhou: Henan renmin chubanshe, 1992), 431–432.
[70] See Smil, *Bad Earth*, 45–47.
[71] Sanmenxia kuzhou yanjianhua qingkuang 三门峡库周盐碱化情况 (The Situation of Salinization around the Sanmenxia Reservoir), 1963. A12-1(1)-31, Yellow River Conservancy Commission Archives, Zhengzhou.

the dam's sluice gates were to be opened during the rainy season. This change of policy helped to relieve silting to a certain degree but, because of the high altitude of the water outlets, more than 60 percent of the sediment in the Yellow River above the dam remained in the reservoir. The reservoir's capacity shrank from 9.84 billion cubic meters of water in 1960 to 5.74 billion cubic meters in 1964. The silting of the Wei River continued. In December 1964, Zhou Enlai hosted a meeting in Beijing to address the problem. After a lengthy discussion, participants agreed to open two tunnels through the dam's north bank and to repurpose four steel tubes to facilitate and accelerate the drainage of water and sediment through the reservoir.[72] This involved sacrificing a certain amount of electricity generation, but it was seen as necessary to mitigate persistent salinization and floods.

The outlet velocity of the dam had doubling by 1969, but the reservoir had still lost half of its capacity because of accumulated sediment. The Wei River and the reservoir above Tongguan were particularly affected, with serious silting persisting despite efforts to reduce it. In 1968, the dikes of the Wei River in Hua County broke, leading to severe flooding that put Xi'an in danger. This revealed that previous reconstruction work had been insufficient. To address the silting issue, when the water level was at 315 meters, it was necessary to reopen eight diversion outlets and lower five penstock inlets from 300 to 287 meters to achieve an outlet velocity of 10,000 cubic meters per second. Also, unlike typical hydropower plants, during flood season all of the generators used the run-of-the-river pattern without impounding water behind the dam, resulting in a capacity of only 250,000 kW. The original design had involved a capacity of 1,000,000 kW. These measures did gradually reduce silting in the Wei River and in the reservoir above Tongguan. After years of experimentation, from November 1973 the reservoir began to store clear water during non-flood seasons and discharge turbid water during flood seasons to improve the efficiency of electricity generation. Finally, then, after fifteen years, the Sanmenxia dam's optimal operating pattern had been found, and it now posed significantly less of a threat to communities in the upper reaches of the Yellow River in Shaanxi and Henan.

CONCLUSION

In his book *Silenced Rivers*, Patrick McCully notes that dam failures are often the result of overoptimistic assumptions made during the planning

---

[72] *Huanghe Sanmenxia shuili shuniuzhi.*

process.⁷³ This is exactly the case with the Sanmenxia project. In the 1950s, the soil and water conservation plan underestimated the challenge of silting in the Sanmenxia reservoir. The result was a high dam design that promised significant hydroelectrical production. As early as 1964, the Water Conservancy and Electricity Ministry convened a conference to draw lessons from the Sanmenxia project. One of the major conclusions was that overemphasizing the dam's function of generating hydroelectric power was to be avoided. Despite the existence of alternatives developed by officials and engineers, the compromise supported by Zhou Enlai, discussed earlier in this chapter, was not enough to turn the tide against making hydroelectric power production one of the dam's key functions.

From very early on, serious silting shattered the illusion that high outputs of electricity could be attained alongside protection from flooding. Thus, given the degradation of riparian ecology and the social disruption caused by reservoir displacement, many people began to question the value of the first mega-dam on the Yellow River. In the official narrative, though, the state continued to proclaim the dam's manifold achievements, thus maintaining and burnishing its own image as "an energetic, determined state capable of taming rivers for the social good."⁷⁴ In this sense, Sanmenxia is both a technological and a political construct.

At the Sanmenxia Dam site now, there is a viewing deck near the former location of Shuzhuangtai, a rock islet that was removed during the dam's construction. On the deck is displayed a poem by He Jingzhi 贺敬之, a major revolutionary poet.

Pangu [one of China's few "creator gods"] gave birth to our new generation! Raising the red flag, opening Earth and Heaven, trampling history underneath our feet, we are writing a new chapter of history. Socialism – here we come! Here we come, here we come, shocking Mang Mountain and Kunlun; showing our river control blueprint, bundling waist of the Yellow River – The gate of the Gods gone. The gate of the Spirits – gone. The gate of Men turning to ashes!

The three gates are not there anymore, and tomorrow the sluice gate will open. Li Bai should revise his poem: the Yellow River comes from our hands!⁷⁵

It is well known that the Tang poet Li Bai's (701–762 CE) original poem includes the line "Can't you see that the Yellow River comes from heaven!" He Jingzhi replaces "Heaven" with "our hands." Written in

---

⁷³ McCully, *Silenced Rivers*, 101.   ⁷⁴ McNeill, *Something New under the Sun*, 157.
⁷⁵ For the whole poem, see He Jingzhi. *He Jingzhi Wenji* 贺敬之文集 (*Collection of He Jingzhi*), vol. 1 (Beijing: Zuojia chubanshe, 2005).

1958, the poem celebrates human domination of nature and the triumph of socialism. In contrast to traditional Chinese poetry, which often emphasized the power of nature, poetry in the twentieth century, and particularly during the Mao era, stressed the power of human beings.[76] He Jingzhi's poem reflects the ideology of the Great Leap Forward, which targeted the Yellow River as a major obstacle to the success of the Communist project.[77] The connection between politics and nature in China, however, dates back to the myth of Yu the Great's successful control of a great flood and the belief, common in the imperial period, that a clear Yellow River was an auspicious omen.[78] Today, the water of the Yellow River is controlled by the Sanmenxia Dam, flowing through its turbines and irrigating the central plain.

Despite its grandeur, official propaganda about the Sanmenxia hydropower project masked the social inequalities and environmental injustices it created. Rather than this representing a simple triumph over the Yellow River, the river was transformed into a stage, and the project also functioned as a way for the Communist Party to assert and reinforce its authority through forced relocation and indoctrination. Whether we see it as an improper intervention of human "hands" or as brute force technology, the dam had to be reengineered to align more closely with the river's nature. The Yellow River's high concentration of sediment is not entirely "natural": The river has been intertwined with human history for a very long time. It has lost its "unmade" qualities.[79] While Europeans transformed the Rhine into a channel for industrial navigation, people in China built thousands of miles of dikes to regulate the flow of the Yellow River on the North China Plain. Technological hubris in twentieth-century China had to be reconciled with the river's natural silt deposits. For sustainable use to be achieved, both the river and the dam had to be transformed. The hydropower nation, that is to say, was formidable but not omnipotent, and it had to conform to environmental and social realities in the process of its ambitious expansion.

---

[76] See Elvin, *Retreat of the Elephants*.   [77] Pietz, *Yellow River*.
[78] Mark Elvin, "Who Was Responsible for the Weather? Moral Meteorology in Late Imperial China," *Osiris* 13 (1998), 213–237.
[79] White, *Organic Machine*.

6

The Human Cost

In February 1956, a sense of excitement filled the village of Shijiatan 史家滩 as its inhabitants busied themselves with preparations for the upcoming lunar new year celebrations. A village cadre returning from a district meeting, however, brought a disturbing message that quickly spread through the community, replacing the festive atmosphere with one of anxiety and unrest. Other village cadres took to adobe walls, scrawling slogans like "Dismantle" and "Sacrifice individual interests for national reconstruction," signaling to the villagers that this would be their final celebration in the home they had long known. In just a few months, they would be forcibly uprooted and relocated elsewhere.

During the 1950s and early 1960s, hundreds of massive dams were built in China as part of a campaign for water conservation and hydro-electric power. In the process, more than 7.8 million people were displaced from their homes, forming one of the largest "unimagined communities" in recent Chinese history.[1] Diana Lary's brief discussion of the reservoir displacement process in Shandong highlights the immense

---

[1] Li Heming, Paul Waley, and Phil Rees, "Reservoir Resettlement in China: Past Experience and the Three Gorges Dam," *The Geographical Journal* 167, no. 3 (2001), 195–212, 197; Also see Jun Jing, "Rural Resettlement: Past Lessons for the Three Gorges Project," *The China Journal*, 38 (1997), 65–92; Mou Mo and Cai Wenmei, "Resettlement in the Xin'an River Power Station Project," in *The River Dragon Has Come! The Three Gorges Dam and the Fate of China's Yangtze River and Its People* ed. Dai Qing (Armonk, NY: M. E. Sharpe, 1998), 104–123. "Unimagined communities" refers to groups of people whose vigorously unimagined condition becomes indispensable to the maintenance of a highly selective discourse of national development worldwide. See Rob Nixon, *Slow Violence and the Environmentalism of the Poor* (Cambridge, MA: Harvard University Press, 2011), 150.

difficulties these migrants faced. They were forced to leave behind not just their homes and livelihoods, but also their ancestors, graves, and sense of belonging to the place.[2] But their experiences were systematically erased from state media and from the sanctioned images meant to showcase the apparent grandeur of the socialist state and its projects.

Alongside the technostructural and statist goals involved, the formation of the hydropower nation came at a tremendous human cost. Political leaders and engineers alike were fixated on their statist aspirations and the technological sublime and showed little regard for the suffering endured by the people displaced by the construction of these dams. Reservoir migrants' struggles and resistance were effectively silenced by state censorship, their stories lost in the grand narrative of the state's achievements. The Sanmenxia project, as the first concrete dam on the Yellow River, was mired in controversy from its very inception. The grand scale of the proposed mega-hydropower project entailed the displacement of countless inhabitants and the submerging of a significant amount of land. If a vast reservoir for flood control and electricity generation was to be built – the original proposal called for a reservoir with a water line 360 meters above sea level – this would displace more than 870,000 residents in the region. The scale of such a displacement was so staggering that even hydraulic specialists like Huang Wanli, discussed in Chapter 5, vehemently opposed the project, together with many officials from Shaanxi province. As a compromise, the planned water level was lowered to 335 meters, resulting in the displacement of officially 37,3678 inhabitants living above the dam in Shaanxi, Shanxi, and Henan provinces.[3]

Most scholarship on the Sanmenxia project has focused on its technical aspects, such as its use of the idea of multipurpose river development, and the effects of radical Maoism on the repression of intellectual freedom and military-style assaults on the natural world.[4] The human and social cost of the mega-dam, particularly the displacement of local residents, has

---

[2] Lary, *Chinese Migrations*, chapter 11; Also see Jun Jing, *The Temple of Memories: History, Power, and Morality in a Chinese Village* (Stanford, CA: Stanford University Press, 1998); Jun Jing, "Villages Dammed, Villages Repossessed: A Memorial Movement in Northwest China," *American Ethnologist* 26, no. 2 (1999), 324–343.

[3] *Huanghe Sanmenxia shuili shuniuzhi*, 172. On the number of displaced people: Because of the remodeling of the dam and the change of the reservoir's mode of operation, riverbank collapse, and population growth after the completion of the dam, the number varies in different sources. The number cited here is the official number released in 1965.

[4] Shapiro, *Mao's War against Nature*.

been largely ignored. Judith Shapiro's study of the project centers on the experience of Huang Wanli, while David Pietz's recent book on the Yellow River discusses the construction and reconstruction of the Sanmenxia Dam but neglects to examine the project's local impact and social consequences.⁵ In contrast, Hu Yingze's work in Chinese describes the impact of the Sanmenxia reservoir on the shoaly lands shaped by the Yellow River and the villagers living along the river between Longmen and Tongguan. Hu concludes that the construction of the Sanmenxia reservoir caused the deterioration of the ecological system of the river valley from Shaanxi to Shanxi by destabilizing the shoaly lands along it.⁶ In this chapter, my aim is to supplement the existing scholarship by highlighting the human and social cost of the Sanmenxia hydropower project.

The majority of the 373,678 official reservoir migrants were resettled to higher-altitude locations near their original homes. However, 32,380 migrants from Shaanxi and 7,879 migrants from Henan were resettled considerably farther away in Ningxia and Gansu, respectively.⁷ In the 1990s, Leng Meng, a freelance writer, conducted interviews with Sanmenxia reservoir migrants and local cadres in Shaanxi, bringing their struggles to the public's attention for the first time.⁸ Following in Leng's footsteps, the writer Xie Zhaoping investigated corruption among local officials in Shaanxi and recorded the suffering and poverty of reservoir migrants from a grassroots perspective.⁹

Both Leng and Xie documented the experiences of people displaced by the Sanmenxia project, particularly those from Shaanxi. In the 1980s, when it became clear that their farmlands would not be inundated because of the remodeling of the dam and a change in its mode of

---

⁵ Pietz, *Yellow River*; An anonymous author using the pseudonym "Wei Shang" introduces the debate over the Soviet design, construction, and reconstruction of the Sanmenxia Dam. See "A Lamentation for the Yellow River: The Three Gate Gorge Dam," in *The River Dragon Has Come! Three Gorges Dam and the Fate of China's Yangtze River and Its People* ed. Dai Qing (Armonk, NY: M. E. Sharpe, 1998), 143–159.
⁶ Hu Yingze 胡英泽, *Liudong de Tudi: Ming Qing yilai Huanghe Xiaobeiganliu quyu shehui yanjiu* 流动的土地：明清以来黄河小北干流区域社会研究 (*Floating Lands: Regional Social Study of Xiaobeiganliu since the Ming and Qing Dynasties*) (Beijing: Beijing Daxue chubanshe, 2012).
⁷ *Huanghe Sanmenxia shuili shuniuzhi*, 172–173.
⁸ See Leng Meng 冷梦, *Huanghe dayimin* 黄河大移民 (*The Yellow River Great Migration*) (Xi'an: Shannxi lüyou chubanshe, 1998); see also its English translation, Leng Meng, *Battle for Sanmenxia: Population Relocation during the Three Gate Gorge Hydropower Project* (Armonk, NY: M. E. Sharpe, 1999).
⁹ See Xie Zhaoping 谢朝平, *Da Qianxi* 大迁徙 (*Exodus*) (Beijing: Huohua zazhishe, 2010).

operation, these migrants attempted to reclaim their homes. But in the meantime, their lands had been occupied by state-owned farms and military institutes. Even after years of collective struggle, they were only able in the end to reclaim a portion of their lands through a compromise with reluctant state authorities. Leng and Xie did not include the experiences of migrants from Henan in their work. This chapter thus seeks to complement their work and to provide a more complete picture of the human costs of the Sanmenxia project.

Unlike Leng and Xie, this chapter draws on county and provincial archival documents from the 1950s and 1960s to reveal the experiences of Sanmenxia reservoir migrants from Shaanxian and Lingbao 灵宝 counties in Henan province. The fate of these migrants was similar to that of their Shaanxi counterparts. Both groups were involuntarily resettled in Dunhuang 敦煌, a frontier garrison along the Silk Road that historically connected Central Asia and China. Many fled back to Henan, but they were repeatedly repatriated to Dunhuang. They struggled to adjust to the natural environment and the indigenous communities they now lived among, and they were unable to resettle in their native places in Henan.[10] Many migrants thus had no permanent place to call home.

An examination of these migrants' mobilization, resettlement, flight, repatriation, and eventual resettlement near their original homes can illuminate the forgotten human costs of the Sanmenxia project. State-led rural collectivization aimed to transform individual peasants into socialist laborers in much larger enterprises, frontier reclamation efforts aimed to transform vast areas of wasteland into socialist granaries, and the construction of mega-dams sought to transform China's rivers into energy for socialist state building and economic development. As we recognize the locality of socialism and acknowledge our currently limited understanding of the effects of campaigns in rural China, we should not overlook interregional dynamics and tensions inside China, or the importance of infrastructure construction as a carrier of state policy and a direct stimulus of social change in the Maoist period.[11]

This chapter seeks to demonstrate the complexity of the displacement process and people's resilience by examining the implementation of Maoist demographic engineering and the agency of reservoir migrants

---

[10] The Mogao Grottoes near Dunhuang were discovered in 1900 and designated as a UNESCO World Heritage Site in 1987. Since then, Dunhuang has become a world-renowned tourism destination.

[11] Hershatter, *Gender of Memory*.

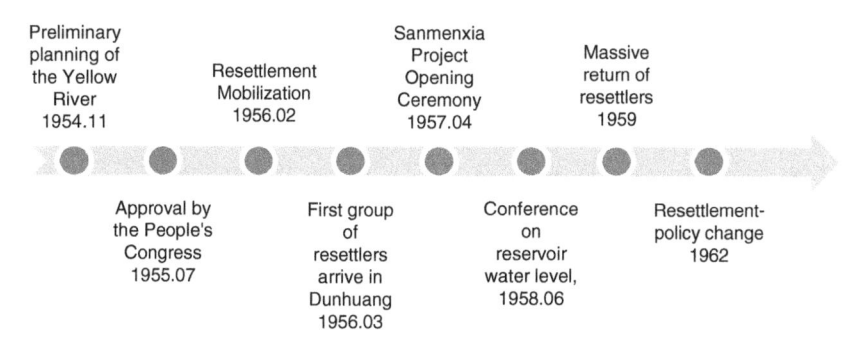

FIGURE 6.1 Timeline of Sanmenxia Project construction and reservoir resettlement in Henan

in the same framework. Although many migrants complied with local cadres as subjects of the state-led resettlement program, many of them also later defied the state-resettlement policy – overtly or covertly – in order to survive.[12] While the silting of the reservoir and the remodeling of the dam represented a technological setback to the state's hydro-engineering ambition, the failure of the long-distance reservoir resettlement scheme exposed the social limits of the hydropower nation (Figure 6.1).

MOBILIZING RESERVOIR INHABITANTS TO MOVE TO DUNHUANG

Shaanxian and Lingbao counties are situated in western Henan, on the edge of the Loess Plateau. In 1952 they had a combined population of 435,100.[13] Unlike the central plains in the east of the province, more than 90 percent of their terrain is hilly or mountainous. The local economy of the two counties was rooted in agriculture, with the most fertile land located along the rivers that flowed through the area, especially the

---

[12] Alongside state-sponsored resettlement programs, Gregory Rohlf discusses voluntary resettlement to Qinghai for better employment opportunities in the 1950s; see Rohlf, "Resettlement Becomes a Frontier Policy, 1955–1956," in *Building New China, Colonizing Kokonor*, chapter 2. On Chinese peasants' counteractions against state policies during the Maoist period, see Gao Wangling 高王凌, *Zhongguo Nongmin Fanxingwei Yanjiu (1950–1980)* 中国农民反行为研究 (1950–1980) (*A Study of Counteractions of Chinese Peasants (1950–1980)*) (Hong Kong: The Chinese University Press, 2013).

[13] Sanmenxiashi difangshizhi bianzuan weiyuanhui 三门峡市地方志编纂委员会, *Sanmenxia shizhi*, vol. 1 三门峡市志 (*Sanmenxia Gazetteer*) (Zhengzhou: Zhongzhou guji chubanshe, 1997).

Yellow River. This land would be submerged by the rising water of the reservoir, displacing people who had lived there for generations, often without ever leaving their villages. According to the gazetteer of Lingbao that was published in the Republican period, local elites complained that most inhabitants lacked consciousness of the importance of nation-building efforts.[14] In the land reform campaign of the early 1950s that redistributed land and other assets among the peasants, the new government effectively dismantled previously existing economic and political orders in rural China centered on landlords and the gentry and replaced them with party branches.[15]

In the 1950s, following the model of the Soviet Union, the new state prioritized rapid industrialization. This involved an extractive agricultural policy, perpetuating an intimate relationship between the state and peasants until hunger killed millions in the late 1950s and early 1960s.[16] For people living in the Sanmenxia reservoir region, though, tensions had already emerged several years earlier. After the State Council approved the Sanmenxia plan in July 1955, central and local administrators began the process of displacing inhabitants in the designated reservoir area in early 1956.

At first, many reservoir residents preferred to move to locations near their old homes but at higher altitudes, which allowed them access to their farmland before the water rose. In response, as a way of minimizing costs, the Shaanxian county secretary Hao Yujiang 郝玉江 and the Lingbao county secretary Zhang Wenying 张文英 proposed that such resettlement at higher altitudes in the two counties could be a general policy.[17]

---

[14] Lingbao xianzhi bianzhuan weiyuanhui 灵宝县志编纂委员会. *Lingbao xianzhi* 灵宝县志 (*Lingbao Gazetteer*) (1936), 94–95.

[15] For classic studies on Mao's rural revolution in the PRC period, see Edward Friedman, Paul Pickowicz, and Mark Selden, *Chinese Village, Socialist State* (New Haven, CT: Yale University Press, 1991); Ralph A. Thaxton Jr., *Catastrophe and Contention in Rural China: Mao's Great Leap Forward Famine and the Origins of Righteous Resistance in Da Fo Village* (Cambridge: Cambridge University Press, 2008). My focus here is on the general trend in the People's Republic.

[16] Robert Ash, "Squeezing the Peasants: Grain Extraction, Food Consumption and Rural Living Standards in Mao's China," *The China Quarterly* 188 (2006), 959–998; Li, *Mao and the Economic Stalinization of China, 1948–1953*; Bernstein and Li, *China Learns from the Soviet Union*; Wemheuer, *Famine Politics in Maoist China and the Soviet Union*; Frank Dikotter, *Mao's Great Famine: The History of China's Most Devastating Catastrophe, 1958–1962* (New York: Walker & Co., 2010); Rohlf, *Building New China, Colonizing Kokonor*.

[17] "Henansheng yimin gongzuo huiyi fenzu taolun ziliao huiji" 河南省移民工作会议分组讨论资料汇集 (1956), J149/05/396 (Zhengzhou, Henan: Henan Provincial Archives).

However, the central government simultaneously embarked upon a national demographic-engineering project, which required that laborers be sent from densely populated provinces to frontier areas like Xinjiang, Gansu, Inner Mongolia, and Heilongjiang for land reclamation.[18] Henan province was one of the most populous in China and, following the call from top party leaders, the Henan provincial leader Wu Zhipu 吴芝圃 decided to send four million people to the frontiers by 1959.[19] The main Henanese candidates for resettlement for land reclamation were people living in poverty, and especially people living in areas that were frequently affected by natural disasters, as well as those affected by hydraulic and hydropower projects or other large public infrastructure projects. According to the quota issued by the provincial government of Henan, Lingbao and Shaanxian were required to send 7,000 people to Dunhuang for land reclamation. Policymakers at both central and provincial levels saw this plan to send residents from the area near the planned Sanmenxia reservoir to Dunhuang as a way of "killing two birds with one stone."

In the twentieth century, technological development and the emergence of high modernism fundamentally transformed the relationship between humanity and nature. The high modernist perspective viewed every person as a valuable source of labor, and every drop of the rivers and every inch of territory as resources to be tapped. Labor, water, and land were all seen as bearers of different forms of energy. Making them legible

---

[18] Henry G. Schwarz, "Chinese Migration to Northwest China and Inner Mongolia 1949–59," *The China Quarterly* 16 (1963), 62–74; Rose Maria Li, "Migration to China's Northern Frontier, 1953–82," *Population and Development Review* 15, no. 3 (1989), 503–538. Transfers of population in China after 1949 did not occur only for land reclamation; they also served political and military purposes. See Shapiro, "War Preparations and Forcible Relocations," in *Mao's War against Nature*, chapter 4. During the second Sino-Japanese War, many wartime refugees had resettled in the northwest for land reclamation. See Zhang Genfu 张根福, *Kangzhan shiqi de renkou qianyi* 抗战时期的人口迁移 (Beijing: Guangming ribao chubanshe, 2006); Micah Muscolino, "Violence against People and the Land: The Environment and Refugee Migration from China's Henan Province, 1938–1945," *Environment and History* 17 (2011), 291–311.

[19] Guanyu 1956 chungengqian yimin kenhuang gongzuode baogao 关于1956春耕前移民垦荒工作的报告 (1956), J149/05/396 (Zhengzhou, Henan: Henan Provincial Archives). Altogether, around 529,000 people from Henan province migrated to frontier regions; of these, around 92,200 migrated to Gansu. See Shen Yimin 沈益民, Tong, Shengzhu 童乘珠, *Zhongguo renkou qianyi* 中国人口迁移 (*The Population Migration in China*) (Beijing: Zhongguo tongji chubanshe, 1992). For studies on land reclamation resettlement in other provinces, see Zhao Rukun 赵入坤, "Ershi shiji wuliushinian daide zhongguo bianjiang yimin" 二十世纪五六十年代的中国边疆移民 (Chinese Frontier Migration in the 1950s and 1960s), *Zhonggong dangshi yanjiu* 中共党史研究 2 (2012), 52–64.

was crucial if the state's capacity to extract energy for modernization was to be maximized. As a mega-project, the Sanmenxia project promised significant growth in energy production, including the generation of electricity and the growth of agricultural output through the expansion of irrigation and the elimination of floods. The construction of the project's reservoir, however, also meant the submerging of farmland, villages, and towns.

To minimize the negative impacts and costs of reservoir displacement, the state decided to send reservoir residents to the northwestern frontier regions for reclamation projects. As a result, in the official narrative of the process of displacement associated with Sanmenxia, moving to create space for the reservoir and transforming the so-called frontier wasteland into fertile farmland for agricultural growth was framed as an "honorable" assignment for these reservoir residents. However, as James Scott observes, "designed or planned social order is necessarily schematic; it always ignores essential features of any real, functioning social order." Thus, this process ignored the fact that the migrants' original residences would be submerged or demolished.[20]

The process of adapting to their new communities also presented the migrants with significant challenges. Without access to their own farmland, it was difficult for them to earn a living. Thus, as the design and modeling of the dam posed technical challenges, the reservoir displacement and resettlement process presented local governments and people with significant political and social challenges.

In the post-Mao period, the standard process of reservoir resettlement can be divided into three phases: Mobilization or education and persuasion aimed at gaining the cooperation of residents of the future reservoir area; relocation or the payment of compensation for old houses and land before the allocation of new land, along with the construction of new houses and site infrastructure; and the development of production activities or providing people with the means to generate improved incomes.[21]

---

[20] Scott, *Seeing Like a State*, 6. Scott develops several helpful analytic frameworks, such as state space versus non-state space, high modernism, and the state's consistent historical efforts at simplification. Drawing on Scott's analysis, the Yellow River valley could be understood as a space created through the interactions of people, nature, and the state over thousands of years. Under the influence of global high modernism, the Chinese Communist state's plans to build dams and reservoirs can be seen as high-modernist state space created through the transformation of natural and agrarian spaces.

[21] Elisabeth J. Croll, "Involuntary Resettlement in Rural China: The Local View," *The China Quarterly* 158 (1999), 468–483, 469.

During the Mao years, though, socialist rural collectivization and the Great Leap Forward made this process much simpler and exacerbated the hardships that Sanmenxia reservoir migrants would face.

Local governments in both origin and destination areas were assigned the political task of making the resettlement program a priority. From the perspective of the Communist cadres, the successful accomplishment of the displacement process required the ideological and political education of the masses.[22] To mobilize the masses, then, the state needed both to organize an efficient propaganda effort and to deploy force. Various county governments gathered village cadres and activists together for meetings where they were informed about the resettlement decision and trained in how to mobilize the people they were responsible for.

The Communist Party had developed an extensive array of party-centered administrative and mobilization tactics in rural China through years of experience in building political revolution and pursuing land reform.[23] Official propaganda highlighted that the project would bring safety and economic prosperity to eighty million people in the region of the lower Yellow River; the "inevitable" cost of the displacement of tens of thousands of residents in the reservoir area was deemed acceptable. In Henan, the local governments of the two counties, together with local cadres, activists, primary-school teachers, students, and amateur theater troupes, mobilized to persuade people of the economic benefits and political significance of both the reclamation–resettlement program and the hydropower project. From the middle of February 1956, they used broadcast technology and group meetings to attempt to persuade people, while cadres from Gansu also came to the counties to assist with the mobilization. To reduce the migrants' concerns and hesitations, they presented a positive image of Dunhuang's agriculture-friendly natural environment and virtuous indigenous residents, claiming that, under the leadership of the Communist Party, the people of Dunhuang had eliminated flood and drought disasters and now used river and spring water for irrigation.

To further persuade the residents of the reservoir area, officially selected migrant delegates went on an inspection tour to Dunhuang, along with local cadres. The tour's report claimed that Dunhuang had over

---

[22] Henansheng yimin weiyuanhui 1956 nian gongzuo zongjie 河南省移民委员会1956年工作总结 J149/05/396 (Zhengzhou, Henan: Henan Provincial Archives).

[23] See Odoric Y. K. Wu, *Mobilizing the Masses: Building Revolution in Henan* (Stanford: Stanford University Press, 1994).

250,000 *mu* of arable land and a population of only 40,000, which suggested that it urgently needed laborers from other places to cultivate that land. The report also concluded that the resettlement of reservoir residents from Lingbao and Shaanxian in Dunhuang would make the reservoir project possible and benefit the expansion of socialist agriculture. The migrant delegation and its report, however, were both part of the government's propaganda effort: The delegates were under intense political pressure to comply with the state resettlement policy and they were prohibited from describing any negative aspects of the designated resettlement place. Some migrant cadres later confessed that they had deceived their fellow villagers, revealing how the Communist Party used both propaganda and coercion to mobilize the masses and promote the reclamation–resettlement program.[24]

After the delegation returned, the Lingbao and Shaanxian county governments immediately summoned all district and commune cadres for meetings where they were enjoined to work hard to persuade reservoir residents to move to Dunhuang. The county secretary and magistrate promoted the positive situation in Dunhuang and proposed slogans like "Migrants are honorable," "One person moves, ten thousand people are safe," "One household moves, ten thousand households are saved," and "Develop the northwest, support the frontier." Unlike "welfare migrants" from disaster-stricken areas, the party secretary of Shaanxian claimed that Sanmenxia reservoir migrants would be resettling in Dunhuang to help the peasants there cultivate cotton. This led many reservoir migrants to believe they would be highly respected when they arrived.[25]

To achieve the quota of migrants assigned by their superiors, many resettlement administrators exaggerated the advantages of Dunhuang and downplayed its challenges. Some cadres lavished praise on the district to which Dunhuang belonged:

There are green mountains, the rivers never stop flowing, many cattle and sheep wander on the grassland. We never worry about food, because we have granaries full of grain; we use tractors to plow the land and electric bulbs to light the darkness. Schools and hospitals are everywhere. Socialism relies on hard work; our happy life shall last for tens of thousands of generations.[26]

---

[24] Xie, "Da Qianxi."
[25] Sanmenxia yimin weiyuanhui 三门峡移民委员会, "Haoshuji zai yimin gongzuo huiyishang de dongyuan baogao" 郝书记在移民工作会议上的动员报告 (1956), folder 91/3 (Sanmenxia, Henan: Sanmenxia Municipal Archives).
[26] "Gansusheng yiminju Wang Xianzhi tongzhi fayan" 甘肃省移民局王宪之同志发言 (1956), J149/05/396 (Zhengzhou, Henan: Henan Provincial Archives).

Local resettlement cadres also realized that material incentives were indispensable in persuading people to move. The potential for the expansion of agricultural production in Dunhuang was limited because of the area's low annual precipitation, but this fact was downplayed in official propaganda, which left many migrants unprepared.

In summary, while the Communist Party used propaganda and material incentives to persuade reservoir residents to move to Dunhuang, the reality was complex. Many migrants arrived ill-equipped to handle the challenges of the area's limited natural resources, highlighting the need for a more complete picture to be shown to people being resettled.

Despite the thoroughness of official propaganda efforts, not all residents of the reservoir area were convinced. At this point, local cadres resorted to coercive methods to suppress dissent. Residents who complained or resisted were labeled ideologically and politically "backward," and complaints were dismissed as counter-revolutionary rumormongering. Stipulations attached to the official resettlement policy meant that certain criteria had to be met in order for a person or household to qualify for frontier migration. The most important criterion was class identity, with only poor or middle peasants deemed suitable for the plan.[27] Seniors and children were often seen as burdens and as a result households with both were not preferred. Some former landlords and members of other politically "backward" categories still asked to be included in the long-distance resettlement program, however, in the hope of starting a new life. In such cases, the county resettlement committee would consult with village cadres in their evaluation of the applicants. Su Yuren 苏玉仁, for example, a former landlord who had become a member of the Shengli agricultural commune, applied to migrate to Dunhuang to be reunited with his family. The commune cadre's letter to the county resettlement committee reported that Su had been passive during political movements and had been sentenced to two years of confinement in the commune from 1956 to 1958. The commune cadres were also skeptical of Su's motivations for resettlement; they suspected that he wanted to move to

---

[27] On the practice of social classification in the PRC, see Yang Kuisong, "How a 'Bad Element' Was Made: The Discovery, Accusation, and Punishment of Zang Qiren" and Jeremy Brown, "Moving Targets: Changing Class Labels in Rural Hebei and Henan, 1960–1979," both in *Maoism at the Grassroots: Everyday Life in China's Era of High Socialism* ed. Jeremy Brown and Matthew Johnson (Cambridge, MA: Harvard University Press, 2015).

escape his identity as a landlord and because his wife was afraid of performing manual labor during his period of confinement.[28]

In the official vetting process, meeting the quota of migrants was local cadres' primary concern, regardless of the applicant's political identity. Su was labeled dishonest, but local cadres in Henan were still happy to let him go, as he was seen as a political burden to the commune. Although this violated the political requirements of the policy and posed a potential threat to the sustainability of the resettlement program in Dunhuang, local cadres did send "politically backward" people to the frontiers for land-reclamation projects. Correspondence between local cadres also suggests, however, that they were cautious about sending members of "hostile" classes to Dunhuang, which suggests that the majority of the long-distance resettlement families were poor or middle peasants.

The county government claimed that more than a thousand people voluntarily submitted applications and pledge letters to migrate to Gansu, and that people were enthusiastic about the resettlement program. But the enthusiasm of the reservoir residents was exaggerated in official reports, and many people were deeply worried about the forthcoming resettlement. Many seniors were reluctant to leave their hometowns. Some started to sell their furniture, while others spent their days drinking and eating with no motivation to work. Some even expressed the desire to die in Shaanxian before they were made to leave.[29] Many younger people, though, were curious about the outside world and in favor of resettlement, which made them easier to mobilize.

To ease migrants' anxiety and to help make their resettlement permanent, the local governments of Shaanxian and Lingbao encouraged migrants' close relatives who were not directly involved in the displacement to join the program. For people who refused to cooperate, local cadres used coercive methods, including demolishing their houses. Despite these challenges, the two counties completed the resettlement registration on time and even exceeded the quota from the provincial government, with 7,879 registered reservoir residents ready to go to Dunhuang by 1956.

---

[28] Lingbaoxian yimin weiyuanhui 灵宝县移民委员会, "1957 nian yimin qinyou waiqian shenqingshu" 1957年移民亲友外迁申请书 (1957), folder 55 (Lingbao, Henan: Lingbao County Archives).

[29] "Guanyu wosheng chungengqian xiangshengwai yimin gongzuo de juti buzhide yijian" 关于我省春耕前向省外移民工作的具体布置的意见 (1956), J149/05/396 (Zhengzhou, Henan: Henan Provincial Archives).

DAILY LIFE AND STRUGGLES IN DUNHUANG

On the night of March 12, 1956, under the surveillance of resettlement cadres, the first group of migrants began their journey to Dunhuang. Despite having pledged to participate in the program, many villagers were reluctant to leave and some even jumped off the train after the journey had begun. The lack of medical services on the long journey led to the deaths of eighteen migrants, eleven adults and seven children.[30]

The journey itself was arduous, with the travelers spending several days on the train from Henan to the Hongliuyuan railway station to the west of Zhangye, and then taking a bus to Dunhuang. Altogether the journey took at least a week to complete. At the Hongliuyuan station, the Gansu government requested that all migrants' clothes and quilts be steamed to prevent the spread of cotton bollworm. They had not been given advance notice of or any explanation for this measure, though, and many felt that they were being discriminated against. They also worried that the steam would damage their belongings. Some older members of the traveling parties were particularly upset when Gansu administrators took their clothing away for steaming.[31] Despite this, all cotton belongings were sanitized before the migrants arrived in Dunhuang.

Despite the fact that the resettlement of migrants to Dunhuang was part of state hydroelectric and land-reclamation projects, both the central and local governments failed to cover specific resettlement costs. These were borne by the inhabitants of Dunhuang and the Henanese migrants themselves. Prior to the migrants' arrival, official reports indicated that the Dunhuang government visited every designated household to encourage them to accommodate the migrants. Local officials emphasized that socialist construction required labor and that people were essential for its success.[32] Gansu officials hailed the Henan migrants as socialist models in their propaganda. They praised the migrants for their high levels of political consciousness and patriotism, arguing that they had responded positively to the call of the nation, leaving their hometowns behind to

---

[30] "Henansheng 1956 nian chunji xiangshengwai yimin gongzuo de zongjie baogao" 河南省1956年春季向省外移民工作的总结报告 (1956), J149/05/396 (Zhengzhou, Henan: Henan Provincial Archives).

[31] Shi Yun 石耘, "Huanghe Sanmenxia kuqu yimin qingku qingkuang gaishu" 黄河三门峡库区移民清库情况概述 (An Overview of the Yellow River Sanmenxia Reservoir Clearance), *Sanmenxia wenshi ziliao* 17 (2007), 106–122, 110.

[32] "Gansusheng yiminju Wang Xianzhi tongzhi fayan" (1956).

support socialist construction.³³ Local cadres certainly did convince some indigenous residents to assist in the resettlement program. Where this failed, those residents were mandatorily assigned to do so.

Cadres in Gansu also organized gatherings of native residents to welcome migrants from Henan at the railway station. However, to spread the costs of resettlement, most migrants were sparsely settled in across a variety of agricultural communes. In one official report, Dunhuang county cadres promised that "eight essentials" – housing, a *kang* (bed), stove, vat, flour, light, oil, and a meal – would be ready for the migrants when they arrived.³⁴ But local accommodations were actually very limited. According to testimonies gathered from migrants to Ningxia, some migrants who arrived at officially designated resettlement sites had to dig caves to provide shelter for themselves.³⁵

In an attempt to foster a sense of community between migrants and local inhabitants, female cadres acted as matchmakers, and in the first year of resettlement several migrants got engaged to or married local people. In Dunhuang county, migrant–indigenous relationships resulted in seven marriages, twenty-four engagements, twenty-one more couples in relationships, and fifty households entering into "sworn kinships."³⁶ According to the official narrative, at least, at the beginning of the resettlement process migrants and indigenous residents got along well.

But this officially cultivated harmoniousness did not last long. Cultural differences, the shortage of housing, unfair treatment, and Dunhuang's harsh, dry, cold, and windy climate contributed to serious homesickness among migrants from Henan. Because of their lack of property, migrants often had to borrow donkeys, mills, and crates from local households, but they were unable to provide any form of payment. Occasionally, these borrowed goods were accidentally damaged or lost. As a result, some local residents stopped sharing their belongings with their new neighbors or, if they did, they added extra charges that made the migrants feel unwelcome.³⁷

---

³³ "Gansusheng yiminju Wang Xianzhi tongzhi fayan" (1956).
³⁴ "Shengyimin weiyuanhui zhuren Jia Xinzhai guanyu Henansheng yinianlai yimin gongzuo de jiben qingkuang de baogao" 省移民委员会主任贾心斋关于河南省一年以来移民工作的基本情况的报告 (November 9, 1956), J149/05/0397 (Zhengzhou, Henan: Henan Provincial Archives).
³⁵ Xie, "Da Qianxi," 87–90.
³⁶ Zhangye zhuanqu 张掖专区, "1956 nian yimin anzhi gongzuo jiancha baogao" 1956年移民安置工作检查报告 (1956), 138/004/2521-2 (Lanzhou, Gansu: Gansu Provincial Archives).
³⁷ Zhangye zhuanqu, "1956 nian yimin anzhi gongzuo jiancha baogao."

On top of everything else, Dunhuang – being located in the middle of the Gobi Desert – had severely limited access to surface water. The scarcity of drinking water only compounded the challenges faced by the Henan migrants. While most local households constructed cellars to collect rainwater for their own and animal consumption, the migrants were made uncomfortable by having to share their drinking water with livestock. For their part, local residents were frustrated with the migrants' use of water cellars to wash clothes, given the immense waste of water this represented in such an arid natural environment. Trees and shrubs were scarce and also precious in Dunhuang, as they played an important role in protecting villages from the threats of sandstorms and moving dunes. Lacking experience of living in a desert environment, though, some Henanese migrants cut trees down for fuel, much to the dismay of the locals. Dunhuang also experienced much colder winters than Sanmenxia. Although the migrants were resettled in Dunhuang for the purpose of expanding farmland, it turned out that most of the so-called wastelands were not arable, and efforts to reclaim that only destroyed the fragile ecology of the meadows and other types of land in the area. In short, the Henan migrants faced numerous difficulties in adapting to Dunhuang's natural environment.

Unable to cultivate new farmland, then, the migrants were compelled to work alongside local residents in the same fields. As new members of the communes they were assigned to, many migrants complained about how local commune cadres recorded work points and deducted payments. They earned fewer work points than local residents for doing the same work. Given that work points were the means by which the commune's revenues were distributed to its members, this entailed disparities in compensation. In the Fengquan commune of Nuanquan district, for example, Liu Yongzhi 刘永治, a migrant, worked in the same field as local commune members removing wheat straws. While native commune members received eighteen points each day, Liu only received twelve points. At harvest time, a team of sixteen migrants reaped nine *mu* of land while a team of twenty local villagers reaped seven *mu*, yet the locals still received eight more work points than the migrants each day.

It should come as no surprise, then, that violent confrontations between migrants and local residents began to occur. In January 1957, a massive brawl between hundreds of migrants from Henan and natives occurred in Minle county, southeast of Dunhuang. Although the reservoir migrants were not directly involved in this conflict specifically, the tension spread through Gansu.³⁸ In Dunhuang, violent confrontations were limited to the individual level, but tensions persisted despite the Gansu government's attempts to ameliorate the situation.

As Dunhuang county began its transformation into a People's Commune in 1958, the county's party secretary, Ma Jiying 马骥英, demanded that all villages contribute half of their cattle and sheep to the commune. To avoid this, many villagers slaughtered their livestock. Many of the cattle and sheep that were successfully requisitioned then died due to mismanagement.³⁹ As a result, local villagers lost an important sideline production to the collectivization process. With the subsequent lowering of living standards compounded by the emergence of starvation in the following years, many local villagers began to hold the migrants responsible for their difficulties.

The tension between the migrants and the local inhabitants was not limited to ordinary people. It also involved the cadres from both groups. In order to help with the management of migrants from Henan, village cadres from Henan were appointed as community leaders in Dunhuang. But local cadres were concerned that the migrant cadres would undermine their authority, so they excluded them from the policymaking process in the communes. Migrant cadres were meant to act as moderators between the migrants and local cadres in the event of confrontations but, because they had been excluded by the local cadres, they had limited success helping the migrants settle in Dunhuang. One migrant cadre, Liu Shuanzhou 刘栓周, expressed his frustration: "Local cadres don't trust me. Many migrants complain about the difficulties of living in this place. I'd better leave all these issues behind. If the migrants choose to go back to Henan, so will I."⁴⁰ As I mentioned earlier, local cadres viewed the migrants as a burden. Wang Yukuan 王玉宽, the accountant of the fifth production team of Hongdun 红墩 Commune, for instance, yelled at a

---

[38] "Guanyu Minlexian yimin zinao shijian de chuli baogao" 关于民乐县移民滋闹事件的处理报告 (April 12, 1957), 138/004/2521-2 (Lanzhou, Gansu: Gansu Provincial Archives). To avoid potential ethnic or religious conflicts, the provincial government avoided placing Han resettlers in Hui Muslim communities. Thus, despite the ethno-religious differences between Henan and Gansu, religion and ethnicity only played a minor role in the tensions between Dunhuang inhabitants and Henan settlers.

[39] Yang Jisheng, "The Three Red Banners: Source of the Famine," in *Tombstone: The Great Chinese Famine, 1958–1962* (New York: Farrar, Straus and Giroux, 2012), chapter 2.

[40] Dunhuangxian renmin weiyuanhui minzhengju 敦煌县人民委员会民政局 (Dunhuang county People's Committee Bureau of Civil Affairs), "Guanyu 1956–1959 nian yimin gongzuo de anpai, zongjie, baogao" 关于1956–1959 年移民工作的安排, 总结, 报告 (1959), folder 34 (Dunhuang, Gansu, Dunhuang Municipal Archives).

migrant who had complained about the difficulties they were facing, saying, "Get out of here if you don't want to stay! Who the hell invited you here?"[41]

The Anti-Rightist Movement of 1957 added further complications to the politicization of the resettlement issue. The historian Cao Shuji has observed that "in rural China, county leaders were the ones who set traps. The Party's attack against rightists at the grassroots was a ruse meticulously planned by the local leaders who carried it out."[42] In Dunhuang, like elsewhere, local cadres exploited the movement to consolidate their power. Complaints and protests from migrant cadres about the resettlement policy were described as anti-party sabotage and as a result some migrant cadres were classified as rightists.

Chen Zhaoxiang 陈兆祥, for instance, who had served as secretary of the Grain Office in Lingbao county before moving to Dunhuang, was appointed as a cadre in the Financial and Trade Department of the Dunhuang county government. Facing unfair treatment and difficulties in finding housing, many Henan migrants sought Chen's assistance. Chen then wrote a letter to the vice secretary of Zhangye district, urging the district government to intervene and find a solution. During the Anti-Rightist campaign, Chen's colleagues in the county government denounced the letter and his meetings with Henan migrants as evidence of counter-revolutionary activities. Other Henan cadres were labeled members of an "anti-party clique" based on accusations of "inciting disturbances among the migrants."[43] After a month of public denunciations, Chen was officially kicked out of the party, lost his government position, and was sent to Huangdunzi 黄墩子 state farm for labor reeducation. His wife was also dismissed from her position in the state-run food company.[44] Migrant cadres like Chen were supposed to facilitate the

---

[41] Zhonggong Dunhuangxian weiyuanhui mishushi 中共敦煌县委员会秘书室 (Office of the Chinese Communist Party Dunhuang County Party Committee Secretary), "Liuyue fanji yimin qingkuang tongbao" 六月返籍移民情况通报 (1957), folder 349 (Dunhuang, Gansu, Dunhuang Municipal Archives).
[42] On the Anti-Rightist Movement at the local level, see Cao Shuji, "An Overt Conspiracy: Creating Rightists in Rural Henan, 1957–1958," in *Maoism at the Grassroots: Everyday Life in China's Era of High Socialism* ed. Jeremy Brown and Matthew Johnson (Cambridge, MA: Harvard University Press, 2015), 77.
[43] Wang, *Dunhuang yishi*.
[44] On the experience of rightists in labor camps in Gansu province, see He Fengming 和凤鸣, *Jingli-wode 1957* 经历-我的1957 (*Experience: My 1957*) (Dunhuang: Dunhuang wenyi chubanshe, 2001); Yang Xianhui 杨显惠, *Jiabiangou Jishi* 夹边沟纪事 (Jiabiangou chronicle) (Guangzhou: Huacheng chubanshe, 2008); see also the English

resettlement program, but political attacks from their counterparts in Dunhuang during the Anti-Rightist Movement eroded their determination to establish a permanent settlement in Dunhuang. Thus, in the following years, many of them would leave.

Promises from both the Henan and Gansu provincial governments to provide new housing for the migrants turned out to be empty, largely because of budgetary limitations. With the beginning of the Great Leap Forward, provincial and county government officials shirked their responsibility to provide residences for the migrants, asserting instead that agricultural cooperatives, communes, and the migrants themselves should solve the housing problem on their own. As I mentioned, most migrants relied on the generosity of their host families in Dunhuang for temporary shelter. The very few shelters for migrants that were actually built were in deplorable condition, with some lacking doors and with walls too thin to keep the occupants warm. Many migrants from Henan had not been informed about the cold winters in Dunhuang and were thus totally unprepared for the freezing weather. Most local people had fur coats to keep warm during the winter; few migrants had anything like that. Fan Shouren's 范守仁 household, for example, consisted of five people but they only had one thin quilt. His children were so cold that they cried and screamed through the winter nights.[45]

In order to build shelter, people needed timber, but Dunhuang was situated in the heart of the Gobi Desert, where trees were scarce and held great ecological importance, as I mentioned. So people had to purchase timber from other parts of the country, which, because of the long distance involved in transporting it, resulted in a much higher price than the national average. Without financial assistance from the state, the migrants simply could not afford it. Despite the fact that the central government provided some limited funding to the Gansu provincial government for reservoir resettlement, county and commune cadres misused the money for administrative expenses. Consequently, many migrants still lived in temporary cellars two years after they arrived. By 1959, roughly 30 percent of the migrants were still forced to move frequently because of the lack of permanent shelter.

---

translation, *Woman from Shanghai: Tales of Survival from a Chinese Labor Camp* (New York: Anchor, 2009).
[45] "Henansheng 1956 dongji weiwen yimin heqing kenduiyuan gongzuo zongjie baogao" 河南省1956冬季慰问移民和青垦队员工作总结报告 (January 1957), J149/05/03967 (Zhengzhou, Henan: Henan Provincial Archives).

In 1956, the year the migrants arrived, Dunhuang experienced a severe drought that caused a dramatic decline in agricultural output. On state instruction, some cultivated land and most of the land that had been newly reclaimed was converted to planting cotton, which increased the demand for food while the food supply remained stagnant. As newcomers, migrant families only received grain rations for two or three days a week. The demand for food outstripped the supply, and most people could only prepare two meals per day, one in the morning and one in the evening. During the Great Famine, Dunhuang's situation was made considerably worse because of the radical policies of the local party secretary An Zhen 安振; it was the worst hit area in Gansu province.[46] "Severe famine has become widespread, and people have been eating tree bark and grass roots to satisfy their hunger."[47] This led to cannibalism in Gansu. According to documents collected by Zhou Xun, in April 1961, a local peasant named Zhao Yuyin bought meat and a pair of leather shoes at the Zhangye railway station but soon discovered that the meat was in fact human flesh, with identifiable nose and ears.[48] While no official record of the death toll among migrants from Henan due to starvation appears to exist, the Dunhuang County Archives contain a document that states that around 110 migrant children and seniors died of measles and typhoid fever in the winter of 1958. Some sick migrants, too weak to work in the fields because of malnutrition, had their already inadequate food rations reduced instead of receiving extra supplies. Commune cadres defended this practice by arguing that those migrants did not deserve equal rations because they did not contribute equally to agricultural production.[49]

In retrospect, pressure from local cadres to support socialist demographic-engineering efforts in Dunhuang led many local residents to share their food, dwellings, and resources with migrants from Henan, at least at first. However, they were unable to sustain their selfless contributions indefinitely. Limited government funding and the agricultural requisition policies of the Great Leap Forward exacerbated their economic hardships. Complaints and antagonism gradually replaced respect and positive

---

[46] Yang, "Three Red Banners."
[47] Zhou Xun (ed.), *The Great Famine in China, 1958–1962: A Documentary History* (New Haven, CT: Yale University Press, 2012), 12.
[48] Zhou, *Great Famine in China, 1958–1962*, 67. On cannibalism in Gansu, see also Yang, *Tombstone*.
[49] "Dunhuang renmin gongshe yimin gongzuo zongjie baogao" 敦煌人民公社移民工作总结报告 (January 5, 1959), folder 34 (Dunhuang, Gansu: Dunhuang Municipal Archives).

expectations between natives and migrants. For their part, many migrants were unprepared for the harsh environment of the northwest and were deeply disappointed by their places of resettlement. This alienation, combined with the Great Famine, ultimately resulted in a mass flight back to Henan not long after the migrants' arrival in Gansu.

FLEEING DUNHUANG

To get back to Henan, some migrants falsely claimed to commune cadres that they needed to visit their families or attend the funerals of their relatives. Many simply left without informing anyone. But the journey back to Henan was treacherous. To reinforce the idea that the population resettlement policy was a "success," the Dunhuang government prohibited the local bus station from selling tickets to fleeing migrants. To reach the nearest railway station, which was over 200 kilometers away, Henan migrants had to navigate the Gobi Desert without any assistance, which was incredibly perilous. Many migrants from Shaanxi lost their lives attempting to cross the deserts in Ningxia to get back to Shaanxi, and it is impossible to determine the number of Henanese migrants who perished in the Gobi Desert.[50] For people who did manage to survive, government officials closely monitored railway stations at both departure and arrival points. If they could be identified by officials and apprehended, fleeing migrants were returned to the resettlement locations from which they had escaped.

Meanwhile, most residents of Dunhuang were pleased to see the Henan migrants leave. The party secretary of the Xiaotun production team in Shahe Commune stated, "Henan migrants are politically backwards and difficult to discipline and educate. Their departure to Henan removes our troubles." Li Jimao 李吉茂, the director of the seventh branch of the Xiangqian production team, added, "Migrants are beggars. They demanded everything from us. Their return to Henan alleviates our burden."[51] Unlike such local cadres, however, officials at higher levels recognized that the success of the resettlement policy was a political responsibility imposed by the central government. A massive exodus of migrants would undermine the authority of the Communist Party.

To prevent migrants from fleeing, the Gansu provincial government directed county and commune officials to respond to their needs swiftly

[50] Xie, "Da Qianxi," 93.
[51] "Guanyu yimin daoliu qingkuang de baogao" 关于移民倒流情况的报告 (August 20, 1959), 138/004/1721 (Lanzhou, Gansu: Gansu Provincial Archives).

and effectively. However, because of the scarcity of food and resources at the village level, such injunctions were unable to stem the flow. Many native commune members complained that the migrants were consuming what little remained of their already depleted resources. "How shameless of them to return! They know how to flee but not how to labor. We shouldn't give them anything to eat!"[52]

Even if they did make it back to their native Henan, many migrants were excluded from joining the people's communes just established in the collectivization of rural China in the late 1950s. The Henan government specifically prohibited communes from providing food and other essential items to returned migrants. As a result, many of them had to rely on friends or family for assistance. Some commune officials sympathized with their situation, though, and arranged for them to take up employment in communes. Da'an Commune, for example, accepted the returned migrant Zhang Xiaowa 张小娃 and assigned him a job feeding livestock. Through personal connections, many returnees began to work on the construction of the Sanmenxia hydropower project itself, earning daily wages to support themselves for a brief period.[53]

As the number of returning migrants increased, they organized themselves to appeal for acceptance by the Henan provincial government. In response, the government put pressure on local officials to locate the migrants and convince them to return to Dunhuang. When confronted by officials, though, returnees employed a variety of tactics to resist repatriation. Some chose to act collectively, criticizing the government for misleading them about conditions in Dunhuang. Others resorted to more extreme measures, like threatening to commit suicide. Others simply refused to cooperate, crying or remaining silent in the face of attempts at persuasion. Some male returnees cited familial resistance, saying that their mothers or wives refused to leave. Many returned migrants hid in their relatives' homes for long periods to avoid detection and repatriation.[54]

In 1958, Vice Premier Peng Dehuai was made aware of the difficulties faced by reservoir migrants because of their long-distance resettlement.

---

[52] Zhonggong Dunhuangxian weiyuanhui mishushi, "Liuyue fanji yimin qiangkuang tongbao."

[53] Sanmenxiashi yimin weiyuanhui 三门峡市移民委员会, "Guanyu dongyuan kuqu yimin fandun de gongzuo anpai" 关于动员库区移民返敦的工作安排 (1956), 91/1 (Sanmenxia, Henan: Sanmenxia Municipal Archives).

[54] "Duifanji yimin qingkuangde fenxi renshi he jinhou gongzuo yijian" 对返籍移民情况的分析认识和今后工作意见 (1957), 91/4 (Sanmenxia, Henan: Sanmenxia Municipal Archives).

He suggested that resettlement to locations closer to their original homes should be the preferred option. This would enable the migrants to harvest their crops before the flood season.⁵⁵ But Mao Zedong and other leaders were not receptive to Peng's proposal. Then, following his removal during the Lushan Plenum in 1959, no one dared to challenge the long-distance reservoir resettlement program, for fear of being labeled a "rightist" or part of an "anti-Party clique."

Communist officials felt that migrants who returned were deliberately undermining state policy and portrayed them as selfish and politically backward individuals who lacked passion and loyalty to the socialist enterprise. Those who voiced concerns about the situation in Dunhuang risked being labeled counterrevolutionaries. Meanwhile, local officials were fully aware that the lack of material support in Dunhuang was a major driver of the mass exodus. But the government took no effective measures to address the dire circumstances during a period of severe famine.

Under pressure from local authorities, many communes denied returned migrants access to food rations, forcing them to resort to stealing to feed themselves and their families. Recognizing that the widespread hunger and resentment among the migrants might lead to riots, the Henan provincial government established a temporary relief system. However, the plan required migrants to apply for grain aid, while the head of each household was required to appear in person to receive the food. Households were also required to resubmit their applications once or twice a month, also in person, which provided local cadres with opportunities to "persuade" them to return to Dunhuang.⁵⁶

As the famine spread, the Henan government tightened its enforcement of the *hukou* (household registration) system to restrict the mobility of starving peasants.⁵⁷ People without local *hukou* would be deported before a specific deadline or imprisoned. This intimidation tactic forced

---

⁵⁵ Zhou Enlai 周恩来, "1958 nian Sanmenxia huiyi jianghua" 1958年三门峡会议讲话 (Zhou Enlai's Talk at the Sanmenxia Meeting) (Sanmenxia, Henan: Sanmenxia Gazetteer Office).

⁵⁶ "Sanmenxiashi yimin bangongshi guanyu dongyuan yimin chongfan anzhi qude yijian" 三门峡移民办公室关于动员移民重返安置区的意见 (1957), 91/5 (Sanmenxia, Henan: Sanmenxia Municipal Archives).

⁵⁷ Tiejun Cheng and Mark Selden, "The Origins and Social Consequences of China's Hukou System," *The China Quarterly* 139 (1994), 644–668; This validates scholarship that has suggested that the household registration system was gradually implemented between 1953 and 1958. See Brown, *City Versus Countryside in Mao's China*; Anthony Garnaut, "The Geography of the Great Leap Famine," *Modern China* 40, no. 3 (2014), 315–348.

many returned migrants to return to Dunhuang, but as the situation there continued to deteriorate, many of them left again. Some migrants made multiple trips between Henan and Dunhuang, while others slept in the wild and begged for food along the Longhai Railway. Some even resorted to robbery.[58] In 1960, with starvation at its worst, nearly all the migrants left Dunhuang. With the adjustment of the party's rural policy in 1962, though, the Henan government began to reconsider its approach to returned migrants, now allowing communes to accept these homeless people. By 1965, around 7,500 of the original 7,869 reservoir migrants had returned to Henan permanently.[59]

But returning to Henan did not mean the end of the challenges associated with resettlement for many migrants. The homes they had left behind had been demolished, and the adjusted resettlement policy prohibited resettlement in the reservoir area, particularly below the designated water level of 350 meters, or in cities.[60] Despite these restrictions, many of the migrants built temporary shelters or caves beneath the 350 meter line to access arable land. In Dawang Commune of Lingbao County, for example, nearly half of the migrants' shelters were located below the 350 meter line. To evade official clearance efforts, the returned settlers would leave their shelters during the day and return at night. Although officials attempted to remove them from the area from time to time, many migrants continued to live in the designated reservoir area. This was possible because of the silting of the reservoir and the remodeling of the Sanmenxia Dam in the 1960s, discussed earlier, which prevented the water level from reaching the 350 meter line. Studies of these migrants' experiences during the post-Mao years have yet to be undertaken.

COMPENSATION FOR PROPERTY LOSS

Jun Jing's research into the Yanguoxia reservoir resettlement process has revealed that many residents of the reservoir area hid portions of their property during the official resettlement survey because of fears of a second land reform campaign and the associated confiscation of property, which could result in the payment of inadequate

---

[58] "Duidangqian yimin gongzuo de jinji zhishi" 对当前移民工作的紧急指示 (July 28, 1959), 1/261 (Dunhuang, Gansu: Dunhuang Municipal Archives).
[59] Shi, "Huanghe Sanmenxia kuqu yimin qingku qingkuang gaisu," 120.
[60] "Lingbaoxian yimin weiyuanhui guayu chengguan gongshe jieshou Dunhuang fanji yimin juzhu chengshi de baogao" 灵宝县移民委员会关于城关公社接受敦煌返籍移民居住城市的报告 (February 2, 1964) 55 (Lingbao, Henan: Lingbao County Archives).

compensation.⁶¹ A similar situation may have occurred during the displacement for the Sanmenxia reservoir. According to official policy in 1956, compensation for long-distance resettlement was set at 790 *yuan* per person, while the compensation for short-distance resettlement was 610 *yuan*. This was supposed to cover the reconstruction of housing, recovery of lost means of production, and any other moving expenses.

As limited as it was, this compensation policy came under attack during the Great Leap Forward: Radical Maoists saw it as "rightist" in the sense that it recognized private ownership and deviated from the principle of self-reliance. As a result, the Henan provincial government reduced the standard compensation amount to 350 *yuan* per person for short-distance resettlement. In practice, though, in 1959 the average compensation was only 290 *yuan* per person.⁶² Furthermore, rather than being given directly to the reservoir migrants, the money was distributed to district- and commune-level administrations to cover collective construction of resettlement housing and other administrative expenses.

Although local corruption was apparently not widespread, it did undermine the compensation policy in practice. Yu Youcai 虞有才, for instance, the chief of the Da'an production team, was accused of embezzlement. The relief fund he administered provided five *yuan* to each returned villager, but Yu appropriated 0.3 *yuan* from each payment, justifying his actions by claiming that the migrants should be grateful to him for the relief funds.⁶³ In this way he embezzled 42 *yuan* from the production team's relief fund. Yu also took advantage of his position and appropriated migrants' funds for housing construction under the pretext of "borrowing" them. As the head of the production team, Yu was responsible for distributing all funds from higher administrative levels, and the villagers were afraid to refuse his "borrowing" requests. In total, then, Yu embezzled 286 *yuan* from the production team's resettlement compensation fund.

Despite the initial compensation rate determined at the beginning of the resettlement process, a combination of political movements, bureaucratic embezzlement, and abuses of power meant that only 60 percent of a single room of residential space were compensated for each migrant.

---

⁶¹ Jing, *Temple of Memories*.
⁶² Shi, "Huanghe Sanmenxia kuqu yimin qingku qingkuang gaishu," 117.
⁶³ Sanmenxiashi yimin weiyuanhui 三门峡市移民委员会, "Guanyu Gaomiao gongshe Da'an dadui duizhang Yu Youcai tanwu yiminkuan he shehui jiujikuan de diaochao baogao" 关于高庙公社大安大队长虞有才贪污移民款和社会救济款的调查报告91/52 (Sanmenxia, Henan: Sanmenxia Municipal Archives).

Because of the lack of funds and construction materials – especially timber, as I mentioned – the collectively built residences (houses and caves) were insufficient in quantity and poor in quality.

Among the reservoir migrants, the belief that "resettlement leads to three years of poverty, and it's hard to recover even after a decade" was widely reported.[64] The loss of their land without fair compensation meant that many migrants' standard of living dropped dramatically. Shijiatan, one of the wealthier villages in Shaanxian, was located one mile from the dam site and was renowned for its fertile farmland along the river. The village director recalled that a farmer who started plowing his land with cattle in the early morning would not return to the place he had started from until noon. This statement attests to the richness of the arable land along the river. But the building of the Sanmenxia Dam submerged these lands, and the reservoir's water level determined whether these residents could harvest their crops. To feed their families, many returned migrants began cultivating terraces on the fragile loess hills, which then exacerbated erosion in the area. This practice was necessary to ensure a sufficient supply of food, but it also had negative ecological consequences.

In the 1950s, hydraulic engineers boasted that the Sanmenxia reservoir would be the largest man-made lake in China. Ironically, though, it became even more difficult for residents who moved uphill to avoid being submerged to obtain drinking water. Many wells in Sanmenxia collapsed because of the rising water level in the reservoir and thus people had to walk for miles to collect water from the river. Many newly built cave houses also collapsed. In December 1959, for example, five cave houses in Lüjiayao collapsed during construction, burying thirteen people alive and injuring another. The builders had failed to examine the geological conditions of the site carefully.[65] People who lived above the planned waterline but along the reservoir's banks were also at risk. Their caves or semi-brick houses were prone to collapse after the reservoir began retaining water.

To make room for workers' dwellings near the Sanmenxia project site, in 1957 Da'an and five other villages were moved. Five years after this resettlement, more than 30 percent of the displaced villagers still had no home of their own. The government-built cave houses began to collapse

---

[64] Shi, "Huanghe Sanmenxia kuqu yimin qingku qingkuang gaishu," 116.
[65] Sanmenxiashi minzhengju 三门峡市民政局, "Sanmenxiashi 1959 nian yimin gongzuo zongjie baogao" 三门峡市1959年移民工作总结报告, 42/1 (Sanmenxia, Henan: Sanmenxia Municipal Archives).

before people even moved in, and many returned migrants refused to put their families at risk by moving into them.⁶⁶ The instability of their lives, economic hardships, and negative rumors about their experiences in Dunhuang eroded their officially cultivated sense of honor as reservoir migrants. In the years following their return from Gansu, few reservoir migrants were willing to identify themselves as such. It had become a synonym for poverty and disgrace.

CONCLUSION

The Sanmenxia hydropower project left almost 400,000 inhabitants of the planned reservoir area homeless and impoverished for decades, and this chapter has examined only a small portion of their experiences. As has been the case in other reservoir displacements around the world, the stress of the Sanmenxia resettlement can be seen as a complex insult with psychological, physiological, and socio-cultural aspects.⁶⁷ The state assumed in this case that reservoir displacement and frontier reclamation were perfectly complementary projects, but this proved misguided. It failed to account for how factors like differences in natural environments, bonds to native places, and grassroots political tensions would affect migrants' resettlement strategies. Frontier reclamation as a developmental and resettlement option proved to be a failure, which further compounded the suffering of people displaced by the Sanmenxia project.

James Scott and Judith Shapiro have argued that Maoist attempts to conquer nature were met with resistance by nature itself. Compelled by the force of the state, thousands of migrants moved to Dunhuang, a place with a fragile oasis ecological system, to expand farmland, resulting in significant desertification of once lush meadows. When they returned to the Sanmenxia region, the migrants' struggle to survive meant that they gave little thought to soil conservation. They cleared vegetation from hillsides to create agricultural land and built cave houses, which then aggravated erosion along the Yellow River and accelerated the silting of the reservoir.

The various levels of the Communist bureaucratic system employed both persuasive and coercive methods to make way for the Sanmenxia

---

[66] Sanmenxiashi minzhengju 三门峡市民政局, "Guanyu jiejue 57 nian neiqian Cizhong yimin zhufang yiliu wenti de qingshi" 关于解决57年内迁磁钟移民住房遗留问题的请示, 91/40 (Sanmenxia, Henan: Sanmenxia Municipal Archives).
[67] Thayer Schudder, "The Human Ecology of Big Projects: River Basin Development and Resettlement," *Annual Review of Anthropology* 2 (1973), 45–55, 51.

Dam. Because the central government had not provided sufficient resources to support the daunting resettlement program, the hierarchical bureaucratic structure meant that local officials were enabled, and even pressured, to act ruthlessly when migrants defied state policy. Most large-scale hydropower projects in China, including the Three Gorges project on the Yangtze River, have relied heavily on state coercion to be completed. In the case of the Sanmenxia project, the initial high dam design caused serious silting in the reservoir and an increased threat of flooding to the Wei River Valley. After it was completed in the early 1960s, the dam had to be remodeled several times. In the end, its total hydroelectric generation capacity was 250,000 kW – a fraction of the original design's capacity of 1,000,000 kW. In other words, the sacrifices that so many migrants endured might have been avoided if political leaders and engineers had conducted more rational and scientific studies on the dam's design and potential environmental challenges.

In the face of the immense challenges faced by tens of thousands of reservoir migrants, the state's attempt to combine dam construction and frontier resettlement ultimately failed. This failure supports Matthew Johnson's and Jeremy Brown's claim that in the early PRC "state control was not always total and centralized but at times appeared limited and tenuous."[68] These migrants, as "development refugees," did not just experience physical displacement; they also experienced a form of imposed invisibility. Despite the fact that official records are dominated by the state's own rhetoric and the censorship of migrant voices, the examination of these reports in this chapter provides a glimpse into their struggles. The drive for increased electricity and agricultural production created a productivist hydropower nation. As the Sanmenxia case demonstrates, the realization of this vision required the collaboration and, in many cases, the sacrifice of its inhabitants. While many complied with state persuasion and coercion, their compliance was never unconditional. Despite the overwhelming force of the state and its hydropower aspirations, the Sanmenxia reservoir resettlement highlights the role of grassroots dynamics and tensions in how Communist social and environmental-engineering processes play out in China.[69]

---

[68] Brown and Johnson, *Maoism at the Grassroots*, 1.
[69] We can see continuity and growth in grassroot actions in protest against and in negotiation with state policies related to large hydropower projects in contemporary China. See Andrew C. Mertha, *China's Water Warriors: Citizen Action and Policy Change* (Ithaca: Cornell University Press, 2008).

7

The Environmental Saga

On May 18, 1999, at the People's Court of Shaanxian in Henan province, two men, Zhao Shishi, a resident of the area, and Zhu Zhongping from Weishan 微山 county in Shandong province, were handed sentences of thirteen and twelve years in prison, respectively. They were charged with killing twenty-seven whooper swans, a species of "second-class" protected wildlife living in the Sanmenxia Reservoir area. In early January of that year, driven by a desire to make some extra money, the two men had purchased 40 kilograms of carbofuran, a highly toxic carbamate pesticide, with the intention of hunting wild birds and waterfowl in the area. They mixed the carbofuran with corn and scattered the mixture in areas where the birds congregated. Before they could collect the poisoned swans and ducks, though, they were reported by witnesses and arrested.

The case created a sensation in the Sanmenxia area because of the harsh punishment meted out for illegal hunting activities. Many local residents were unaware of the classification of state-protected wildlife and the criminal sanctions that accompanied it. The case served as a valuable lesson to the local population about the importance of wildlife protection, and it also brought attention to the thriving population of whooper swans in the Sanmenxia area, which had previously been known only as the site of the first concrete dam on the Yellow River.

The focus of this chapter shifts from the human aspect explored in Chapter 6 to the relationships between nonhuman elements and human society in Sanmenxia. My aim is to provide a more comprehensive understanding of the hydropower project and to emphasize the importance of both biodiversity and environmental justice. This chapter will review the environmental changes brought about by the construction of the

Sanmenxia Dam and examine the dynamics between top-down conservation efforts and the resistance or acquiescence of local residents, many of whom had been displaced by the formation of the reservoir in the 1950s. It also delves into the political ecology of the Sanmenxia Reservoir, particularly in light of the recent arrival of whooper swans. The case of Sanmenxia shows that the environmental aspect of the hydropower nation is more complex than the narrative of decline suggests. While certain consequences, like the decline of fish populations and the degradation of agricultural land and riverbanks, can be seen as negative, others, like the expansion of a wetland system in a human-made water complex, do not necessarily fit into the category of ecological hazards. In the age of the Anthropocene, the formation of a hydropower nation involves more than just the relationship between humans and rivers; it encompasses the interactions and histories of other species as well.

THE GEOGRAPHIC SETTING

Before the construction of the Sanmenxia Dam, the Yellow River was not consistently constrained by the mountains and hills along its middle course before it reached the North China Plain. As it flowed from Longmen to Tongguan, the width of the river channel expanded from a narrow V-shape of only 1 km to a much wider channel of between 4 and 19 km. At this point, the Yellow River converges with the Wei River from the west and the Fen River from the east. In the 1960s, the wide and flat river valley below Longmen and the fertile eastern part of the Guanzhong Plain – which was central to the development of the first empire in Chinese history, the Qin dynasty – was threatened by floods as a result of the construction of the dam.[1]

Throughout recorded history, the Yellow River has been prone to both flooding and silting, which meant that its path below Longmen frequently changed. This resulted in a wide, shallow, and divided river course that was sometimes difficult to identify with certainty. Between the lower regions of Shaanxi and Shanxi provinces, the river formed a mini flood plain. For generations, local farmers have made use of the fertile alluvial lands, which were covered in sediment carried by the river from the Loess Plateau, for agriculture. Despite the persistent threat of flooding, these lands provided a valuable supplement to farmland, and in some cases

---

[1] See Brian Lander, *The King's Harvest: A Political Ecology of China from the First Farmers to the First Empire* (New Haven: Yale University Press, 2021).

were an important source of income for households in subsistence economies.

To deal with the Yellow River's seasonal flooding and the sediment it deposited, local peasants used the alluvial lands to plant crops like cotton, beans, and peanuts. Not all of these lands were suitable for agriculture, though. The sediment contained a mixture of minerals that were essential for crop cultivation, but it also contained salt, which was harmful to plants. To make use of the barren saline lands, local people would dig shallow holes to collect brine, which was then evaporated to produce salt and alkali.[2] The annual flooding of the Yellow River had also resulted in severe erosion of its riverbanks between Longmen and Tongguan, which caused the river to widen its course. This had forced people to seek higher ground, known as *yuan* 塬, as riverbank collapses and course shifts increased in frequency.[3] Not all households could afford to move their homes and farms to higher ground, though, leaving many poor families dependent on the fertile alluvial lands along the river for their livelihoods.

The Yellow River has claimed the lives of many, and many others who fell victim to floods but survived had to migrate as environmental refugees. In his book, *Liudong de tudi* (Floating Lands), published in 2012, Hu Yingze has shed light on the environmental changes and social tensions surrounding the alluvial lands along the Yellow River between Longmen and Tongguan. This chapter expands his inquiry into the latter half of the twentieth century and attempts to encompass the experiences of nonhuman species as well. The reservoir region between Tongguan and the Sanmenxia Dam will be the central focus, while the area between Longmen and Tongguan will also be frequently referred to.

Upon reaching Tongguan, the Yellow River is met by the imposing Qinling Mountains, forcing it to turn eastward. The river channel between Tongguan and Sanmenxia was narrow, only 3–8 km wide, because of the combined presence of the Qinling and Zhongtiao Mountains. Despite the narrowness of this channel, the river's depth and rapid flow were a result of the surrounding topography and geology. People living along this stretch of the Yellow River were thus rarely troubled by silting or flooding. But the construction of the Sanmenxia Dam would bring about a radical transformation of the landscape and of

---

[2] Hu, *Liudong de tudi*, 50–52.
[3] Shi Nianhai 史念海, *Huangtu gaoyuan lishi dili yanjiu* 黄土高原历史地理研究 (*Study of the Historical Geography of the Loess Plateau*) (Zhengzhou: Huanghe shuili chubanshe, 2001).

the local community's relationship with the river. Its impact would be far-reaching and significant, irrevocably changing the lives of people who lived along the banks of the Yellow River for generations.

A LAND IN TRANSFORMATION

According to the original plan for the Sanmenxia project, all residents in the designated reservoir area were to be relocated before the area was flooded. This resulted in the moving of six county towns, including the historic Tongguan, and 373,408 residents. On September 14, 1960, Chinese engineers successfully shut the bottom sluice gates of the Sanmenxia Dam, triggering a series of unprecedented environmental changes in the Sanmenxia Reservoir region. In a single day, the water level in the reservoir rose by 3.5 meters and then continued to rise. Over the next two years, the level of this section of the Yellow River was maintained at an unprecedentedly high level of 332.58 meters above sea level. The consequences of this project were immense: nearly 900,000 *mu* of fertile farmland and 356 villages were swallowed by this massive man-made lake.

The first concrete dam on the Yellow River was intended not just to impound water but also to trap a significant amount of sediment. To the surprise of many engineers and many of the political leaders involved in the project, within just a year and a half, the river had deposited approximately 153 million tons of sediment in the reservoir. While the retention of sediment was part of the design, the speed at which this was happening was both surprising and a matter of concern. Unfortunately, the Soviet engineers who had designed the dam were no longer in China as a result of the Sino-Soviet split, which left the hundreds of Chinese hydraulic engineers and bureaucrats involved to find a solution on their own. By October 1964, the amount of sediment in the reservoir had risen to 470 million tons. This made the dam an unprecedented threat to the upper reaches of the Yellow River.[4]

The deposited sediment formed a "door blockade" (*lanmensha* 拦门沙) where the Wei River converged with the Yellow River near Tongguan. This caused the flow of the Wei River to reverse, which in turn created a threat of flooding in the Guanzhong Plain, where the Shaanxi provincial capital of Xi'an was located. In response, the party center and the central

---

[4] *Huanghe Sanmenxia shuili shuniuzhi*, 6.

government authorized engineers to sacrifice part of the dam's power-generation capabilities and to begin remodeling the dam. This included a fundamental reorientation of the way the reservoir operated.

For the ambitious Communist bureaucrats and engineers, this was clearly a major setback. But for the hundreds of thousands of people displaced by the construction of the reservoir, the effects were mixed. They had been forcibly relocated from their homes and farms to make way for the dam. But, due to the heavy sediment deposits carried by the Yellow River from the Loess Plateau in the north, the dam had to be remodeled and its operational principles changed. Much of the land that had been initially submerged would thus no longer be or would only be inundated seasonally.[5] Facing challenges to feeding their families in the resettlement areas to which they had been sent by the government, many displaced people sought to reclaim their former lands. Sadly, this led to a deterioration of the reservoir's environment. The concrete dam, sediment deposits, and the altered flow of the reservoir which was no longer in sync with the natural rhythms of the Yellow River created a new set of environmental dynamics in the area. This created a wide range of problems for both human populations and wildlife.

RIVERBANK COLLAPSE

In Shaanxi, migrants who returned to their homes found themselves resettling on dry alluvial lands along the Yellow River, which were at increased risk of flooding. Despite ongoing land disputes after they had resettled from sites in Ningxia, Gansu, and northern Shaanxi, many of them quickly built earth dikes to defend their limited lands.[6] Between Henan and Shanxi, the Yellow River winds through loess hills, unlike the fertile basin lands where the Wei and the Yellow Rivers converge in Shaanxi, which left flat land along the river between Tongguan and the

---

[5] After 1973, the dam opened its sluice gate without impounding water unless the lower reaches of the Yellow River faced a serious threat of flood. It would impound water during the non-rainy season, usually from November to June. This mode of operation significantly lowered the water level in the reservoir compared to pre-1973. Between 1970 and 1990, the reservoir's highest water level was 318.5 meters. In seventeen of those years the highest water level was below 316 meters during rainy seasons. During non-rainy seasons, when the reservoir impounded water, the highest water level was 326 meters and in eighteen of those years the water level was below 325 meters. This meant that many previously cleared lands below 335 meters were accessible for at least part of the year.

[6] For more details, see Leng, *Huanghe dayimin*; Hu, *Liudong de Tudi*.

site of the dam in short supply. Faced with the reality that they could not stop the influx of migrants, local officials mandated that all returning migrants settle on hills above the 335-meter line, while still allowing them access to farmland below that line. This decision aimed to achieve the dual goal of keeping the reservoir clear while also minimizing social unrest and economic losses caused by the rising water. With no other options, the returning migrants had to accept the separation of altitude between their homes and their farmland.[7]

Even settling on higher ground along the river, though, was not without its dangers. The man-made rise of the water level in the reservoir led to a dramatic surge in life-threatening riverbank collapses, particularly between 1960 and 1962 when the water level was at its highest. Prior to the construction of the dam, the Yellow River flowed mostly eastward between Tongguan and Sanmenxia, without significant turns or twists, but the dam and the subsequent silting fundamentally changed the river's course, causing many sections to turn north-to-south. This resulted in an unruly, twisting river that now flowed directly against the unprotected loess riverbanks.[8]

The artificial ebb and flow of the reservoir water level also destabilized the soil structure, leading to the collapse of numerous sections of riverbank. The steep topography of the river valley and heavy loess coverage made the section of the river between Tongguan and the dam particularly susceptible to riverbank collapse. By the late 1980s, the accumulated width of riverbank collapse had reached between 200 and 300 meters, with the most severe cases, at Yangjiawan and Xiguyi in Lingbao county, surpassing 800 meters. In the early 1960s, riverbank collapse affected a staggering 179 million square meters over a length of 201 kilometers.[9]

It would be a mistake to assume on the basis of the above that people living above the 335-meter line, the state-designated water level of the reservoir, were immune to danger. In fact, riverbank collapse was a new and devastating environmental hazard for many of the migrants who had resettled on the upper hills, as well as for residents who had not been displaced by the reservoir. Their farmlands, homes, and lives were all at risk. A number of villages had to be hastily evacuated due to sudden

---

[7] Yang Anqing 杨庆安 and Luo Qimin 罗启民, "Huanghe Sanmenxia shuikuqu de zhili jiqi jingyan" 黄河三门峡水库区的治理及其经验 (The Yellow River Sanmenxia Reservoir Area Management and Experience), *Renmin Huanghe*人民黄河 5 (1986), 29–32.
[8] Shanxisheng Sanmenxia kuqu guanliju 山西省三门峡库区管理局 (ed.), *Shanxisheng Sanmenxia kuquzhi* 山西省三门峡库区志 (Zhengzhou: Huanghe shuili chubanshe, 2007).
[9] *Huanghe Sanmenxia shuili shuniuzhi*, 192.

collapses, and the number of people who lost their lives during the period of the highest water levels in the early 1960s is unknown. Official data shows that, by 1987, riverbank collapse had claimed 25 lives and nearly 100 head of livestock and required more than 30 villages to be resettled.[10] To address this threat and to protect the communities living along the river, the reservoir management authority implemented engineering measures to reinforce the riverbanks.

A recent investigation of riverbank collapse between Tongguan and the dam site on the Henan side from 1996 to 2014, however, revealed a staggering impact. The length of the collapsed riverbank reached 864 kilometers, affecting 128,700 *mu* of land, which included 67,000 *mu* of irrigated farmland. In total, 12,658 residents were forced to relocate, and 29 lost their lives or suffered serious injury.[11] One example is that of Zhang Jianye, a resident of Houdi Village in Lingbao county, who at one point owned twenty *mu* of orchards. Only three *mu* of his land remained after various riverbank collapses.[12]

Riverbank collapse has not only caused loses of land and property; it has also created a life-threatening environmental hazard for the communities along the reservoir. The combination of the area's steep topography, its loess geology, and the fluctuation of the reservoir's water level continues to this day to wreak havoc on the landscape and those communities.

GROUNDWATER AND LAND DETERIORATION

In addition to riverbank collapse, the reservoir's rising water also had a negative effect on the surrounding groundwater. Although this was unseen by local residents, the reservoir directly caused the general groundwater level in the region to rise. By the end of 1961, just a year into its full operation, this rise caused 1,296 wells across Shaanxi, Shanxi, and Henan along the Yellow River to collapse. In areas located on lower

---

[10] *Huanghe Sanmenxia shuili shuniuzhi*, 192.

[11] Henan Jianghe huanjing keji youxian gongsi 河南江河环境科技有限公司, "Huanghe Tongguan zhi Sanmenxia daba heduan 'shisanwu' zhili gongcheng huanjing yingxiang baogaoshu" 黄河潼关至三门峡大坝河段 "十三五" 治理工程环境影响报告书 (Environmental Assessment Report of the "Thirteenth Five-Year Plan" Management Project of the Section of the Yellow River from Tongguan to Sanmenxia Dam), March 2018, 23–24.

[12] Yang Lijuan 杨利娟, "Shinian xinku buxunchang: Sanmenxia kuqu yimin yiliu wenti chuli gongzuo chengji feiran 十年辛苦不寻常—三门峡库区移民遗留问题处理工作成绩斐然 (A Decade of Hard Work: Achievements in Solving Remaining Issues in the Sanmenxia Reservoir Displacement), *Henan qingnianbao* 河南青年报, May 21, 1997.

ground along the river, the groundwater also became salty and bitter and was thus no longer suitable for drinking. As a result, many villages faced severe water shortages, which required them to spend a great deal of time and effort carrying water from farther away.[13] This shortage and other deterioration issues were eventually alleviated when the water level in the reservoir began to be lowered regularly and electricity pumps were introduced, making deeper wells accessible again.

In some areas, the rise of the groundwater level also made the ground soft and muddy. This made twenty-eight villages west of Tongguan uninhabitable. In the most severe cases, rural homes and caves collapsed. In Wangzhuang village in 1962, for example, over 200 homes crumbled. Those households were not even considered reservoir migrants at the time.[14]

During the process of displacement, reservoir migrants lost their rights to the farmlands below the 335-meter line which were to be flooded. Sadly, much farmland above the line was also impacted by deterioration from below. The rising water elevated the groundwater, bringing salt and other harmful minerals to the surface soil, causing salinization and significantly reducing the fertility of the land, resulting in lower agricultural yields. By 1963, the amount of salinized land in Huayin and Huaxian counties had increased to 389,000 *mu*.[15] Compared to yields in 1958, yields of cotton and grain had decreased by 73 percent and 53 percent, respectively. Since most of the salinized land was located above the 335-meter line, local farmers were not compensated for the degradation of their land caused by the dam. It was not until the 1970s, after more than a decade of struggle, that soil fertility in the region improved. This was due to a reduction in the reservoir water level and the creation of discharge ditches to drain the water away from the salinized lands.[16]

LAND USE CHANGE

In Xiaobeiganliu, the stretch of the Yellow River between Longmen and Tongguan provided vast shorelands because of its topography and V-shaped river course. The availability of shorelands was limited between

---

[13] *Huanghe Sanmenxia shuili shuniuzhi*, 197.
[14] *Huanghe Sanmenxia shuili shuniuzhi*, 197.
[15] "Sanmenxia kuzhou yianjianhua qingkuang" 三门峡库周盐碱化情况 (Salinization near the Sanmenxia Reservoir), A12-1(1)-31, 1964, The Yellow River Conservancy Commission Archives, Zhengzhou.
[16] *Huanghe Sanmenxia shuili shuniuzhi*, 196.

FIGURE 7.1 Sanmenxia Reservoir at low water level (Photo by Jie Zhao/Corbis News via Getty Images)

Tongguan and Sanmenxia before the construction of the dam, though.[17] The dam was designed to retain most of the sediment carried by the Yellow River within the reservoir, but in just a few years the V-shaped course of the river was completely flattened by the deposited sediment. As a result, a vast area of previously stable farmland was transformed into seasonally flooded shoreland, competing with local communities for land and space.

As a result of modifications to the dam's operation, vast areas of shoreland along the Yellow River to the west of Tongguan would not be submerged as originally planned. Meanwhile, about 200,000 *mu* of land to the east of Tongguan between the 315-meter and 326-meter lines was now subject to seasonal floods, which limited peasants to only one crop during the non-flood months each year rather than the two crops that they produced prior to the dam's construction (see Figure 7.1).[18]

Layers of sediment carried by the Yellow River have been deposited on the flat lands in the river valley, creating new land and reshaping the landscape. Ling Zhang has discussed the seasonal changes in the

---

[17] The section of the Yellow River between Longmen and Tongguan is traditionally called Xiaobeiganliu 小北干流. See Hu, *Liudong de Tudi*.
[18] *Huanghe Sanmenxia shuili shuniuzhi*, 202.

composition of river silt on the North China Plain insightfully.[19] However, there may be a different interpretation of the source she cites regarding the decrease in silt fertility. According to the compilers of the official *History of the Song Dynasty* (*Songshi*), the loss of fertility experienced at the time was a result of changes in the sediment carried and deposited by the river, rather than necessarily a gradual loss of fertility of a particular patch of silt, which may be true. In the summer, the river carries and deposits fertile glutinous soil, as heavy rainfall from the Loess Plateau picks up organic topsoil and carries it downstream. As the river's velocity and flow capacity decrease in the autumn, however, only smaller sediment and fine sands are carried downstream.[20] Although the deposited sediments often improve the land's fertility, some consist mainly of sands low in the fertilizers nitrogen and phosphate. Because of the seasonal flooding, local peasants also hesitated to invest in these lands, for example by applying fertilizer. All of these factors have made these lands less productive than before.

Crops like wheat, which require a relatively long growing season, can no longer be cultivated on the seasonally flooded lands. Instead, people switched to growing soybeans, peanuts, and corn. The Reservoir Management Bureau informed local peasants that they would not be responsible for any crops that were flooded below the 335-meter line. This meant that there would be no compensation for this land, as it was considered state-designated reservoir space and not suitable for agriculture. For a long time, people in Sanmenxia had relied on precipitation, as there was a lack of irrigation infrastructure. After the dam was built, though, the state-engineered hydropower project became the land's new "common dominator." While the flow of the Yellow River changes each year, the flood and non-flood seasons are now predictable and the reservoir can impound water at any time to prevent flooding in the river's lower reaches on the North China Plain. This means that the land in the reservoir is destined to be sacrificed to a largely human-made environmental system, and residents around the reservoir must adapt their daily lives to this artificial unpredictability.

Most reservoir migrants in Henan and Shanxi were forced to move to higher ground along the Yellow River, where the availability of arable land was limited. In many resettled villages, the amount of arable land per

---

[19] Zhang, *River, Plain, and State*, 260–261.
[20] Tuoteo (ed.), *Songshi, Hequ, Huanghe* 宋史,河渠一, 黄河 (*The History of the Song Dynasty, Rivers, the Yellow River*), vol. 91 (Beijing: Zhonghua shuju, 1985).

capita was as little as 0.3 *mu*. To sustain themselves, many peasants had no choice but to use the alluvial lands, which had been relatively stable and accessible but were now subject to seasonal flooding without compensation. The experimental modifications of the reservoir's mode of operation and the problem of heavy silting caused the river's course to meander more frequently than before. In addition to the threat of riverbank collapse, residents now had to deal with the collapse of alluvial lands as well. In Henan, 97,300 *mu* of alluvial lands collapsed in three decades following the completion of the dam. Although new alluvial lands did eventually form, it took three to five years for them to be usable for agriculture.[21]

According to estimates from 2003, 204,306 *mu* of alluvial land lay below the 335 meters line in the Sanmenxia reservoir. Despite controversies surrounding the resettlement process in the 1950s, the Sanmenxia Yellow River and Migrants Management Bureau maintains the legal right to use these state-owned lands, which were often claimed by resettled villages along the river.[22] To address this, the bureau decided to lease the lands to the villages, distributing them to individual households through a contracting system.

However, because of the poor quality of the topsoil and a lack of irrigation infrastructure, much of the land was relatively unproductive for grain crop cultivation. In Houdi Village in Lingbao County, for example, the alluvial land was so poor that nobody was willing to lease them at a price of five *yuan* per *mu*. It wasn't until the late 1980s that the local government began to offer financial subsidies and technical assistance to help improve the productivity of the land.[23] After that, the land gradually became more productive.

According to official data, as of the end of 1996, 84,600 migrants in Sanmenxia still struggled to feed themselves.[24] To address this problem, the Reservoir and Migrants Management Bureau encouraged them to develop aquaculture, pig farming, and other cash-earning industries. With the Bureau's assistance, local farmers began to plant apple and

---

[21] Sanmenxiashi Huanghehewu yiminguanlijuzhi bianzuan weiyuanhui 三门峡市黄河河务移民管理局志编纂委员会, *Sanmenxiashi Huanghe hewu yimin guanlijuzhi* 三门峡市黄河河务移民管理局志 (Beijing: Fangzhi chubanshe, 2006), 156.
[22] *Sanmenxiashi Huanghe hewu yimin guanlijuzhi*, 245.
[23] Yang, "Shinian xinku buxunchang."
[24] Li Chunlei 李春雷, "Tuanjie fenjin, zaizhan hongtu: qianjinzhongde Henansheng Sanmenxiashi Huanghekuqu guanliju" 团结奋进, 再展宏图: 前进中的河南省三门峡市黄河库区管理局, *Zhongguo shuilibao* 中国水利报, December 9, 1997.

jujube trees, and also converted a large area of alluvial lands near the river into lotus root and fishponds. While these efforts were limited in scope, they did help to connect many migrants to the growing urban market and to diversify local agriculture. Since its establishment, the management of the Sanmenxia reservoir has focused on two major tasks: The alleviation of poverty and maintenance of the reservoir. In the present century, though, with top-down enforcement of wetland conservation and an increasing appreciation of biodiversity, new environmental dynamics have emerged in the reservoir area. Thus, the pursuit of the preservation of biodiversity and wetland conservation may outweigh the subsistence needs of reservoir migrants and the poverty alleviation efforts undertaken by the reservoir management authority.

THE RISE OF WETLAND CONSERVATION

Since the late 1950s, the "grain as the key line" policy and the rise of radical Maoism had shaped the relationship between humans and so-called wasteland across China.[25] The ecological value of wetlands and their pivotal role in preserving biodiversity were not widely recognized either by the Chinese state or by the people. While the first national nature reserve was established in Guangdong in 1956, it was intended primarily for scientific research rather than conservation and tourism.[26] In recent decades, in response to domestic and international pressure, the government of China has started to integrate environmental protection and sustainability into its laws, national planning, and policies.[27] China joined the Ramsar Convention on Wetlands in 1992 and officially placed wetland conservation on its agenda.[28] The government then conducted two nationwide surveys between 1995 and 2013 to obtain a systematic overview of the status of China's wetlands. The second survey showed that the country has 53.60 million *ha* of wetlands, 46.67 million *ha* of which are natural wetlands while the rest is human-made via dam construction.[29]

---

[25] Shapiro, *Mao's War against Nature*.
[26] Robert P. Weller, *Discovering Nature: Globalization and Environmental Culture in China and Taiwan* (Cambridge: Cambridge University Press, 2006), 77.
[27] Shapiro, *China's Environmental Challenges*, 59–70.
[28] The Ramsar Convention is an international convention on wetlands conservation and sustainable development. Its first meeting took place at Ramsar, Iran, 1971.
[29] A hectare (symbol: ha) is a unit of area that is accepted in the International System of Units (SI). It is primarily used to measure land area. One hectare is equal to 10,000 square meters and is equivalent to approximately 2.471 acres.

Because of urbanization and related human activities, though, natural wetlands have suffered a dramatic rate of loss of 9.33 percent over a decade, while large dam construction has increased the water surface of many rivers and thus, to some extent, "offset" the loss of natural wetlands.[30] From an ecological perspective, wetlands are seen as the "kidneys of the earth": They play a crucial role in preserving biodiversity by providing habitats for waterfowl, such as migratory birds, cranes, wild geese, and ducks. Riverine wetlands also serve as buffer zones during flood seasons.

The disastrous ecological consequences of reckless reclamation and deforestation were exposed by the devastating flood of the Yangtze River in 1998. As part of its response to this crisis, the central government initiated the National Wetland Conservation Action Plan in 2000 and the National Wetland Conservation Program in 2003, both of which aimed to preserve natural wetland and to restore low-yield croplands to wetland. Since then, at national and regional administrative levels, China has established more than 200 new wetland nature preserves.[31]

In 1995, as part of national efforts toward wetland conservation, the government of Henan Province sponsored the establishment of the Yellow River Wetland Provincial Level Natural Preserve at Sanmenxia and Luoyang. Then, in June 2003, with the approval of the central government, the preserve was promoted from provincial to the state level. The Sanmenxia Reservoir wetland, along with other sections of the Yellow River in Henan, is an essential component of this wetland preserve.

---

[30] Zhigao Sun, Wenguang Sun, Chuan Tong, Congsheng Zeng, Xiang Yu, and Xiaojie Mou. "China's Coastal Wetlands: Conservation History, Implementation Efforts, Existing Issues and Strategies for Future Improvement," *Environment International* 79 (2015), 25–41.

China's wetlands are unevenly distributed among seven wetland regions:

1. The northeast region, dominated by freshwater marshes.
2. The Xinjiang-Inner Mongolia region, dominated by saline lakes and swamps in dry climate.
3. The Qinghai-Tibet plateau region, dominated by alpine lakes and marshes.
4. The Yunnan-Guizhou plateau, dominated by subalpine lakes.
5. The middle-lower Yellow River region, dominated by rivers and coastal wetlands.
6. The middle-lower Yangtze River region, dominated by lakes, rivers, and coastal wetlands.
7. The south and southeast regions, dominated by rivers and coastal wetlands.

[31] Weihua Xu, Xinyue Fan, Jungai Ma, et al., "Hidden Loss of Wetlands in China," *Current Biology* 29 (2019), 3065–3071.

The wetland's promotion from provincial to state level brought not only a title change but also an increase to its budget and administrative expansion. In September 2003, the Sanmenxia municipal government created the Wetland Management Committee, which later established the Wetland Management Office with twenty permanent employees. To better manage the condition of the wetland, the office created four local management stations at Lingbao, Shaanxian, Hubin, and Mianchi.[32] Although the Reservoir and Migrants Management Bureau has served as the managing authority of the reservoir for decades, the Wetland Management Office is independent of it. Despite what would appear to be bureaucratic redundancy, these two bodies represent two different kinds of management policies: One represents the traditional mode centered on economic exploitation, while the other represents the more recent state-led environmentalist principle of nature conservation. The tensions and dynamics between these two modes have become a central new line of narrative in the history of the reservoir.

Although the Sanmenxia wetland is designated as a "natural" preserve, its status as "pristine" nature is difficult to discern. The long history of interaction between the river and agrarian societies, now coupled with the effects of modern concrete-dam technology, have altered the riverine environment significantly. This particular national wetland preserve is, therefore, a product of the bringing together of the Yellow River and state-led hydro-engineering mechanisms. The rhythms of the wetland system at Sanmenxia are directed by the ebb and flow of the reservoir's water level. Every November, the reservoir begins to impound water, with the water level reaching its highest point in April. After that, it discharges water until July or August, when the water level reaches its lowest point of the year. This mode of operation is central to the size and ecology of the wetland at Sanmenxia, as shown by Table 7.1. A low water level in the reservoir can lead to a significant shrinkage of the wetland it determines, resulting in a corresponding lowering of its ecological vitality.

Despite its artificial origins, the Sanmenxia wetland is nonetheless an essential habitat for a number of species, including waterfowl, migratory birds, and cranes. During flood seasons, the wetland also serves as a buffer zone, mitigating the risk of flooding in the surrounding areas.

---

[32] Yang Yuqiu 杨玉秋 and Luo Song 罗松, "Henan Huanghe shidi guojia ziran baohuqu Sanmenxiaduan jiben qingkuang jianjie" 河南黄河湿地国家级自然保护区三门峡段基本情况简介 (Brief Introduction of the National Yellow River Wetland Conservation in Sanmenxia, Henan), *Sanmenxia wenshiziliao* 三门峡文史资料 17 (2007), 431–435.

TABLE 7.1 *Sanmenxia Reservoir water level and wetland size*[33]

Reservoir Water Level (meters)	320	315	310	305
Wetland Size (square meters)	274.7	92.3	38	13.5

As a result, it is imperative that the reservoir water level be managed carefully to maintain the wetland's ecological balance and vitality.

WHOOPER SWAN: AN ENVIRONMENTAL WINDFALL

A thriving wetlands system is essential for maintaining biodiversity in the region. In recent years, the arrival of large flocks of whooper swans during winter has garnered widespread recognition for the wetland and the city of Sanmenxia as a champion of environmental sustainability. But the effects of the dam that has facilitated the development of this wetland on local species have been mixed. On the one hand, the dam has created a vast expanse of water that has become an ideal habitat for a diverse range of waterfowl and other birds in the reservoir area. On the other, the fish population in the middle and lower streams of the Yellow River has undergone a dramatic decline. This is principally due to the absence of fish ladders, which has cut off the path that many fish species take to their spawning grounds in the river's upper stream or its tributaries. In combination with overfishing, the dam has contributed to the extinction of thirteen species of fish.[34]

In the 1950s, the average annual yield of fish in the Gangkou area in Tongguan was 30,000–40,000 kilograms. By the 1970s, this number had plummeted to less than 10,000 kilograms. By the 1980s, it had fallen further to less than 5,000 kilograms per year. Prior to the dam's construction, migratory fish like carp comprised around 72 percent of the annual

---

[33] Mao Zhanpo 毛战坡, Peng Wenqi 彭文启, Wang Shiyan 王世岩, and Zhou Huaidong 周怀东. "Sanmenxia shuiku yunxing shuiwei dui shidi shuiwen guocheng yingxiang yanjiu" 三门峡水库运行水位对湿地水文过程影响研究 (Study of the Sanmenxia Reservoir Water Level's Impact on Wetland Hydrology), *Zhongguo shuilishuidian keyuan yanjiuyuan xuebao* 中国水利水电科学研究院学报 (*Journal of the China Institute of Water Resources and Hydropower Research*) 4, no. 1 (2006), 36–41, 38.

[34] Guo Qiaoyu 郭乔羽 and Yang Zhifeng 杨志峰, "Sanmenxia shuili shuniu gongcheng shengtai yingxianghou pingjia" 三门峡水利枢纽工程生态影响后评价 (Assessment of the Ecological Impact of the Sanmenxia Hydropower Project), *Huanjing kexue xuebao* 环境科学学报 25, no. 5 (2005), 580–585.

catch in Tongguan. After its completion, this proportion dropped to less than 20 percent.[35]

Despite these challenges, the Sanmenxia Reservoir has nonetheless experienced a remarkable increase in its whooper swan population in recent years.[36] As the area was once just a temporary stop for most swans on their way south in the 1970s and 1980s, only a small number would stay for the entire winter season.[37] This has changed dramatically, though, with the number of swans present surging from 410 in 2010 to a staggering 7,858 in 2014. Unfortunately, by 2018, because of an outbreak of a highly pathogenic avian influenza, H_5N_1, during the winter of 2014, the population had fallen to 4,437.[38] The question remains: What draws such a vast number of whooper swans to the Sanmenxia Reservoir as their new wintering ground? This section explores the factors that contribute to this ecological boon as well as new challenges that have emerged in the reservoir as a result.

When choosing a wintering ground, whooper swans exhibit a preference for wetlands that are abundant in flora and the smaller organisms they often eat. Recent scientific research has suggested that swans can tolerate a certain degree of human disturbance in their selected habitats, provided that an ample food supply is present.[39] The Sanmenxia

---

[35] Huanghe shuixi yuyeziyuan diaochaxie zuozu 黄河水系渔业资源调查协作组, Huanghe shuixi yuye ziyuan (Fishery Resources of the Yellow River Basin) (Shenyang: Liaoning keji chubanshe, 1986), 66.

[36] The whooper swan's scientific name is *Cygnus*. It can be found in many parts of the northern hemisphere. In Asia, they breed in Siberia, Mongolia, Northern Manchuria, and Central Asia. Their wintering grounds in China include Qinghai, Shandong, Henan, and the lower reaches of the Yangtze River. In the late 1990s, the estimated population in China was about 5,000 in breeding grounds and 15,000 in wintering grounds. Throughout the second half of the twentieth century, under the dual pressures of hunting and loss of habitat, the general population was declining. For more detail, see Yuan Li, Li Xiaomin, Yu Hongxian, Guo Libin, Zhang Taizhong, Gao Qingjun, and Zeng Daihua, "The status and conservation of whooper swans (*Cygnus cygnus*) in China," *Journal of Forestry Research* 8, no. 4 (1997), 235-239.

[37] Baitian'e daidong yizuocheng 白天鹅带动一座城, March 24, 2014, available at www.shidicn.com/sf_44C6D2602D2B474BB2BAEF944384BD66_151_smxhhsd.html, last accessed March 13, 2024.

[38] Ru Jia, Shu-Hong Li, Wei-Yue Meng, et al., "Wintering Home Range and Habitat Use of the Whooper Awans in Sanmenxia Wetland, China," *Ecological Research* 34 (2019), 637-643, 638.

[39] Yuhong Liu, Ying Lu, Cheng Chen, et al., "Behavioural Responses of the Whooper Swans *Cygnus cygnus* to Human Disturbance and Their Adaptability to the Different Habitats in the Rongcheng Lagoon of China" *Ecohydrology* 11 (2018), 1-9, 2.

Reservoir offers both open water and an abundance of food, making it a particularly attractive wintering ground.

As I discussed previously, the hydraulic engineering force behind the rise of the Sanmenxia Reservoir's water level during the non-flood season is water retention. Not all reservoirs provide ideal habitats for wild swans, however. In some cases, the heavy weight of the sediment deposited during flood season impedes the flourishing of the plankton and other organisms that serve as food for these birds.[40] Fortunately, the enormous alluvial lands and flatter higher grounds that have been submerged in the Sanmenxia Reservoir have created a large and shallow water area that is perfect for waterfowl. In addition, these alluvial lands had until recently been cultivated by local farmers, meaning that plenty of plant stalks, soybeans, corn, and peanuts have been left behind. These then become crucial sources of nutrition for the swans during the long winter months. The Sanmenxia Reservoir has been transformed into a seasonal wetland habitat for whooper swans by a fortuitous combination of the river, the dam, alluvial lands, and the local history of agriculture undertaken by local farmers.

China introduced its first law for the protection of wildlife in 1989. In its early period, unfortunately, the enforcement of this law was lax. Despite the new legal prohibitions on hunting and consuming wild animals, a high level of demand existed for swan meat and other waterfowl, which thus fetched a good price in local markets. Despite the clear violation of the new wildlife protection laws, demand persisted. Although swan meat was not publicly advertised on the menus of local restaurant, many waiters and waitresses discreetly recommended it, with each dish costing hundreds of *yuan*.[41] This issue was certainly not limited to the Sanmenxia area. Across China, the survival of wildlife was threatened not only by loss of habitat but also by the lucrative game meat industry.

THE CRACKDOWN ON ILLEGAL HUNTING

It was only after the swans at Sanmenxia became a significant presence that the local government started to take wildlife protection laws more

---

[40] Shuilibu Huangweihui kance guihua shejiyuan 水利部黄委会勘测规划设计院, *Sanmenxia shuiku jianku qianhou ludi, shuisheng shengwu de bianhua* 三门峡水库建库前后陆地, 水生生物的变化 (*Changes of Terrestrial and Aquatic Species before and after the Construction of the Sanmenxia Reservoir*) (Zhengzhou, 1988), 146.

[41] "Baohuqunei bushefang, Tongguan you jiudian jingchi tian'e rou" 保护区内不设防, 潼关有酒店竟吃天鹅肉 (Defenseless in the Conservation Area, a Hotel in Tongguan Sells Swan Meat," *Huashangbao* 华商报, May 10, 2002.

seriously. Under the banner of the central slogan "building an ecological civilization," the government recognized the new mass congregation of swans as a stroke of ecological fortune and presented it as a testament to the success of its environmental protection and sustainable development strategy.[42] It also viewed the fortuitous situation as an opportunity to develop an eco-tourism industry centered on the swans. Local authorities were galvanized into taking action to protect these creatures.

To start, the local government undertook to tackle the problem of illegal hunting. While the high-profile poisoning incident discussed at the beginning of this chapter and the subsequent harsh penalties imposed in 1999 did serve as a warning to the public about how seriously the state now took wildlife protection, this was just the tip of the iceberg. Illegal hunting continued and many restaurants still offered swan and other game meat to their customers. It became clear that more effective measures were needed to protect the swans.

As a first step, surveillance of the swan habitats was increased. The Wetland Management Bureau hired more than twenty unemployed city residents in 2011 to patrol Wangguan and other wetland areas at Sanmenxia. The bureau also installed surveillance cameras in the Wangguan and Shuanglonghu wetlands.[43] The following year, the patrol team grew by several orders of magnitude, expanding to 4,050 members. They were now able to provide 24/7 surveillance of the wetland areas through rotating shifts.[44] Effective surveillance is an essential element of wildlife protection, and the measures taken by the local government at Sanmenxia have been instrumental in safeguarding the swan habitats from illegal hunting.

As well as increasing its law enforcement capacity, the Wetland Management Office at Sanmenxia also took steps to raise public awareness of the importance of wildlife protection and biodiversity. It set up exhibition booths and banners throughout the city and dispatched propaganda teams to villages around the reservoir to educate residents about

---

[42] See Yifei Li and Judith Shapiro, *China Goes Green: Coercive Environmentalism for a Troubled Planet* (New York: Polity, 2020).

[43] "Woshi baitian'e baohu gongzuo zhashi youxiao" 我市白天鹅保护工作扎实有效 (Our City's Whooper Swan Protection Work Is Solid and Effective) *Sanmenxia Huanghe shidi guojiaji ziran baohuqu guanwang* 三门峡黄河湿地国家级自然保护区官网, November 17, 2011.

[44] "Zhoumi bushu, tianqian zuohao shidi baohu gongzuo anpai" 周密部署, 提前做好湿地保护工作安排 (Be Prepared for the Wetland Protection Work Arrangement), *Sanmenxia Huanghe shidi guojiaji ziran baohuqu guanwang*, October 30, 2012.

the law and the consequences of violating it. In previous decades, organized hunting squads had existed in many villages along the Yellow River, and hunting was an established side business. In the late 1970s, such hunting squads could harvest around 150 kilograms of bird and waterfowl feathers per year.[45] The establishment of the Natural Wetland Preserve and the enforcement of wildlife protection laws have, however, criminalized such activities, thus eliminating one of the supplementary sources of income peasants relied on. In order to transform people's traditional perceptions of wildlife and its legitimate uses, local authorities distributed pamphlets and signed wildlife protection agreements with town-level administrative bodies. This was to encourage the lowest level of the administrative system to participate in the collective effort to maintain a harmonious habitat for the swans. These efforts have been crucial in raising public awareness and increasing the sense of responsibility toward the protection of wildlife, particularly swans, in Sanmenxia.

The crackdown on illegal hunting in Sanmenxia has been largely successful, although incidents of individual hunting do occasionally still occur. In 2012, three men were caught hunting waterfowl in the Wangguan wetland. Even though no animals were found to have been killed, according to China's wildlife preservation law they faced a fine of 2,000 *yuan*. In their defense, the men claimed ignorance of the law and argued that they only hunted for personal consumption. Given that the three of them were poor local peasants in their seventies and in poor health, the authorities just confiscated their tools and warned them not to hunt wild animals again.[46] While this particular case did not result in a major confrontation, it highlights the continuing tension between traditional local assumptions and practices of free access to natural resources and the criminalization of such practices by state authorities.

Unlike in Caohai, in Guizhou province, where collective resistance to the new legal regime occurred, the majority of villagers along the Sanmenxia Reservoir do not rely on fishing or hunting as their primary source of livelihood.[47] As a result, local residents have largely complied

---

[45] Shuilibu Huangweihui kance guihua shejiyuan, *Sanmenxia shuiku jianku qianhou*, 143.
[46] "Shidi baohuqu yancha luanbu lanlie houniao xingwei" 湿地保护区严查乱捕滥猎候鸟行为 (Crackdown on Illegal Bird Hunting in the Wetland Conservation Area), *Sanmenxia Huanghe shidi guojiaji ziran baohuqu guanwang*, November 2, 2012.
[47] Melinda Herrold, "The Cranes of Caohai and Other Incidents of Fieldwork in Southwestern China," *Geographical Review* 89, no. 3 (1999), 440–448; "Peasant Resistance against Nature Reserves," in *Reclaiming Chinese Society: The New Socialist Activism*, ed. You-tien Hsing and Ching Kwan Lee (New York: Routledge, 2010), 83–98.

with the government's crackdown on illegal hunting. Although resistance to the state's frontier resettlement in the 1950s might be defined as rightful, traditional notions of free access to wildlife are less justifiable in the twenty-first century. The government's efforts to protect wildlife have been largely successful and local residents have generally complied with the wildlife preservation code.

FEEDING THE SWANS

Initially, the soybeans, peanuts, corn, and plant stalks left behind by local peasants served as a major source of nutrition for the swans in the reservoir, which contributed to their growth in numbers. However, precisely as a result of this growth, these sources of food began to be exhausted. Organisms present in the water were not sufficient to sustain thousands of swans throughout the winter, so many began to forage in agricultural fields. This created problems for local farmers, particularly in Pinglu county, Shanxi province, on the northern bank of the Sanmenxia Reservoir. Farmers there were annoyed that swans were eating their wheat seedlings, causing a dramatic decrease in their yield. Here, the demands of wildlife protection were at odds with local agricultural production, and farmers demanded compensation.

To satisfy the disgruntled farmers while accommodating the swans, the county government agreed to compensate local farmers along the reservoir at a rate of 500 *yuan* per *mu* in exchange for not driving away the swans. The government also purchased 50 tons of corn and 5 tons of Chinese cabbage to supplement the swans' food sources. Given that swans prefer soybeans, the Pinglu government even cultivated more than 100 *mu* of soybeans for the sole use of their wild guests.[48] These measures reflect the need to strike a balance between wildlife protection policies and local agricultural production.

The municipal government of Sanmenxia, recognizing that the presence of swans had become a valuable eco-tourism resource, took similar measures but on an even larger scale (see Figure 7.2).[49] In this case, their actions were not solely intended to satisfy local farmers. The government

---

[48] Shanxi Pinglu Huanghe shidi cheng baitian'e meili jiayuan 山西平陆黄河湿地成白天鹅"美丽家园, November 28, 2012, 中华人民共和国生态环境部, available at www.mee.gov.cn/xxgk/gzdt/201211/t20121128_242788.shtml, last accessed March 13, 2024.
[49] "Woshi baitian'e baohu gongzuo zhashi youxiao."

FIGURE 7.2 Feeding swans at the Sanmenxia Wetland Preserve (Photo by Visual China Group via Getty Images)

was more concerned about the possibility that the swans would migrate somewhere else for the winter.

Unfortunately, the feeding program organized by the government caused unexpected problems. For tourism purposes, the wetland management office dispensed food at specific locations where swans typically gathered, particularly at Shuanglong Lake. This practice drew swans that would typically have dispersed throughout the reservoir to these specific locations, leading to a densification of the swan population. Unfortunately, this high density created an environment well-suited to the spread of viruses.

In January 2015, a highly fatal H5N1 influenza outbreak occurred among waterfowl in the Sanmenxia area. It was suggested that the high density of swans in the area contributed to the outbreak.[50] This incident highlights the need for more careful consideration of the potential risks and consequences of large-scale ecological initiatives. While efforts to protect wildlife and to promote tourism are surely important, they must be balanced with careful consideration of potential consequences.

[50] Zhang Guogang 张国钢, "Huanghe Sanmenxia kuqu yuedong datian'e de zhongqun xianzhuang" 黄河三门峡库区越冬大天鹅的种群现状 (The Current Status of Wintering Population of Whooper Swans at Sanmenxia Reservoir Region," *Dongwuxue zazhi* 动物学杂志 (*Chinese Journal of Zoology*) 51, no. 2 (2016), 190–197, 195.

BUILDING A BENIGN ENVIRONMENT

The rise of environmental consciousness and the growth of eco-tourism have spurred local governments around the Sanmenxia reservoir to take action to improve the swans' habitat. As Robert Weller's work has shown, the implementation of environmental policies at the intersections of various levels of government has involved a mix of successes and failures.[51] While it is still too early to undertake a definitive assessment of the policies implemented by local governments in Sanmenxia, they have a significant stake in wetland and wildlife conservation as a means of promoting economic development and transformation.

Unlike the giant panda, which has needed both captive breeding and the conservation of its habitat to survive, whooper swans do not require human intervention in their reproduction.[52] However, as is the case for almost all species, humans must refrain from encroaching on or destroying their habitats to ensure their survival. The actions taken by local governments to protect the swans' habitat reflect the need to strike a balance between environmental protection and economic development.

In 2012, the Sanmenxia government recognized the need to protect the swans' habitats by relocating its wastewater plant to a location farther away. At the same time, it initiated a series of measures to address water pollution in the region.[53] The following year, the Sanmenxia Management Bureau of the Yellow River Wetland Reserve mobilized its entire staff to clear the stalks left behind by local farmers in the Wangguan wetland. Normally, farmers in the area would plant a single crop, like peanuts or soybeans, from late June to mid-October. In 2013, though, they planted sunflowers to produce oil, which left a large area of tall stalks in the fields after the harvest. Concerned that the presence of these stalks would make it harder for swans to land or take off, the Wetland Management Bureau mobilized its employees to remove them, thereby creating a more accommodating habitat for the swans.[54] The office also employed hydro-engineering techniques: It built sluice gates to impound

---

[51] Weller, *Discovering Nature*, 155.
[52] Elena Songster, *Panda Nation: The Construction and Conservation of China's Modern Icon* (Oxford: Oxford University Press, 2018).
[53] "Baitian'e qianghua shengtaiguan" 白天鹅强化生态观 (Swans strengthen ecological perspective), *Sanmenxia Huanghe shidi guojiaji ziran baohuqu guanwang*, March 14, 2014.
[54] "Baohu baitian'e cong qingli gaogao zuowu kaishi" 保护白天鹅 从清理高杆作物开始 (Swan Protection Starts with Clearing Tall Stalk Crops), *Sanmenxia Huanghe shidi guojiaji ziran baohuqu guanwang*, October 12, 2013.

or pump Yellow River water to the "wetlands" on higher ground if the reservoir water level was too high or too low.

On December 27, 2013, in an attempt to restore the original wetland landscape, the Wetland Management Office shut down a chicken farm in the Wangguan wetland area. Local fisher households were also prohibited from fishing in key preserve areas.[55] The following year, the Sanmenxia People's Congress approved rigorous regulations to protect whooper swans, declaring October to March every year as a swan-wintering protection period. The regulations prohibit all activities that would disturb the swans, including fishing, hunting, foraging for eggs, agriculture, construction, and fireworks. To better protect the wetland, it was divided into three categories, following the criteria of the UNESCO biosphere reserve program: Preserve, key preserve, and buffer zone.[56] In the key preserve areas, crops with tall stalks are now prohibited during summer and autumn to ensure that they do not impede the whooper swans' landing, taking off, and foraging activities. In the buffer zones, tall buildings and brightly colored paint are not permitted, to prevent the swans being scared away.[57] To better accommodate the swans, then, the local government required both rural and urban residents near the wetland preserve to follow rigorous rules, regardless of their economic needs.

MARKETING "SWAN CITY"

The presence of whooper swans in the Sanmenxia Reservoir has become a central motivation behind the local government's commitment to environmental protection and sustainable development. With millions of *yuan* invested in protecting and accommodating these creatures, the government has manifested its determination to maintain a harmonious

---

[55] "Lianhe kaizhan zhuanxiang xingdong daji pohuai shidi xingwei" 联合开展专项行动 打击破坏湿地资源行为 (Take Collective Action against Wetland Sabotage Activities), *Sanmenxia Huanghe shidi guojiaji ziran baohuqu guanwang*, December 27, 2013.

[56] UNESCO (The United Nations Educational, Scientific, and Cultural Organization) demarcates three biosphere zones: A core, where preservation is the overriding goal and there should be no human activity beyond scientific monitoring and research; a buffer zone, which protects the core area but that can be used for tourism, experiments in ecological rehabilitation, or research on economic productivity consistent with conservation; and an outer transition area, where local inhabitants, governments, and others work together for sustainable development. See Weller, *Discovering Nature*, 77.

[57] "Sanmenxia chutai wosheng shoubu baohu baitian'e guiding" 三门峡出台我省首部保护白天鹅规定 (Sanmenxia Promulgates the Province's First Set of Rules on the Protection of Swans), *Sanmenxia Huanghe shidi guojiaji ziran baohuqu guanwang*, January 8, 2014.

relationship with the natural world. However, this commitment goes beyond a simple desire to protect the swans. The Sanmenxia government has recognized that ecological restoration and improvement are essential components of its response to the central government's call for the building of an ecological civilization, and it sees the development of a swan-centered tourism industry as a new driver of local economic transformation. The protection of the swans is thus not only a moral imperative but also an economic opportunity, inspiring the city to reshape its environment and economy.[58]

Alongside the measures taken to safeguard the swans, the government of Sanmenxia also made a concerted effort to elevate the birds to the status of the city's new cultural emblem. In 2013, the film channel of China Central Television broadcast the movie "Love Story in Swan City," which recounts a romantic story set in Sanmenxia in the late 1950s. Despite its historical setting during the construction of the first large dam on the Yellow River, the film centers on the swans that bring the two main characters together. The movie begins with Suyun, a journalist from Beijing, attempting to prevent local farmers from hunting the swans. The farmers argue that they are merely hunting for sustenance, but Siyuan, the male lead, intervenes and accuses them of threatening the friendship between the Soviet Union and China by killing the birds that had migrated to the area from Siberia. Intimidated by this accusation, the farmers retreat quietly.

While the political discourse and love story in the film may not be particularly credible, the sheer number of swans present in the film in Sanmenxia in the 1950s stands out. It becomes apparent that the main sponsor of the film was the Sanmenxia municipal government, and that it was not intended to be an accurate historical reflection of the construction of the Sanmenxia Dam, but a promotional tool for the city's burgeoning swan tourism industry.

To further enhance the city's image as a haven for swans, the Sanmenxia Radio and Television Station produced a documentary about a pair of injured whooper swans who were compelled to spend the breeding season in Sanmenxia, when most of the swans had migrated

---

[58] "Baitian'e daidong Sanmenxia dongyure" 白天鹅带动三门峡"冬游热 (Swans Bring Winter Tourist Wave to Sanmenxia), *Sanmenxia Huanghe shidi guojiaji ziran baohuqu guanwang*, March 10, 2014. According to official data, from November 2013 to February 2014, the city of Sanmenxia received 3.8011 million tourists who generated a total tourism income of 3.71 billion *yuan*. The Swan Wetland Park alone received 619,900 visitors.

north to their usual breeding grounds. Together with a series of news reports carried by state media, this documentary showcased the beauty of the Sanmenxia wetland to a national audience, with the intention of transforming the perception of Sanmenxia from a "dam city" to a "Swan City."

CONCLUSION

The intricate relationships between nature and society can be better understood through a careful examination of the ways in which access to and control over resources are shaped by social and political forces.[59] In the 1950s and 1960s, the socialist state forcibly relocated the residents of the reservoir area, depriving them of access to their historical farmlands. Subsequent modifications to the dam, however, allowed many of them to regain access to those lands, at least seasonally, despite some adverse environmental impacts.

For a long time, the primary focus of the Reservoir Management Bureau and local government was the alleviation of poverty among the reservoir migrants. They provided support and incentives for the development of aquaculture, agriculture, and other income-generating activities. But the sudden surge in the swan population altered the priorities of the local government. With the establishment of the Yellow River Wetland Preserve, many alluvial lands, previously open to local farmers, were converted into restricted zones. While the economic struggles of reservoir migrants are rarely highlighted in official narratives of development, they are deeply ingrained in the broader political-economic framework. This also reveals the resilience of these communities, who have persisted despite these struggles. Unlike the imposing dam and the graceful swans, the economic hardships faced by reservoir migrants remain largely unseen.

The creation of a nature preserve and the implementation of wetland protection laws have outlawed hunting and other economic activities that had long been established parts of local life. The state's confiscation of farmland along the Yellow River in the 1950s was, from the farmers' perspective, inadequately compensated. In the decades following the completion of the dam, widespread poverty and hardship characterized

---

[59] Melinda Herrold-Menzies, "Peasant Resistance against Nature Reserves," in *Reclaiming Chinese Society: The New Social Activism*, ed. You-tien Hsing and Ching Kwan Lee (New York: Routledge, 2009), 85.

these communities, making access to their former farmland during periods of low water levels an important means of sustaining themselves and their families.

For people displaced by the dam, the prohibition of economically productive activities in the newly created "natural" wetland reserves is seen as unjust. The official narrative of preservation often fails to acknowledge the human factor and the specific historical circumstances that gave rise to the wetlands near Sanmenxia. These residents feel that their perspectives and experiences have been disregarded and marginalized in the broader discourse around the nature reserve. The relationships between nature, society, and the political forces that shape them are complex and multifaceted. The struggles of local farmers and formerly displaced residents testify to the ways in which the interactions of these factors can have profound and lasting impacts on communities and the environment.

The story of the Sanmenxia wetland illustrates the complex interplays between social and environmental factors in the development of a hydropower nation. Despite the pressing need for energy in response to the demands of a modern society and the challenges posed by climate change, the construction of large dams often has profoundly negative consequences for the environment and local communities. In official discourse, dams are portrayed as "natural" solutions to the problem of energy production, but this characterization belies the impact they have on surrounding ecosystems. The Chinese state's pursuit of an "ecological civilization" and a growing environmental consciousness among the public have challenged the traditional productivist and declensionist narratives associated with the hydropower nation.

The broader global critique of large concrete dams in terms of their ecological impact on rivers is well-known. The extinction of the Yangtze paddlefish caused by the construction of the Gezhouba Dam is an especially clear example of this.[60] The extinction of Yangtze dolphins should have been a wake-up call to the potentially devastating effects of human activities on the environment. The Sanmenxia project, however, presents a perhaps surprising outcome. The massive hydro-system has become an ideal winter habitat for thousands of whooper swans and other migratory birds from the north. This has been an ecological windfall for the local government, which has successfully rebranded the city from "the Pearl on

---

[60] Samuel Turvey, *Witness to Extinction: How We Failed to Save the Yangtze River Dolphins* (Oxford: Oxford University Press, 2008), 202.

the Yellow River" to "Swan City," reflecting its new emphasis on its harmonious relationship with nature. As the urban middle class grows and becomes increasingly environmentally conscious, more and more people visit the wetland preserve for leisure. Despite the government's significant investment in maintaining the wetland, though, the legacy of the displacement and struggles of reservoir migrants from decades ago is still present.

The visibility of the swans and the related invisibility of the reservoir migrants reveals a problematic political ecology that needs to be addressed. While the beauty of the whooper swans is certainly worthy of admiration, it is also important to recognize the environmental degradation that has taken place elsewhere. As a historian, I believe that acknowledging the history of displacement by the reservoir and the sacrifices of local peasants is a crucial step toward creating a more just and sustainable future. As visitors and local decisionmakers stroll through the wetland park, I hope they will be mindful of the complex dynamics and tensions behind this human-made nature preserve.

Epilogue

In 1992, amid intense controversy among both hydraulic engineers and the general public, the proposal for the Three Gorges project was submitted to the National People's Congress for a vote. Li Peng, the Premier of the PRC, who had received training in hydroelectric engineering in the Soviet Union in the early 1950s, oversaw the submission. Despite the fact that proposals submitted to the Congress typically receive unanimous support, the Three Gorges project proposal was passed with vocal oppositions. Despite this, the launch of the massive project on the Yangtze River was seen as a triumph for individuals like Li Peng who were proponents of a technology-centered approach to development. It was broadly perceived as a statement of the Communist Party's firm control over the Chinese people as well as the country's land and waterways. Viewing through the lens of environmental history, water control is a crucial aspect of human societies' relationships with the natural world. In the context of China's long history of water management, the Three Gorges project elevated the country's status as a major player in hydropower. From its inception, though, the project has been plagued with social conflicts and environmental difficulties. The full story of the interplay between this massive dam, the Yangtze River, and the Chinese people is yet to be told.

In other regions of the world, the trend toward constructing large dams appears to be declining, despite the fact that many existing ones continue to function. The United States was once a prime mover in the realm of large hydroelectric projects. From the 1930s, the Tennessee Valley Authority and the Hoover Dam served as symbols of a new epoch in which human beings used rivers for economic expansion, and these structures became sources of inspiration for other nations. However, with the emergence of environmental consciousness in the 1960s and a lack of suitable locations for building new large dams in the country, the construction of new large-scale hydropower projects came to a halt in the 1970s. In 1994, the commissioner of the US Bureau of Reclamation stated confidently that "the dam building era in the United States is now over."[1] The focus of energy production in the United States and other industrialized nations in the global North has shifted away from maximizing electrical output toward prioritizing the conservation of biodiversity conservation and preventing the displacement of communities. Increasing recognition of the reality of climate change and its impacts on both the environment and humanity has brought the necessity for more sustainable and equitable practices in energy development and use into focus.

In stark contrast, many countries in the global South, including China, persist in building large dams both within and beyond their borders. This pursuit of energy has elicited criticism both domestically and internationally, yet proponents of such projects argue that they are necessary for the greater good of the majority. Such opposition has been met with skepticism by those who view it as a form of "full stomach environmentalism." The question is asked: How can we expect people to prioritize conservation when millions still struggle to meet their most basic needs, such as food security?[2] One cannot deny that hydropower projects, especially those that are on a smaller scale and that are considered less environmentally harmful, can provide electricity to remote communities and have a meaningful impact in relieving poverty. Unfortunately, the construction of large dams often brings negative consequences like community displacement, loss of biodiversity and cultural heritage, the degradation of water quality, increased seismic activity, and a multitude of other unanticipated risks. Despite these challenges, the official narrative in China downplays or ignores these problems. Non-government organizations have led protests against the construction of large dams, but public perception of these projects remains deeply influenced by their perceived economic benefits, rather than by environmental considerations.[3] In 2000, Pan Jiazheng, a hydraulic engineer who was involved in designing the Three Gorges and several other large dams in China,

---

[1] Tilt, *Dams and Development in China*, 203.
[2] Bryan Tilt, *The Struggle for Sustainability in Rural China: Environmental Values and Civil Society* (New York: Columbia University Press, 2010), 6.
[3] Mertha, *China's Water Warriors*.

responded to criticism from domestic and foreign environmentalists with a resolute statement:

The Chinese people will build more large dams, including record-breaking high dams. We will tackle the threat of floods and droughts and harness the abundant hydropower reserves gifted to us by nature. Essentially, we will have control over every drop of water for our use.[4]

The unapologetic tone of this declaration reflects the determination of the Chinese state to pursue its goals in the face of opposition. Here, technological hubris defends itself under the shield of the people without acknowledging the human cost of its own actions. In the twenty-first century, China has embraced hydropower as a key source of energy, driven by the expanding technostructure, an almost exclusively utilitarian view of rivers, and rising energy demand. Despite the human and environmental costs associated with hydropower projects, the concept of the hydropower nation will likely remain important in China for the foreseeable future. As dam sites become depleted and as pressure from environmental movements grows, however, the growth of hydropower may slow down and policies may shift toward more sustainable forms of renewable energy. Still, existing hydropower projects will continue to play a significant role in China's energy mix and the legacy of the hydropower nation will endure.

As global demand for energy continues to rise, China has emerged as a leading player in the deployment of hydropower worldwide. China's investments in infrastructure projects like hydropower facilities in Africa have attracted international attention. Despite the persistent and well-documented issues surrounding large dams, hydropower is often presented as a renewable source of energy with a relatively low carbon footprint. This has led to a proliferation of large dams across the global South, as proponents of development seek to address the pressing issue of climate change caused by the consumption of fossil fuels.

Small hydropower projects are recognized everywhere as being more environmentally friendly, and China is well-known for building many of these projects. It is worth noting, though, that a staggering 80 percent of the hydropower facilities built in Africa and Asia by Chinese investors and companies are large dams.[5]

---

[4] Pan Jiazheng 潘家铮, *Qianqiu gongzui huashuiba* 千秋功罪话水坝 (On Hydropower Dams) (Beijing: Qinghua daxue chubanshe, 2000), 202.

[5] Giuseppina Siciliano and Frauke Urban (eds.), *Chinese Hydropower Development in Africa and Asia: Challenges and Opportunities for Sustainable Global Dam-building* (New York: Routledge, 2017), 5.

It is clear that large-scale hydropower projects are the preferred choice of policymakers and companies. Despite concerns over the potentially devastating impact of large dams on rivers, as well as criticism of them as a form of "brute force technology," their economic productivity and the profits that can be derived from them appear to take precedence.[6] To the most powerful stakeholders, the preservation of biodiversity and the wellbeing of marginalized communities are of secondary concern when compared to the production of millions of watts of electricity and the potential for economic growth that entails. Collaboration between state policymakers, hydropower companies, and investors will continue to transform rivers and communities in the global South, with both positive and negative consequences.

The concept of the hydropower nation is rooted in the political changes and energy demand in China. If it were to be applied to other countries, the definition would necessarily vary according to the unique political, social, and geographical factors present in those places. Despite this, the desire for increased energy production and the utilitarian view of rivers remains prevalent in many places.

Can the history of hydropower, then, shed light on the challenge of sustainability? In his reflections on the Three Gorges project, Donald Worster argues that there is nothing new about China's efforts in hydropower construction. "It emerges out of two thousand years of Chinese history."[7] The use of hydropower certainly does date back to the earliest stages of Chinese history, but the technologies of turbines, generators, and concrete dams were not developed within the country. In the twentieth century, the United States and the Soviet Union were seen as leaders in the field of hydropower and their influence can be seen in large-scale projects across the globe. Moreover, according to Bryan Tilt, in a longer historical perspective – particularly since the Industrial Revolution – "this technocratic drive to harness the power of nature in the service of human needs constitutes more of a continuity with the past than a radical break from it."[8]

The core of the issue lies in a pattern of intensive energy consumption that has both shaped and been perpetuated by the modern world. Despite their benefits, large dams and the people who support their construction

---

[6] Josephson, *Industrialized Nature*.
[7] Donald Worster, "The Flow of Empire: Comparing Water Control in the United States and China," *RCC Perspectives* 2011, no. 5 (2011), 1–23, 5.
[8] Tilt, *Dams and Development in China*, 46.

have rightfully faced criticism for their adverse effects, such as the displacement of communities, loss of biodiversity, and environmental degradation. Over a century ago, Winston Churchill famously claimed that any drop of water from the Nile that flowed freely into the Mediterranean was wasted. Unfortunately, in the present, many countries, including those in the developed global North, still cannot afford to let their rivers run freely to the sea. In the American West, rivers are controlled in order to sustain a high standard of living, while, in Africa and other regions, they are harnessed to facilitate the development of production beyond basic subsistence needs. These projects often result in environmental injustices and regional imbalances that disproportionately affect local communities.

Policymakers with political influence, technicians with scientific knowledge, and investors with financial resources have combined to integrate the natural elements of flora and fauna, landscape, hydrosphere, geosphere, and atmosphere into an anthropocentric system: the hydropower nation. Despite the broad recognition of small hydropower as a more sustainable and renewable source of energy, the history of small hydropower in China since the 1950s highlights the challenges and obstacles that can arise in its pursuit of energy production.

It is important to recognize the limitations of our human-constructed water regime. And it is as important to realize that it is influenced by both environmental and technological systems. The notion of the Anthropocene – a new geological era defined by human activity – suggests that the natural world is subservient to human desires. However, this does not mean that the natural world is, or ever will be, fully subordinated to human wants and needs.[9]

The examination of China's small and large-scale hydropower initiatives in the twentieth century that I have undertaken in this book highlights the interplay between hydraulic technology and the complex natural and artificial elements of rivers. Failure to consider this interplay can result in catastrophic consequences for communities. It can also mean that the expected increased productivity fails to materialize. Rivers have a far longer history than human civilization, and their fluctuating waters have formed and sustained diverse riparian ecosystems. These ecosystems have become some of the most fertile habitats for human habitation, shaped by their floodplains and deltas. The hydropower nation, despite

---

[9] Jeremy Davis, *The Birth of the Anthropocene* (Berkeley: University of California Press, 2016), 7.

its immense size and technological sophistication, is still at the mercy of larger ecological processes and forces.

The hydropower nation is a product of human ingenuity and ambition, but it must also be guided by a sense of environmental stewardship and social responsibility. The long-term viability of concrete dams, the cornerstone of large hydropower projects, remains uncertain despite them having been around for less than a century. With the risk of dams collapsing, the topic of decommissioning has become an important one in North America, while in China the problem of silting has prompted policymakers to consider decommissioning the Sanmenxia Dam. However, simply "blowing it up" will probably not solve all the problems, and in any case raises further questions about the intrinsic value of nature, technologies, and infrastructure. The scale of dam failures is also a cause for concern. In its first forty years, the People's Republic of China experienced an average of 110 dam collapses per year, with a high point of 554 collapses in 1973.[10] In 1975, a combination of heavy typhoon rainfall and poor reservoir design and management led to the collapse of sixty-two dams on the upper stream of the Huai River in Henan, resulting in the loss of between 100,000 and 240,000 lives.[11] What is more, according to a recent Bloomberg report, nearly 40,000 hydropower plants in China are at risk of being shut down because of low productivity and the drying up of rivers.[12]

For its part, the Chinese government has shifted its focus from its earlier ideology of "conquering nature" toward recognizing the ecological significance of rivers. Despite this shift, though, internal pressure to achieve the goal of carbon-neutrality by 2060 has resulted in a continued reliance on hydropower projects, specifically large dams, as essential to the state's objectives.

The Sanmenxia hydropower project proved to be an unexpected boon for wildlife when it created a thriving wetland along its reservoir. It has

---

[10] Chetham, *Before the Deluge*, 186.
[11] Yi Si, "The World's Most Catastrophic Dam Failures: The August 1975 Collapse of the Banqiao and Shimantan Dams," in *The River Dragon Has Come: Three Gorges Dam and the Fate of the Yangtze River and Its People*, ed. Dai Qing (New York: M. E. Sharpe, 1998), 25–38; Henansheng shuiliting 河南省水利厅 (ed.), *Henan 75.8 teda hongshui zaihai* 河南75.8 特大洪水灾害 (The Henan 75.8 Catastrophic Flood Disaster) (Zhengzhou: Huanghe shuili chubanshe, 2005).
[12] Bloomberg News, "China Has Thousands of Hydropower Projects It Doesn't Want," available at www.bloomberg.com/news/features/2021-08-14/china-wants-to-shut-down-thousands-of-dams, last accessed March 13, 2024.

become a destination for migratory birds from Mongolia and Siberia, who flock to the area for its abundant food and shelter. The dam operates by holding water back during the non-flooding season (usually winter and early spring) and releasing it through its sluice gates during the flooding season. This elevates the water level in the adjacent valley, creating a seasonal wetland ecosystem that has flourished over the years. The surprising emergence of this ecosystem again highlights the complex interplay between human infrastructure and the natural world. The arrival of tens of thousands of migratory birds, particularly whooper swans, has led local governments to advocate a narrative of harmony between humans and nature. The construction of dams has produced vastly different outcomes for different species. While it has had catastrophic effects on fish populations, it has provided new habitats for whooper swans and other waterfowl.

The displacement and impoverishment of reservoir resettlers in the process of building China's dams created "unimagined communities," while the electricity thus generated was channeled to cities and factories in support of nation-building and industrialization efforts. In stark contrast, many reservoirs, such as the Thousand Islands Lake on the upper stream of the Xin'anjiang hydropower project, have become popular destinations for ecological tourism and recreation for China's growing urban middle class. Concrete dams have become an integral part of the lived experience of many species, offering tragic consequences for some, and entertainment for others.

When I visited Sanmenxia in January 2015, I was taken aback to find that the wetland park near the city was off-limits to visitors. The swans floating on the water were cordoned off and closely guarded. A watchful passerby noted my confusion and informed me that H_5N_1 influenza had been discovered among the swans in the area, but that thankfully, no human infections had been reported. Fast forward to late 2019 and the world was grappling with a new virus, COVID-19, which has upended daily life around the world. The possibility that the virus had been transmitted from pangolins to humans in a wet market in Wuhan led to a ban on consuming terrestrial wildlife in China.[13] Though I have not explored this issue in this book, it is important to note that ecological systems transformed by human activity, such as those created by reservoirs, have been linked in various parts of the world to the spread of

---

[13] Lingyun Xiao, Zhi Lu, Xueyang Li, Xiang Zhao, and Binbin V. Li, "Why do We Need a Wildlife Consumption Ban in China?" *Current Biology* 31 (2021), R161–R185.

diseases like malaria.[14] Lockdown, isolation, and outbreaks of infectious diseases are unfamiliar experiences to much of the human population. According to David Morens and Anthony Fauci:

> As human societies grow in size and complexity, we create an endless variety of opportunities for genetically unstable infectious agents to emerge into the unfilled ecological niches we continue to create. There is nothing new about this situation, except that we now live in a human-dominated world in which our increasingly extreme alterations of the environment induce increasingly extreme backlashes from nature.[15]

In the grand scheme of things, dams may simply represent another chapter in the history of humanity's relationship with nature, but it is an important chapter. While rivers may no longer flow uninterrupted and free in a world dominated by human influence, they will not be silenced entirely.

---

[14] United States Public Health Service and Tennessee Valley Authority Health and Safety Department, *Malaria Control on Impounded Water* (Washington DC: US Government Printing Office, 1947); McCully, *Silenced Rivers*.

[15] David M. Morens and Anthony S. Fauci, "Emerging Pandemic Diseases: How We Got to COVID-19," *Cell* 182, no. 5 (2020), 1077–1092.

Bibliography

ARCHIVES AND SPECIAL COLLECTIONS

Chongqing Municipal Archives, Chongqing.
Dunhuang Municipal Archives, Dunhuang, Gansu.
Gansu Provincial Archives, Lanzhou, Gansu.
Henan Provincial Archives, Zhengzhou, Henan.
Institute of Modern History Archives, Academia Sinica, Taipei.
Lingbao County Archives, Lingbao, Henan.
The National Archives at Atlanta, GA.
Percy Othus Collection on Chinese Hydrology, Special Collection Library, The Pennsylvania State University, College Park, PA.
Sanmenxia Municipal Archives, Sanmenxia, Henan.
The Second Historical Archives of China, Nanjing.
Wilson Center Digital Archives, Washington, DC.
The Yellow River Conservancy Commission Archives, Zhengzhou, Henan.

SECONDARY SOURCES

Anderson, Benedict. *Imagined Communities: Reflection on the Origin and Spread of Nationalism* (rev. ed.). London: Verso, 2016.
Andreas, Joel. *Rise of the Red Engineers: The Cultural Revolution and the Origins of China's New Class*. Stanford: Stanford University Press, 2009.
Ash, Robert. "Squeezing the Peasants: Grain Extraction, Food Consumption and Rural Living Standards in Mao's China." *China Quarterly* 188 (2006), 959–998.
Ba, Mu 巴牧. *Sanmenxia de chuanshuo* 三门峡的传说 (*Folktales of Sanmenxia*). Beijing: Zuojia chubanshe, 1958.
Bailes, Kendall. *Technology and Society under Lenin and Stalin*. Princeton: Princeton University Press, 1978.

"Baitian'e daidong Sanmenxia dongyure" 白天鹅带动三门峡"冬游热 (Swans Bring Winter Tourist Wave to Sanmenxia). *Sanmenxia Huanghe shidi guojiaji ziran baohuqu guanwang*, March 10, 2014.

"Baitian'e qianghua shengtaiguan" 白天鹅强化生态观 (Swans Strengthen Ecological Perspective). *Sanmenxia Huanghe shidi guojiaji ziran baohuqu guanwang*, March 14, 2014.

Bao, Heping 包和平. *Gongchengde shehuiyanjiu: Sanmenxia gongchengzhongde zhenglunyu jiejue* 工程的社会研究：三门峡工程中的争论与解决 (*Social Study of Engineering: Debates and Solution of the Sanmenxia Project*). Hohhot: Neimenggu jiaoyu chubanshe, 2007.

"Baohu baitian'e cong qingli gaogao zuowu kaishi" 保护白天鹅 从清理高杆作物开始 (Swan Protection Starts with Clearing Tall Stalk Crops). *Sanmenxia Huanghe shidi guojiaji ziran baohuqu guanwang*, October 12, 2013.

"Baohuqunei bushefang, Tongguan you jiudian jingchi tian'e rou" 保护区内不设防，潼关有酒店竟吃天鹅肉 (Defenseless in the Conservation Area, a Hotel in Tongguan Sells Swan Meat), *Huashangbao* 华商报, May 10, 2002.

Barnes, Nicole. *Intimate Communities: Wartime Healthcare and the Birth of Modern China, 1937–1945*. Berkeley: University of California Press, 2018.

Bennett, Gordon. *Yundong: Mass Campaigns in Chinese Communist Leadership*. Berkeley: University of California Press, 1976.

Bernstein, Thomas and Hua-Yu Li (eds.). *China Learns from the Soviet Union, 1949–Present*. Lanham, MD: Lexington Books, 2010.

Bian, Morris L. *The Making of the State Enterprise System in Modern China: The Dynamics of Institutional Change*. Cambridge, MA: Harvard University Press, 2005.

Blackbourn, David. *The Conquest of Nature: Water, Landscape, and the Making of Modern Germany*. New York: W. W. Norton & Company, 2006.

Bloch, Marc, translated by J. E. Anderson. *Land and Work in Medieval Europe*. Berkeley: University of California Press, 1967.

Bray, Francesca. *Technology and Gender: Fabrics of Power in Late Imperial China*. Berkeley: University of California Press, 1997.

"Science, Technique, Technology: Passages between Matter and Knowledge in Imperial Chinese Agriculture." *British Journal for the History of Science* 41, no. 3 (2008), 319–344.

Brazelton, Mary Augusta. *Mass Vaccination: Citizen's Bodies and State Power in Modern China*. Ithaca: Cornell University Press, 2019.

Brown, Jeremy. *City Versus Countryside in Mao's China: Negotiating the Divide*. Cambridge: Cambridge University Press, 2012.

Brown, Jeremy and Matthew Johnson (eds.). *Maoism at The Grassroots: Everyday Life in China's Era of High Socialism*. Cambridge, MA: Harvard University Press, 2015.

Brown, Jeremy and Paul Pickowicz (eds.). *Dilemmas of Victory: The Early Years of the People's Republic of China*. Cambridge, MA: Harvard University Press, 2010.

Buck, John Lossing. *Land Utilization in China*. New York: Paragon Book Reprint Corp, 1964.

Buoye, Thomas M. *Manslaughter, Markets, and Moral Economy: Violent Disputes over Property Rights in Eighteenth-Century China*. Cambridge: Cambridge University Press, 2000.
Burt, Sally. *At the President's Pleasure: FDR's Leadership of Wartime Sino-US Relations*. Leiden: Brill, 2015.
Cao, Shuji. "An Overt Conspiracy: Creating Rightists in Rural Henan, 1957-1958," in Jeremy Brown and Matthew Johnson (eds.), *Maoism at the Grassroots: Everyday Life in China's Era of High Socialism*. Cambridge, MA: Harvard University Press, 2015.
Carin, Robert. "Rural Electrification," in E. Stuart Kirby (ed.), *Contemporary China, VI 1962-1964*. Hong Kong: Hong Kong University Press, 1968.
Cen, Zhongmian 岑仲勉. *Huanghe bianqianshi* 黄河变迁史 (*History of the Yellow River*). Beijing: Renmin chubanshe, 1957.
Chang, Jue 常珏. *Shuili Fadian* 水力发电 (*Hydropower*). Beijing: Tongshuduwu chubanshe, 1954.
Chao, Xiaohong 钞晓鸿. "Guangai, huanjing yu shuili gongtongti – jiyu Qingdai Guanzhong zhongbu de fenxi" 灌溉, 环境与水利共同体-基于清代关中中部的分析 (Irrigation, Environment and Hydraulic Community: An Analysis based on the Central Part of the Guanzhong Plain). *Zhongguo Shehui Kexue* 中国社会科学 4 (2006), 190-204.
Chen, Jian. *Mao's China and the Cold War*. Chapel Hill: The University of North Carolina Press, 2001.
Chen, Shaoming 陈少明. "Zuowei wentide Zhongguo zhishifenzi" 作为问题的中国知识分子 (Chinese Intellectuals as a Problem). *Kaifang shidai* 开放时代 5 (2013), 6-26.
Chen, Zhang 陈章. "Duiyu shuili fadian yingyoude renshi" 对于水力发电应有的认识 (What We Should Know about Hydroelectric Power). *Xinminzu* 新民族 3, no. 14 (1939), 4-7.
Chen, Zudong 陈祖东. "Su'e jianshe zuida shuilifadianchang" 苏俄建设最大水力发电厂 (Soviet Union Builds the Largest Hydroelectricity Plant), *Su'e Pinglun* 苏俄评论 5 (1932), 616.
  "Cong dianlishuili shuodao Sulianjianguo yu Zhongguojianguo" 从电力水力说到苏联建国与中国建国 (From Electricity, Hydropower to the Construction of Soviet Union and China). *Xinjingji* 新经济 2, no. 4 (1939), 88.
Cheng, Li and Lynn White. "Elite Transformation and Modern Change in Mainland China and Taiwan: Empirical Data and the Theory of Technocracy." *China Quarterly* no. 121 (1990), 1-35.
Cheng, Tiejun and Mark Selden. "The Origins and Social Consequences of China's Hukou System." *The China Quarterly* 139 (1994), 644-668.
Cheng, Yufeng 程玉凤 and Cheng Yuhuang 程玉凰 (eds.). *Ziyuan weiyuanhui jishurenyuan fumei shishi shiliao, 1942* 资源委员会技术人员赴美实习史料, 1942 (*Archives on the National Resources Commission Technicians' Training in the United States, 1942*). Taipei: Academia Historica, 1988.
Chetham, Deirdre. *Before the Deluge: The Vanishing World of the Yangtze's Three Gorges*. New York: Palgrave Macmillan, 2002.

Chi, Ch'ao-ting. *Key Economic Areas in Chinese History: As Revealed in the Development of Public Works for Water Control*. London: George Allen & Unwin Ltd., 1936.

Chongqingshi nongji shuidianju 重庆市农机水电局. *Chongqingshi shuilizhi* 重庆市水利志 (*Water Conservancy Gazetteer of Chongqing*). Chongqing: Chongqing chubanshe, 1996.

Cochran, Sherman. *Encountering Chinese Networks: Western, Japanese, and Chinese Corporations in China, 1880–1937*. Berkeley: University of California Press, 2000.

Coopersmith, Jonathan. *The Electrification of Russia, 1880–1926*. Ithaca: Cornell University Press, 1992.

Courtney, Chris. *The Nature of Disaster in China: The 1931 Yangzi River Flood*. Cambridge: Cambridge University Press, 2018.

Croll, Elisabeth J. "Involuntary Resettlement in Rural China: The Local View." *China Quarterly* 158 (1999), 468–483.

Crook, Isabel B., Christina K. Gilmartin, Xiji Yu, Gail Hershatter, and Emily Honig. *Prosperity's Predicament: Identity, Reform, and Resistance in Rural Wartime China*. New York: Rowman & Littlefield, 2013.

Crosby, Alfred W. *Children of the Sun: A History of Humanity's Unappeasable Appetite for Energy*. New York: W. W. Norton & Company, 2006.

Dagao qunzhong yundong quanmin ban shuidian 大搞群众运动全民办水电 (*The Mass Campaign of Hydropower Construction*). Beijing: Shuili shuidian chubanshe, 1959.

Dai, Qing (ed.). *The River Dragon Has Come! Three Gorges Dam and the Fate of China's Yangtze River and Its People*. New York: Routledge, 1998.

Davis, Jeremy. *The Birth of the Anthropocene*. Berkeley: University of California Press, 2016.

Des Forges, Roger V. *Cultural Centrality and Political Changes in Chinese History: Northeast Henan in the Fall of Ming*. Stanford: Stanford University Press, 2003.

Dharmadhikary, Shripad. *Unraveling Bhakra: Assessing the Temple of Resurgent India: Report of a Study of the Bhakra Nangal Project* (Badwani: Manthan Adhyayan Kendra, 2005).

Dikötter, Frank. *Exotic Commodities: Modern Objects and Everyday Life in China*. New York: Columbia University Press, 2006.

Mao's Great Famine: The History of China's Most Devastating Catastrophe, 1958–1962. New York: Walker & Co., 2010.

Dinmore, Eric. "Concrete Result? The TVA and the Appeal of Large Dams in Occupation-Era Japan." *The Journal of Japanese Studies* 39, no. 1 (2013), 1–38.

Dodgen, Randall A. *Controlling the Dragon: Confucian Engineers and the Yellow River in Late Imperial China*. Honolulu: University of Hawai'i Press, 2001.

Domes, Jurgen. *Socialism in the Chinese Countryside: Rural Societal Policies in the People's Republic of China, 1949–1979*, translated by Margitta Wending. London: C. Hurst & Co. Ltd., 1980.

"Dongyang diyi fadian gongshe (shuli fadian)" 东洋第一发电工事 (水力发电) (The Largest Hydroelectricity Plant in East Asia). *Diqigongye Zhazhi* 电气工业杂志 2, no. 2 (1924), 74.

D'Souza, Rohan. "Damming the Mahanadi River: The Emergence of Multi-Purpose River Valley Development in India (1943–46)." *Indian Economic Social History Review* 40, no. 1 (2003), 81–105.

Duara, Prasenjit. *Culture, Power, and the State: Rural North China, 1900–1940*. Stanford: Stanford University Press, 1991.

Rescuing History from the Nation: Questioning Narratives of Modern China. Chicago: University of Chicago Press, 1997.

Editorial, "Dajia lai zhiyuan Sanmenxia a!" (Everybody Come to Support Sanmenxia!). *Renmin ribao*, April 14, 1957.

Eisenman, Joshua. *Red China's Green Revolution: Technological Innovation, Institutional Change, and Economic Development under the Commune*. New York: Columbia University Press, 2018.

Ekbladh, David. "Meeting the Challenge from Totalitarianism: The Tennessee Valley Authority as a Global Model for Liberal Development, 1933–1945." *The International History Review* 32, no. 1 (2010), 47–67.

Elman, Benjamin A. *On Their Own Terms: Science in China, 1550–1900*. Cambridge, MA: Harvard University Press, 2005.

Elvin, Mark. "Who Was Responsible for the Weather? Moral Meteorology in Late Imperial China." *Osiris* 13 (1998), 213–237.

The Retreat of the Elephant: An Environmental History of China. New Haven: Yale University Press, 2004.

Entenmann, Robert. "Migration and Settlement in Sichuan 1644–1796." PhD Dissertation, Harvard University, 1982.

Espy, Willard R. "Dams for the Floods of War." *New York Times Magazine* 27 (1946), 12–13.

Evenden, Matthew. *Allied Power: Mobilizing Hydro-Electricity during Canada's Second World War*. Toronto: University of Toronto Press, 2015.

Fan, Rongkang 范荣康. "Guanche Sulianzhuanjia Jianyide Fanli" 贯彻苏联专家建议的范例 (An Example of Following Soviet Experts' Advice). *Renmin ribao*, January 23, 1954.

Fang, Wanpeng 方万鹏. "Xiangdi zuomo: Mingqing yilai Hebei Jingjing de shuilijiagongye – jiyu huangjingshi shijiao de kaocha" 相地作磨：明清以来河北井陉的水力加工业 – 基于环境史视角的考察 (Install Watermills According to the Terrain: The Hydraulic Machining in Jingxing of Hebei Province since the Ming and Qing Dynasties: Based on the Perspective of Environmental History Studies). *Zhongguo Nongshi* 中国农史 (*Chinese Agricultural History*) 3 (2014), 51–58.

Feng, Yan 冯艳 (dir.). Bing'Ai 秉爱. Beijing: Beisen Films, 2007.

Feuerwerker, Albert. "Industrial Enterprise in Twentieth-Century China: The Chee Hsin Cement Co," in Albert Feuerwerker, Rhoads Murphey, and Mary C. Wright (eds.), *Approaches to Modern Chinese History*. Berkeley: University of California Press, 1967, 304–342.

Friedman, Edward, Paul Pickowicz, and Mark Selden. *Chinese Village, Socialist State*. New Haven, CT: Yale University Press, 1991.

Fu, Zuoyi. "Sulian dui Zhongguo shuili shiye de bangzhu" (Soviet Assistance to Chinese Hydraulic Enterprise). *Renmin ribao*, October 22, 1957.

Fujiansheng difangzhi bianzuan weiyuanhui 福建省地方志编纂委员会. *Fujian shengzhi: shuilizhi* 福建省志:水利志 (*Fujian Provincial Gazetteer: Water Conservancy*). Beijing: Zhongguo shehui kexue chubanshe, 1999.

Gao, Jun 高峻. "Ershi shiji wushi niandai Fujian xiaoshuidian jianshe de xingqi" 二十世纪五十年代福建小水电建设的兴起 (The Rise of Small Hydropower in Fujian in the 1950s). *Dangshi Yanjiu yu Jiaoxue* 党史研究与教学 6 (2009), 67–75.

Gao, Mobo. *Gao Village: Rural Life in Modern China*. Honolulu: University of Hawai'i Press, 2007.

Gao, Wangling 高王凌. *Zhongguo Nongmin Fanxingwei Yanjiu (1950–1980)* 中国农民反行为研究 (1950–1980) (*A Study of Counteractions of Chinese Peasants (1950–1980)*). Hong Kong: The Chinese University Press, 2013.

Gao, Yan. *Yangzi Waters: Transforming the Water Regime of the Jianghan Plain in Late Imperial China*. Leiden: Brill, 2022.

"Gaosuduo xiang dianqihua maijin, Jin Xuecheng weiyuan de fayan" 高速度向电气化进军, 金学成委员的发言 (Marching toward Electrification with High Speed: Speech by Jin Xuecheng). *Renmin ribao*, April 10, 1960.

Garnaut, Anthony. "The Geography of the Great Leap Famine." *Modern China* 40, no. 3 (2014), 315–348.

Gerth, Karl. *China Made: Consumer Culture and the Creation of the Nation*. Cambridge, MA: Harvard University Asia Center, 2003.

"Geshuini gangcaihe mucai deming" 革水泥, 钢材和木材的命 (Revolutionize Cement, Steel, and Timber). *Shuili fadian* 水力发电 18 (1958), 3–5.

Gezhi xinbao (*Scientific Review*) 格致新报 (1898) 近代中国史料丛刊三编 24, Taibei: Wenhai chubanshe, 1987.

Ghosh, Arunabh. *Making It Count: Statistics and Statecraft in the Early People's Republic of China*. Princeton: Princeton University Press, 2020.

"Multiple Makings at China's First Hydroelectric Power Station at Shilongba, 1908–1912." *History and Technology* 38 (2022), 167–185.

Gimpel, Jean. *The Medieval Machine: The Industrial Revolution of the Middle Ages*. London: Penguin Books, 1976.

Graham, Loren R. *The Ghost of the Executed Engineer: Technology and the Fall of the Soviet Union*. Cambridge, MA: Harvard University Press, 1993.

Greene, Megan J. *The Origin of the Developmental State in Taiwan: Science Policy and the Quest for Modernization*. Cambridge, MA: Harvard University Press, 2008.

Gu, Yuxiu 顾毓秀. "Dianqi yu jianshe" (Electricity and Construction) 电气与建设, *Xinmin* 新民 8 (1931), 10.

"Guangfan yongdianli daiti laoli, quanguo jiangxianqi nongcun dianqihua gaochao" 广泛用电力代替劳力, 全国将掀起农村电气化高潮 (Widely Using Electrical Power to Replace Labor Force, a Nationwide High Tide of Rural Electrification Is Coming). *Renmin ribao*, August 18, 1958.

Guanggongsheng shuilizhi 广东省水利志 (*Water Conservancy Gazetteer of Guangdong Province*). Guangzhou: Guangdongsheng shuili dianliting, 1994.

Guo, Qiaoyu 郭乔羽 and Zhifeng Yang 杨志峰. "Sanmenxia shuili shuniu gongcheng shengtai yingxianghou pingjia" 三门峡水利枢纽工程生态影响后评价

(Assessment of the Ecological Impact of the Sanmenxia Hydropower Project). *Huanjing kexue xuebao* 环境科学学报 25, no. 5 (2005), 580–585.

Guo, Yi 郭仪. "Keke mingzhu fangguangcai: caise kejiaopian 'Shancun xiaoshuidian' guanhou" 颗颗明珠放光彩—彩色科教片 '山村小水电' 观后 (Every Pearl Shines: Review of the Color Scientific Education Film "Small Hydroelectricity in the Mountainous Countryside"). *Renmin ribao*, May 9, 1975.

Hao, Ping. "A Study of the Construction of Terraced Fields in Liulin County, Shaanxi Province in the Era of Collectivization," in Thomas DuBois and Huaiyin Li (eds.), *Agricultural Reform and Rural Transformation in China Since 1949*. Leiden: Brill, 2016, 101–114.

He, Fengming 和凤鸣. *Jingli-wode 1957* 经历-我的1957 (*Experience: My 1957*). Dunhuang: Dunhuang wenyi chubanshe, 2001.

He, Jingzhi. *He Jingzhi Wenji* 贺敬之文集 (*Collection of He Jingzhi*) Vol.1. Beijing: Zuojia chubanshe, 2005.

Henan Jianghe huanjing keji youxian gongsi 河南江河环境科技有限公司. "Huanghe Tongguan zhi Sanmenxia dabaheduan 'shisanwu' zhiligongcheng huanjingyingxiang baogaoshu" 黄河潼关至三门峡大坝河段 "十三五" 治理工程环境影响报告书 (Environmental Assessment Report of the "Thirteenth Five-Year Plan" Management Project of the Section of the Yellow River from Tongguan to Sanmenxia Dam), 2018.

Henansheng shuiliting 河南省水利厅 (ed.). *Henan 75.8 teda hongshui zaihai* 河南 75.8 特大洪水灾害 (*Henan 75.8 Catastrophic Flood Disaster*). Zhengzhou: Huanghe shuili chubanshe, 2005.

Herrold, Melinda. "The Cranes of Caohai and Other Incidents of Fieldwork in Southwestern China." *Geographical Review* 89, no. 3 (1999), 440–448.

Herrold-Menzies, Melinda. "Peasant Resistance against Nature Reserves," in You-tien Hsing and Ching Kwan Lee (eds.), *Reclaiming Chinese Society: The New Social Activism*. New York: Routledge, 2009.

Hershatter, Gail. *The Gender of Memory: Rural Women and China's Collective Past*. Berkeley: University of California Press, 2011.

Hinton, William. *Hundred Days War: The Cultural Revolution at Tsinghua University*. New York: Monthly Review Press, 1972.

Ho, Denise Y. *Curating Revolution: Politics on Display in Mao's China*. Cambridge: Cambridge University Press, 2018.

Hodgen, Margaret T. "Domesday Water Mills." *Antiquity* 13 (1939), 261.

Hou, Li. *Building for Oil: Daqing and the Formation of the Chinese Socialist State*. Cambridge, MA: Harvard University Asia Center, 2021.

Hsing, You-tien and Ching Kwan Lee (eds.). *Reclaiming Chinese Society: The New Socialist Activism*. New York: Routledge, 2010.

Hsu, Robert C. *Food for One Billion: China's Agriculture since 1949*. Boulder: Westview Press, 1982.

Hu, Ming 胡明. "Qunzhong bandian" 群众办电 (The Masses Produce Electricity). *Renmin ribao*, November 15, 1958.

Hu, Yingze 胡英泽. *Liudong de Tudi: Ming Qing yilai Huanghe Xiaobeiganliu quyu shehui yanjiu* 流动的土地:明清以来黄河小北干流区域社会研究

(*Floating Lands: Regional Social Study of Xiaobeiganliu since the Ming and Qing Dynasties*). Beijing: Beijing daxue chubanshe, 2012.

Hua, Shan. "Sanmenxia: yige gaibian Huanghe ziran mianmao de weida jihua xuanbule!" (Sanmenxia: A Great Plan to Change the Yellow River has been Announced). *Renmin ribao*, July 21, 1955.

Huang, Ailian 黄爱莲. *Wo shi Shunma* 我是顺妈 (*I am Shun's Mom*). Shanghai: Shanghai cishuchubanshe, 2012.

Huang, Wanli. "Duiyu Huanghe Sanmenxia shuiku xianxing guihua fangfa de yijian" (Comments on the Current Principles of Operation of the Sanmenxia Reservoir on the Yellow River). *Zhongguo shuili* 8 (1957), 26–29.

"Huacong xiaoyu" 花丛小语 (Murmuring among Flowers), in *Huang Wanli wenji* 黄万里文集 (*Anthology of Huang Wanli*). Beijing: Huang Wanli wenji bianjichuban xiaozu, 2001.

Huang, Wenxi 黄文熙. "Shuili jianshe zouyi" 水力建设刍议 (Comments on Hydropower Construction). *Jingji jianshe jikan* 经济建设季刊 (*Economic Reconstruction Quarterly*) 1 (1942), 154.

Huang, Yuxian 黄育贤. "Zhanhou kaifa woguo shuili ziyuan zhi guanjian" 战后开发我国水力资源之管见. *Jingji jianshe jikan* 经济建设季刊 1, no. 4 (1943), 42–47.

Huanghe Sanmenxia Shuili shuniuzhi 黄河三门峡水利枢纽志 (*Annals of the Yellow River Sanmenxia Hydro Station*). Beijing: Zhongguo dabai kequanshu chubanshe, 1993.

"Huanghe shuili weiyuanhui 1954 nian shuitu baochi gongzuo huiyi jielun" 黄河水利委员会1954年水土保持工作会议结论 (Conclusion of the 1954 Yellow River Conservancy Commission Water and Soil Conservation Working Meeting)." *Kexue tongbao* 科学通报 3 (1955), 19–21.

Huanghe shuixi yuyeziyuan diaochaxie zuozu 黄河水系渔业资源调查协作组, *Huanghe shuixi yuye ziyuan* (*Fishery Resources of the Yellow River Basin*). Shenyang: Liaoning keji chubanshe, 1986.

Hughes, Thomas. *Networks of Power: Electrification in Western Society, 1880–1930*. Baltimore: John Hopkins University Press, 1983.

Hunter, Louis C. *A History of Industrial Power in the United States, 1780–1930, Volume 1: Waterpower in the Century of the Stream Engine*. Charlottesville: University Press of Virginia, 1979.

Hunter, Louis C. and Lynwood Bryant. *A History of Industrial Power in the United States, 1780–1930, Volume 3: The Transmission of Power*. Cambridge, MA: MIT Press, 1991.

Jansson, Roland, Christer Nilsson, and Brigitta Renofalt. "Fragmentation of Riparian Floras in Rivers with Multiple Dams." *Ecology* 81, no. 4 (2000), 899–903.

Ji, Wenxuan 季文选. "Quanguo zhiyuan Sanmenxia" 全国支援三门峡 (The Whole Country Supports Sanmenxia), in Zhongguo renmin zhengzhixieshang huiyi Sanmenxia shi weiyuanhui中国人民政治协商会议三门峡市委员会 and Zhongguo shuilishuidian dishiyi gongchengju中国水利水电第十一工程局 (eds.), *Wanli Huanghe diyiba*. Zhengzhou: Henan renmin chubanshe, 1992, 431–432.

Jia, Ru, Li, Shu-Hong, Meng, Wei-Yue, et al. "Wintering Home Range and Habitat Use of the Whooper Swans in Sanmenxia Wetland, China." *Ecological Research* 34 (2019), 637–643.

Jiang, Daquan 姜达权. "Sanmenxia baduande xuanding" 三门峡坝段的选定 (Selecting the Sanmenxia Dam Site). *Shuiwei dizhi gongcheng dizhi* 水文地质工程地质 no. 12 (1957), 1–3.

Jijifazhan nongcun xiaoxing shuidianzhan 积极发展农村小型水电站 (*Actively Develop Rural Small Hydropower*). Beijing: Renmin chubanshe, 1972.

Jing, Jun. "Rural Resettlement: Past Lessons for the Three Gorges Project." *The China Journal* 38 (1997), 65–92.

  The Temple of Memories: History, Power, and Morality in a Chinese Village. Stanford, CA: Stanford University Press, 1998.

  "Villages Dammed, Villages Repossessed: A Memorial Movement in Northwest China." *American Ethnologist* 26, no. 2 (1999), 324–343.

"Jinnian quanguo jiang xingjian yiqianduo zuo xiaoxing shuidianzhan" 今年全国将兴建一千多座小型水电站 (One Thousand Small Hydropower Stations Will Be Built This Year Nationwide). *Renmin ribao*, February 10, 1956.

"Jishu geming yidingyao fadong qunzhong" 技术革命一定要发动群众 (We Must Mobilize the Masses in Technological Revolution). *Renmin ribao*, June 24, 1968.

Jones, Christopher F. *Routes of Power: Energy and Modern America.* Cambridge, MA: Harvard University Press, 2014.

Josephson, Paul R. *Industrialized Nature: Brute Force Technology and the Transformation of the Natural World.* Washington: Island Press, 2002.

Kander, Astrid, Paolo Malanima, and Paul Warde. *Power to the People: Energy in Europe over the Last Five Centuries.* Princeton: Princeton University Press, 2013.

Kaple, Deborah A. *Dream of a Red Factory: The Legacy of High Stalinism in China.* Oxford: Oxford University Press, 1994.

Kibler, Kelly and Desiree Tullos. "Cumulative Biophysical Impact of Small and Large Hydropower Development in Nu River, China." *Water Resources Research* 49 (2013), 3104–3118.

Kikkawa, Takeo. "The History of Japan's Electric Power Industry before World War II." *Hitotsubashi Journal of Commerce and Management* 46, no. 1 (2012), 10–16.

Kinzley, Judd C. "Crisis and the Development of China's Southwestern Periphery: The Transformation of Panzhihua, 1936–1969." *Modern China* 38, no. 5 (2012), 559–584.

Kirby, William. *Germany and Republican China.* Stanford: Stanford University Press, 1984.

  "The Chinese War Economy," in James C. Hsiung and Steven I. Levine (eds.), *China's Bitter Victory: The War with Japan, 1937–1945.* London: M. E. Sharpe, 1992, 185–212.

  "Engineering China: Birth of the Developmental State, 1928–1937," in Wen-hsin Yeh (ed.), *Becoming Chinese: Passages to Modernity and Beyond.* Berkeley: University of California Press, 2000, 137–160.

Kitchens, Carl. "The Role of Publicly Provided Electricity in Economic Development: The Experience of the Tennessee Valley Authority, 1929–1955." *The Journal of Economic History* 74, no. 2 (2014), 389–419.

Knight, John. "Savior of the East: Chinese Imagination of Soviet Russia during the National Revolution, 1925–1927." *Twentieth Century China* 43, no. 2 (2018), 120–138.

Ko, Humphrey. *The Making of the Modern Chinese State: Cement, Legal Personality, and Industry*. Singapore: Palgrave Macmillan, 2016.

Toa Kenkyujo dai 2 chosa linkai hokushi linkai dai 4 bukai 東亞研究所第二調查委員會內地委員會第四部. *Koga suiryoku hatsuden keikaku hokokusho* 黃河水力發電計劃報告書 (*Report on the Yellow River Hydropower Development Plan*). Tokyo: Toakenkyujo, 1941.

Koll, Elisabeth. *From Cotton Mill to Business Empire: The Emergence of Regional Enterprises in Modern China*. Cambridge, MA: Harvard University Asia Center, 2003.

  Railroads and the Transformation of China. Cambridge, MA: Harvard University Press, 2019.

Kololev, A. A. "Huanghe Sanmenxia shuili shuniu" (Sanmenxia Hydropower Project on the Yellow River). *Zhongguo shuili* 7 (1957), 5.

Koss, Daniel. *Where the Party Rules: The Rank and File of China's Communist State*. Cambridge: Cambridge University Press, 2018.

Kung, James Kai-Sing and Shuo Chen. "The Tragedy of the Nomenklatura: Career Incentives and Political Radicalism during China's Great Leap Famine." *American Political Science Review* 105, no. 1 (2011), 27–45.

Kuo, Leslie T. C. *Agriculture in the People's Republic of China: Structural Changes and Technical Transformation*. New York: Praeger Publishers, 1976.

  "A Lamentation for the Yellow River: The Three Gate Gorge Dam," in Dai Qing (ed.), *The River Dragon Has Come! Three Gorges Dam and the Fate of China's Yangtze River and Its People*. Armonk, NY: M. E. Sharpe, 1998, 143–159.

Lander, Brian. *The King's Harvest: A Political Ecology of China from the First Farmers to the First Empire*. New Haven: Yale University Press, 2021.

Landry II, Marc D. "Europe's Battery: The Making of the Alpine Energy Landscape, 1870–1955." PhD Dissertation, Georgetown University, 2013.

Lary, Diana. *The Chinese People at War: Human Suffering and Social Transformation, 1937–1945*. Cambridge: Cambridge University Press, 2010.

  Chinese Migrations: The Movement of People, Goods, and Ideas over Four Millennia. Lanham, MD: Rowman & Littlefield, 2012.

Lavelle, Peter. *The Profits of Nature: Colonial Development and the Quest for Resources in Nineteenth-Century China*. New York: Columbia University Press, 2020.

Lean, Eugenia. *Vernacular Industrialism in China: Local Innovation and Translated Technologies in the Making of a Cosmetics Empire, 1900–1940*. New York: Columbia University Press, 2020.

Lee, Hong Yung. *From Revolutionary Cadres to Party Technocrats in Socialist China*. Berkeley: University of California Press 1991.

Leng, Meng 冷梦. *Huanghe dayimin* 黄河大移民 (*The Yellow River Great Migration*). Xi'an: Shanxi luyou chubashe, 1998.
"The Battle for Sanmenxia." *Chinese Sociology and Anthropology* 31, no. 3 (1999), 4–98.
Battle for Sanmenxia: Population Relocation during the Three Gate Gorge Hydropower Project. Armonk, NY: M. E. Sharpe, 1999.
Lewis, C. S. "The Abolition of Man," in *The Complete C. S. Lewis Signature Classics*. Grand Rapids: Zondervan, 2007, 689–730.
Li, Bozhong 李伯重. "Chucai Jinyong: Zhongguo shuizuan dafangche yu Yingguo Arkwright shuilifangshaji" 楚才晋用：中国水转大纺车与英国阿克莱水力纺纱机 (The Talent of Chu Put to Use by Jin: China's Water-Powered Spinning Wheel and Britain's Arkwright Water Frame). *Lishi Yanjiu* 历史研究 (*Historical Studies*) 1 (2002), 62–74.
Li, Chunlei 李春雷. "Tuanjie fenjin, zaizhan hongtu: qianjinzhongde Henansheng Sanmenxiashi Huanghekuqu guanliju" 团结奋进,再展宏图:前进中的河南省三门峡市黄河库区管理局. *Zhongguo Shuilibao* 中国水利报. December 9, 1997.
Li, Daigeng 李代耕. *Xinzhongguo dianli gongye fazhan shilue* 新中国电力工业发展史略 (*Historical Outline of the Electrical Industry in New China*). Beijing: Qiyeguanli chubanshe, 1984.
Li, Heming, Paul Waley, and Phil Rees. "Reservoir Resettlement in China: Past Experience and the Three Gorges Dam." *The Geographical Journal* 167, no. 3 (2001), 195–212.
Li, Hua-yu. *Mao and the Economic Stalinization of China, 1948–1953*. Lanham: Rowman & Littlefield, 2006.
Li, Huaiyin. *Village China under Socialism and Reform: A Micro-History, 1948–2008*. Stanford: Stanford University Press, 2009.
Li, Peng 李鹏. *Li Peng Huiyilu,1928–1983* 李鹏回忆录 (*Memoir of Li Peng*). Beijing: Zhongguo dianli chubanshe, 2014.
Li, Rose Maria. "Migration to China's Northern Frontier, 1953–82." *Population and Development Review* 15, no. 3 (1989), 503–538.
Li, Rui 李锐. *Li Rui Riji:1946–1979* 李锐日记 (*Diary of Li Rui*). Fort Worth, TX: Fellow Press of America, 2008.
"Sulian shuili fadian jianshe de jiben qingkuang he zhuyao jingyan" 苏联水力发电建设的基本情况和主要经验 (The Basic Situation and Major Experiences of Soviet Hydroelectric Construction), in *Li Rui wenji* 李锐文集 (*Collected Writings of Li Rui*). Hong Kong: Zhongguo shehui jiaoyu chubanshe, 2009, 275–317.
Li Rui koushu wangshi 李锐口述往事 (*Memoir of Li Rui*). Hong Kong: Dashan wenhuachubanshe, 2016.
Li, Xinchun 李欣春 and Baiyuan Zhou 周百源. "Woguo shuili fadian shiyede xianquzhe Huang Yuxian xiansheng" 我国水力发电事业的先驱者黄育贤先生 (Mr. Huang Yuxian, Pioneer of China's Hydropower Enterprise). *Chongren wenshi ziliao* 崇仁文史资料 3 (1991), 67–73.
Li, Yifei and Judith Shapiro. *China Goes Green: Coercive Environmentalism for a Troubled Planet*. New York: Polity, 2020.
Li, Yizhi 李仪祉. *Huanghe gaikuang ji zhiben tantao* 黄河概况及治本探讨 (*Overview of the Yellow River and Exploration of Fundamental Solutions*). Kaifeng: Huanghe shuili weiyuanhui, 1935.

Li, Zili 李自力. "Sannian zanshi kunnan zai Yongchun chengguan gongshe" 三年暂时困难在永春城关公社 (Three difficult Years in Chengguan Commune, Yongchun). *Yongchun wenshi ziliao* 永春文史资料 21 (1994), 35-43.

"Lianhe kaizhan zhuanxiang xingdong daji pohuai shidi xingwei" 联合开展专项行动 打击破坏湿地资源行为 (Take Collective Action against Wetland Sabotage Activities). *Sanmenxia Huanghe shidi guojiaji ziran baohuqu guanwang*, December 27, 2013.

Lieberthal, Kenneth and Michel Oksenberg. *Policy Making in China: Leaders, Structures, and Processes*. Princeton: Princeton University Press, 1988.

Lilienthal, David. *TVA: Democracy on March*. New York: Harper & Brothers, 1944.

Lin, Adet, Anor Lin, and Meimei Lin. *Dawn over Chungking*. New York: John Day, 1941.

Lin, Feng 林风. "Hongqi qupan" 红旗渠畔 (On the Banks of the Hongqi Canal). *Renmin ribao*, December 9, 1973.

Lingbao xianzhi bianzhuan weiyuanhui 灵宝县志编纂委员会. Lingbao Xianzhi 灵宝县志 (*Lingbao gazetteer*), 1936.

Liu, Lanbo 刘澜波. "Wei quanguo chubu dianqihua er fendou" 为全国初步电气化而奋斗 (Fight for National Preliminary Electrification). *Renmin ribao*, June 21, 1958.

Liu, Shengyuan 刘盛源. *Shui Xiheng zhuan* 税西恒传 (*Biography of Shui Xiheng*). Beijing: Tuanjie chubanshe, 2016.

Liu, Xiaoping 刘小平. "Tangdai shiyuan de shuinianai jingying" 唐代寺院的水碾硙经营 (The Management of Buddhist Temples' Watermills during the Tang Dynasty). *Zhongguo nongshi* 中国农史 (*Chinese Agricultural History*) 4 (2005), 44-50.

Liu, Yanwen 刘彦文. *Gongsi shehui: Yintao shangshan shuili gongchengde geming, jitizhuyi yu xiandaihua* 工地社会:引洮上山水利工程的革命集体主义与现代化 (*Revolution, Collectivism and Modernization in China: A Case Study of the Yintao Water Conservancy Project in Gansu Province*). Beijing: Shehui kexue wenxian chubanshe, 2018.

Liu, Yuhong, Ying Lu, Cheng Chen, et al. "Behavioural Responses of the Whooper Swans *Cygnus cygnus* to Human Disturbance and Their Adaptability to the Different Habitats in the Rongcheng Lagoon of China." *Ecohydrology* 11 (2018), 1-9.

Liujiaxia shuidian changzhi 刘家峡水电厂志 (*Gazetteer of Liujiaxia Hydropower Plant*). Lanzhou: Gansu renmin chubanshe, 1999.

"Lizheng gaosudu" 力争高速度 (Work for High Speed). *Renmin ribao*, June 21, 1958.

Lowdermilk, Walter C. *Soil, Forest, and Water Conservation and Reclamation in China, Israel, Africa, and the United States*, vols. 1 and 2. Berkeley: University of California, Regional Oral History Office, 1969.

Lu, Qixun 盧起勳 and Junxi Liu 劉君錫, *Changshou Xian zhi* 长寿县志: *16 juan*. Taipei: Chengwen chubanshe, 1976.

Lu, Shiqian 卢世钤. "Shuili liyong yu guomin jingji" 水力利用与国民经济 (Hydropower Exploitation and National Economy). *Zhonghua Yubao* 中华月报 5 (1935), 6.

Lu, Weizhen 陆为震. "Zhongguo weilai zhi shuili jianshe" 中国未来之水力建设 (China's Hydropower Construction in the Future). *Hankou Shangye Yuekan* 汉口商业月刊 1, no. 8 (1934), 21.

Luo, Xuan 罗漩. "Kan shuidian fadian jianshe zhanlan" 看水力发电建设展览 (Visit the Hydropower Construction Exhibit). *Renmin ribao*, October 13, 1957.

Ma, Junya 马俊亚. *Beixisheng de jubu: Huaibei shehui shengtai bianqian yanjiu, 1680-1949* 被牺牲的局部:淮北社会生态变迁研究 (*The Sacrificed Region: A Study of the Social Ecological Changes in Huaibei*). Beijing: Beijingdaxue chubanshe, 2011.

Ma, Min 马敏. *Guanshang zhi jian: shehui jubianzhong de jindai shenshang* 官商之间:社会巨变中的近代绅商 (*Between Official and Merchant: The Modern Gentry-Merchants amid Drastic Social Change*). Tianjin: Tianjin renmin chubanshe, 1995.

MacKinnon, Stephen R. *Wuhan, 1938: War Refugees, and the Making of Modern China*. Berkeley: University of California Press, 2008.

Magee, Darrin, "Powershed Politics: Yunnan Hydropower under Great Western Development." *The China Quarterly* no. 185 (2006), 23-41.

Mann, Susan. *Local Merchants and the Chinese Bureaucracy, 1750-1950*. Stanford: Stanford University Press, 1987.

Mao, Zedong. "Guanyu daxue jiaoyu gaigede yiduan tanhua" 关于大学教育改革的一段谈话 (A Talk on Higher Education Reform), in Zhonggong zhongyang wenxian yanjiushi 中共中央文献研究室 (ed.), *Jianguo yilai Mao Zedong wengao* 12 建国以来毛泽东文稿 (*Manuscripts of Mao Zedong since 1949*). Beijing: Zhongyang wenxian chubanshe, 1998.

Mao Zedong Sixiang wansui, 1949-1957 毛泽东思想万岁 (*Long Live Mao Zedong Thought*). Neibuziliao.

Mao, Zhanpo 毛战坡, Peng Wenqi 彭文启, Wang Shiyan 王世岩, and Zhou Huaidong 周怀东. "Sanmenxia shuiku yunxing shuiwei dui shidi shuiwen guocheng yingxiang yanjiu" 三门峡水库运行水位对湿地水文过程影响研究 (Study of the Sanmenxia Reservoir Water Level's Impact on Wetland Hydrology). *Zhongguo shuilishuidian keyuan yanjiuyuan xuebao* 中国水利水电科学研究院学报 (*Journal of China Institute of Water Resources and Hydropower Research*) 4, no. 1 (2006), 36-41.

Marks, Robert B. *Tigers, Rice, Silk and Silt: Environment and Economy in Late Imperial South China*. Cambridge: Cambridge University Press, 1998.

Martin, William A. P. 丁韪良. *Gewu Rumen* 格物入门 (*Introduction to Science*). Beijing: Beijing tongwenguan, 1868.

McCully, Patrick. *Silenced Rivers: The Ecology and Politics of Large Dams*. New York: Zed Books, 2001.

McNeill, John. *Something New under the Sun: An Environmental History of the Twentieth-Century World*. New York: W. W. Norton, 2000.

McNeill, John and Peter Engelke. *The Great Acceleration: An Environmental History of the Anthropocene since 1945*. Cambridge: Belknap Press, 2014.

McRoberts, Duncan. *Pleading China*. Grand Rapids, MI: Zondervan, 1946.

Mertha, Andrew C. *China's Water Warriors: Citizen Action and Policy Change.* Ithaca: Cornell University Press, 2008.

Meyskens, Covell. "Labour," in Christian Sorace, Ivan Franceschini, and Nicholas Loubere (eds.), *Afterlives of Chinese Communism*. Canberra: Australian National University Press, 2019.

Mao's Third Front: The Militarization of Cold War China. Cambridge: Cambridge University Press, 2020.

"Rethinking the Political Economy in Mao's China." *Positions: Asia Critique* 29, no. 4 (2021), 809–834.

Mi, Rucheng 宓汝成. *Diguo zhuyi yu Zhongguo tielu, 1847–1949* 帝国主义与中国铁路 (*Imperialism and China's Railway, 1847–1949*). Beijing: Jingji guanli chubanshe, 2007.

Miller, Ian J. *The Nature of the Beasts: Empire and Exhibition at the Tokyo Imperial Zoo.* Berkeley: University of California Press, 2013.

Mitchell, Timothy. *Carbon Democracy: Political Power in the Age of Oil.* New York: Verso, 2011.

Mitter, Rana. *Forgotten Ally: China's World War II, 1937–1945.* Boston: Mariner Books, 2013.

Moore, Aaron. *Constructing East Asia: Technology, Ideology, and Empire in Japan's Wartime Era, 1931–1945.* Stanford: Stanford University Press, 2015.

Morens, David and Anthony S. Fauci. "Emerging Pandemic Diseases: How We Got to COVID-19." *Cell* 182, no. 5 (2020), 1077–1092.

Morita, Akira, translated by Lei Guoshan. *Qingdai shuili yu quyu shehui* 清代水利与区域社会 (*Hydraulic and Local Societies of the Qing Dynasty*). Jinan: Shandong huabao chubanshe, 2008.

Mostern, Ruth. *The Yellow River: A Natural and Unnatural History.* New Haven: Yale University Press, 2021.

Mote, Frederick W. "The Growth of Chinese Despotism: A Critique of Wittfogel's Theory of Oriental Despotism as Applied to China." *Oriens Extremus* 8, no. 1 (1961), 1–41.

Mou, Mo and Cai Wenmei. "Resettlement in the Xin'an River Power Station Project," in Dai Qing (ed.), *The River Dragon Has Come! The Three Gorges Dam and the Fate of China's Yangtze River and Its People.* Armonk, NY: M. E. Sharpe, 1998, 104–123.

Muhlhahn, Klaus. *Making China Modern: From the Great Qing to Xi Jinping.* Cambridge, MA: Belknap Press, 2019.

Muscolino, Micah. "Violence against People and the Land: The Environment and Refugee Migration from China's Henan Province, 1938–1945." *Environment and History* 17 (2011), 291–311.

The Ecology of War in China: Henan Province, the Yellow River, and Beyond, 1938–1950. Cambridge: Cambridge University Press, 2015.

"Energy and Enterprise in Liu Hongsheng's Cement and Coal-Briquette Business, 1920–37." *Twentieth-Century China* 41, no. 2 (2016), 159–179.

"Water Has Aroused the Girls' Hearts: Gendering Water and Soil Conservation in 1950s China." *Past and Present* 255 (2021), 351–387.

Nanpingshi difangzhi bianzhuan weiyuanhui 南平市地方志编纂委员会 (eds.). *Nanping diquzhi* 南平地区志 (*Gazetteer of Nanping*). Beijing: Fangzhi chuabanshe, 2004.

Nayak, Arun Kumar. "The Mahanadi Multipurpose River Valley Development Plan in India." *World Affairs: The Journal of International Issues* 20, no. 4 (2016), 76–93.
Needham, Andrew. *Power Lines: Phoenix and the Making of Modern Southwest.* Princeton: Princeton University Press, 2014.
Needham, Joseph, Ling Wang, and Gwei-djen Lu. *Science and Civilization in China, Vol. 4 Physics and Physical Technology, Part II: Mechanical Engineering.* Cambridge: Cambridge University Press, 1965.
Nelson, Daniel. *Journey to Chungking.* Minneapolis: Augsburg Publishing House, 1945.
"Neng fadian de shuili gongcheng douyao fadian" 能发电的水利工程都要发电 (Every Qualified Hydraulic Project Shall Generate Electricity). *Renmin ribao*, December 4, 1958.
Nield, Robert. *China's Foreign Places: The Foreign Presence in China in the Treaty Port Era, 1840–1943.* Hong Kong: Hong Kong University Press, 2015.
Nien, Chen-ho 粘振和. "Lun Bei Song shuimo chafa" 论北宋水磨茶法 (The Study of the Monopoly System of Tea by Water-Powered Mills in the Northern Song Dynasty). 成大历史学报 (Cheng Kung Journal of Historical Studies) 47 (2014), 1–28.
Nishiijima, Sadao 西嶋定生. *Tyuugoku Keizaisi Kennkyuu* 中國經濟史研究. Tokyo: Tokyo daigaku shuppan-kai, 1966.
  Zhongguo Jingjishi Yanjiu. Beijing: Nongye chubanshe, 1984.
Nongcun xiaoxing shuidianzhan cankao ziliao 农村小型水电站参考资料 (*Materials on Rural Small Hydropower Stations*). Beijing: Shuili chubanshe, 1956.
Nye, David. *Electrifying America: Social Meanings of a Technology, 1880–1940.* Cambridge, MA: MIT Press, 1992.
  Consuming Power: A Social History of American Energies. Cambridge, MA: MIT Press, 1997.
Ogashi, Heriku. 小越平陸. *Koga chisui* 黄河治水. Tokyo: Seikyosha, 1929.
Pan, Jiazheng 潘家铮. *Qianqiu gongzui huashuiba* 千秋功罪话水坝 (*On Hydropower Dams*). Beijing: Qinghua daxue chubanshe, 2000.
Pan, Junxiang 潘君祥. *Zhongguo jindai guohuo yundong* 中国近代国货运动 (*National Products Movements in Modern China*). Beijing: Zhongguo wenshi chubanshe, 1996.
Pang, Mingyue, Lixiao Zhang, Sergio Ulgiati, and Changbo Wang. "Ecological Impacts of Small Hydropower in China: Insights from an Energy Analysis of a Case Plant." *Energy Policy* 76 (2015), 112–122.
Pantsov, Alexander V. and Steven Levine. *Mao: The Real Story.* New York: Simon & Schuster, 2012.
Parrinello, Giacomo. "Systems of Power: A Spatial Envirotechnical Approach to Water Power and Industrialization in the Po Valley of Italy, ca. 1880–1970." *Technology and Culture* 59 (2018), 652–688.
Perkins, Dwight and Shahid Yusuf. *Rural Development in China.* Baltimore, MD: The Johns Hopkins University Press, 1984.
Pietz, David. *Engineering the State: the Huai River and Reconstruction in Nationalist China, 1927–1937.* New York: Routledge, 2002.

The Yellow River: The Problem of Water in Modern China. Cambridge, MA: Harvard University Press, 2015.

Premalatha, M., Tabassum-Abbasi, Tasneem Abbasi, and S. A. Abbasi. "A Critical View on the Eco-Friendliness of Small Hydroelectric Installations." *Science of the Total Environment* 481 (2014), 638–643.

Pritchard, Sara B. *Confluence: The Nature of Technology and the Remaking of the Rhone*. Cambridge, MA: Harvard University Press, 2011.

Ptak, Thomas. "Towards an Ethnography of Small Hydropower in China: Rural Electrification, Socioeconomic Development and Furtive Hydroscapes." *Energy Research & Social Science* 48 (2019), 116–130.

Purcell, Fernando. "Dams and Hydroelectricity: Circulation of Knowledge and Technological Imaginaries in South America, 1945–1970," in Andra Chastain and Timothy Lorek (eds.), *Itineraries of Expertise: Science, Technology, and the Environment in Latin America's Long Cold War*. Pittsburgh: University of Pittsburgh Press, 2020, 217–236.

Purdue, Peter. *Exhausting the Earth: State and Peasant in Hunan, 1500–1850*. Cambridge, MA: Harvard University Asia Center, 1987.

"Is There a Chinese View of Technology and Nature?" in Martin Reuss and Stephen Cutcliffe (eds.), *The Illusory Boundary: Environment and Technology in History*. Charlottesville: University of Virginia Press, 2010, 101–119.

Qu, Zhide and Bai Wenying. "Hanhao wolunji, weiguo zhengguang" (Welding the water turbine well, honoring the Motherland), in *Wanli Huanghe diyiba*. Zhengzhou: Henan renmin chubanshe, 1992, 462.

Quanguo Nongye Zhanlanguan Shuiliguan 全国农业展览馆水利馆 (ed.), *Nongcun Shuidian* 农村水电 (*Rural Hydro*). Beijing: Nongye chubanshe, 1960, 8.

Quanzhoushi shuilishuidianju 泉州市水利水电局. *Quanzhoushi shuilizhi* 泉州市水利志 (*Water Conservancy in Quanzhou*). Beijing: Zhongguo shuilishuidian chubanshe, 1998.

Rassweiler, Anne D. *The Generation of Power: The History of Dneprostroi*. Oxford: Oxford University Press, 1988.

Reuss, Martin. "Seeing Like an Engineer: Water Projects and the Mediation of the Incommensurable." *Technology and Culture* 3 (2008), 531–546.

Rohlf, Gregory. *Building New China: Colonizing Kokonor: Resettlement to Qinghai in the 1950s*. Lanhan, MD: Lexington Books, 2016.

"Sanmenxia chutai wosheng shoubu baohu baitian'e guiding" 三门峡出台我省首部保护白天鹅规定 (Sanmenxia Promulgates the Province's First Set of Rules on the Protection of Swans). *Sanmenxia Huanghe shidi guojiaji ziran baohuqu guanwang*, January 8, 2014.

Sanmenxiashi difangshizhi bianzuan weiyuanhui 三门峡市地方史志编纂委员会. *Sanmenxia shizhi*, vol. 1 三门峡市志 (*Sanmenxia Gazetteer*). Zhengzhou: Zhongzhou guji chubanshe, 1997.

Sanmenxiashi Huanghehewu yiminguanlijuzhi bianzuan weiyuanhui 三门峡市黄河河务移民管理局志编纂委员会. *Sanmenxiashi Huanghe hewu yimin guanlijuzhi* 三门峡市黄河河务移民管理局志. Beijing: Fangzhi chubanshe, 2006.

Schaller, Michael. "FDR and the 'China Question'," in David Woolner, Warren Kimball, and David Reynolds (eds.), *FDR's World: War, Peace, and Legacies*. New York: Palgrave MacMillan, 2008, 145–174.

Schmalzer, Sigrid. *Red Revolution, Green Revolution: Scientific Farming in Socialist China*. Chicago: University of Chicago Press, 2016.

Schoppa, R. Keith. *Xiang Lake: Nine Centuries of Chinese Life*. New Haven: Yale University Press, 1989.

Schudder, Thayer. "The Human Ecology of Big Projects: River Basin Development and Resettlement." *Annual Review of Anthropology* 2 (1973), 45–55.

Schumacher, E. F. *Small is Beautiful: Economics as if People Mattered*. London: Harper & Row, 1975.

Schwarz, Henry G. "Chinese Migration to North-West China and Inner Mongolia 1949–59." *The China Quarterly* 16 (1963), 62–74.

Scott, James. *Seeing Like a State: How Certain Schemes to Improve the Human Condition Have Failed*. New Haven: Yale University Press, 1999.

Seow, Victor. *Carbon Technocracy: Energy Regimes in Modern East Asia*. Chicago: University of Chicago Press, 2022.

Shan, Yubin 单毓斌. "Kaocha Riben Dongjing diandeng huishe guichuan shuilifadiansuo jilue" 考察日本东京电灯会社桂川水力发电所纪略 (Visit of Japan's Tokyo Light Company Guichuan Hydroelectricity Plant). *Dianqixiehui Zhazhi* 电气协会杂志 8 (1914), 54–63.

"Shangban Yunnan Yaolong diandeng gongsi Shilongba gongcheng jilue" 商办云南耀龙电灯公司石龙坝工程纪略 (Brief Records of the Commercial Yunnan Yaolong Company Shilongba Project). *Zhongguo Shuilifadian Shiliao* 中国水力发电史料 (China Historical Materials on Waterpower) 1 (1987), 73.

Shanghai shuili fadian shejiyuan 上海水力发电设计院. "Minzhe liangsheng xiaoxing shuidianzhan diaocha baogao" 闽浙两省小型水电站调查报告 (Report on Small Hydropower in Min and Zhe Provinces). *Shuili fadian* 水力发电 8 (1958), 41–43.

Shangyuan 商原, *Shangwubao* 商务报 31 (1908).

Shanxisheng Sanmenxia kuqu guanliju 山西省三门峡库区管理局 (ed.). *Shanxisheng Sanmenxia kuquzhi* 山西省三门峡库区. Zhengzhou: Huanghe shuili chubanshe, 2007.

Shapiro, Judith. *Mao's War against Nature: Politics and the Environment in Revolutionary China*. Cambridge: Cambridge University Press, 2001.

China's Environmental Challenges. Cambridge: Polity Press, 2019.

Shen, Grace Yen. *Unearthing the Nation: Modern Geology and Nationalism in Republican China*. Chicago: The University of Chicago Press, 2014.

Shen, Yimin 沈益民 and Shengzhu Tong 童乘珠. *Zhongguo renkou qianyi* 中国人口迁移 (The Population Migration in China). Beijing: Zhongguo tongji chubanshe, 1992.

Shen, Zhihua 沈志华. *Sulian Zhuanjia zai Zhongguo,1948–1960* 苏联专家在中国 (*Soviet Advisors in China, 1948–1960*). Beijing: Zhongguo guoji guangbo chubanshe, 2003.

Zhonghua renmin gonghe guoshi, disanjuan, sikao yu xuanze: cong zhishifenzi huiyi dao fanyoupai yundong (1956-1957) 中华人民共和国史, 第三卷 思考

与选择–从知识分子会议到反右派运动 (1956–1957) (*The History of the People's Republic of China, vol. 3, Reflections and Choices: The Consciousness of the Chinese Intellectuals and the Anti-Rightist Campaign, 1956–1957*). Hong Kong: Research Center for Contemporary China, The Chinese University of Hong Kong, 2008.

Shen, Zhihua and Douglas Stiffer (eds.). *Cuiruo de lianmeng: Lengzhan yu Zhongsu guanxi* 脆弱的联盟：冷战与中苏关系 (*Frail Alliance: The Cold War and Sino-Soviet Relations*). Beijing: Shehui kexue wenxian chubanshe, 2010.

Shen, Zhihua, *Mao Zedong, Sidalin yu Chaoxianzhanzheng* 毛泽东，斯大林与朝鲜战争 (*Mao Zedong, Stalin, and the Korean War*). Guangzhou: Guangdong renmin chubanshe, 2013.

Shi, Nianhai 史念海. *Huangtu gaoyuan lishi dili yanjiu* 黄土高原历史地理研究 (*Study of the Historical Geography of the Loess Plateau*). Zhengzhou: Huanghe shuili chubanshe, 2001.

Shi, Yun 石耘. "Huanghe Sanmenxia kuqu yimin qingku qingkuang gaishu" 黄河三门峡库区移民清库情况概述 (An Overview of the Yellow River Sanmenxia Reservoir Clearance). *Sanmenxia wenshi ziliao* 三门峡文史资料 17 (2007), 106–122.

"Qinghua daxue shuilixi zai Sanmenxia" 清华大学水利系在三门峡 (The Hydraulics Department of Tsinghua University at Sanmenxia). *Henan Wenshiziliao* 河南文史资料 1 (2011), 4–20.

"Shidi baohuqu yancha luanbu lanlie houniao xingwei" 湿地保护区严查乱捕滥猎候鸟行为 (Crackdown on Illegal Bird Hunting in the Wetland Conservation Area). *Sanmenxia Huanghe shidi guojiaji ziran baohuqu guanwang*, November 2, 2012.

Shuili dianli jianshe zongju xinan gongzuozu 水利电力建设总局西南工作组. *Dagao qunzhong yundong quanmin banshuidian* 大搞群众运动全民办水电 (*Mobilizing the Masses to Build Hydroeletricity*). Beijing: Shuili dianli chubanshe, 1959.

Shuili Fadian Jianshe Zongju Zhuanjia Gongzuoshi 水力发电建设总局专家工作室. "Sulian zhuanjia dui woguo shuidian jianshede bangzhu" 苏联专家对我国水电建设的帮助 (Soviet Experts' Aid on Our Country's Hydroelectric Construction). *Shuili Fadian* 水力发电 4 (1954), 1.

"Shuili fadian zengjia qushi" 水力发电增加趋势 (Increasing Trend of Hydroelectricity). *Dianqi Gongye Zazhi* 电气工业杂志 1, no. 5 (1923), 64.

Shuilibu Huangweihui kance guihua shejiyuan 水利部黄委会勘测规划设计院. *Sanmenxia shuiku jianku qianhou ludi, shuisheng shengwu de bianhua* 三门峡水库建库前后陆地，水生生物的变化 (*Changes of Terrestrial and Aquatic Species before and after the Construction of the Sanmenxia Reservoir*). Zhengzhou, 1988.

Shuili dianlibu disi gongchengju 水利电力部第四工程局. *Liujiaxia shuidianzhan (tupianji)* 刘家峡水电站(图片集) (*Liujiaxia Hydropower Project, Picture Collection*). Beijing: Renmin meishuchubanshe, 1977.

Siciliano, Giuseppina and Frauke Urban (eds.). *Chinese Hydropower Development in Africa and Asia: Challenges and Opportunities for Sustainable Global Dam-Building*. New York: Routledge, 2017.

Smil, Vaclav. *China's Energy: Achievements, Problems, Prospects*. New York: Praeger, 1976.
  The Bad Earth: Environmental Degradation in China. Armonk: M. E. Sharpe, 1984.
  Energy in China's Modernization: Advances and Limitations. Armonk: M. E. Sharpe, 1988.
  Energy in World History. Boulder: Westview Press, 1994.
  Enriching the Earth: Fritz Haber, Carl Bosch, and the Transformation of World Food Production. Cambridge, MA: MIT Press, 2001
Smith, Aminda. *Thought Reform and China's Dangerous Classes*. New York: Rowman & Littlefield, 2012.
Smith, Norman. *Man and Water: A History of Hydro-Technology*. London: Peter Davies, 1976.
Sneddon, Christopher. *Concrete Revolution: Large Dams, Cold War Geopolitics, and the US Bureau of Reclamation*. Chicago: University of Chicago Press, 2015.
Songster, Elena. *Panda Nation: The Construction and Conservation of China's Modern Icon*. Oxford: Oxford University Press, 2018.
Stavis, Benedict. *The Politics of Agricultural Mechanization in China*. Ithaca: Cornell University Press, 1978.
Steinberg, Theodore. *Nature Incorporated: Industrialization and the Waters of New England*. Cambridge: Cambridge University Press, 1991.
Stine, Jeffery K. and Joel A. Tarr. "At the Intersection of Histories: Technology and the Environment." *Technology and Culture* 39, no. 4 (1998), 601–640.
Sun, Yat-sen. *Sun Yat-sen Quanji* 孙中山全集, Vol. 1, Beijing: Zhonghua shuju, 1981.
Sun, Yusheng 孙玉声. "Kanzhan banianlai zhi dianlishiye" 抗战八年来之电力事业(The Electricity Industry in the Eight Years War of Resistance). *Ziyuan weiyuanhui jikan* 资源委员会季刊 6 (1946), 143.
Sun, Zhaoming 孙肇明 and Han Hai 韩海 (eds.). *Qinglong Manzhu zizhixian shuilizhi* 青龙满族自治县水利志 (*Water Conservancy in Qinglong Manchu Autonomous County*). Tianjin: Tianjin daxue chubanshe, 1993.
Sun, Zhigao, Wenguang Sun, Chuan Tong, Congsheng Zeng, Xiang Yu, and Xiaojie Mou. "China's Coastal Wetlands: Conservation History, Implementation Efforts, Existing Issues and Strategies for Future Improvement." *Environment International* 79 (2015), 25–41.
Tan, Xuming 谭徐明. "Zhongguo shuili jixie de qiyuan, fazhan jiqi zhongxi bijiao"中国水力机械的起源、发展及其中西比较研究 (A Study of the Origin and Development of Hydraulic Machinery in China and Its Comparison with the West). *Ziran kexue shi yanjiu* 自然科学史研究 (Studies in the History of Natural Sciences) 14, no. 1 (1995), 83–95.
Tan, Ying Jia. *Recharging China in War and Revolution, 1882–1955*. Ithaca: Cornell University Press, 2021.
Thaxton Jr., Ralph A. *Catastrophe and Contention in Rural China: Mao's Great Leap Forward Famine and the Origins of Righteous Resistance in Da Fo Village*. Cambridge: Cambridge University Press, 2008.
Tian, Dongkui 田东奎. "Zhongguo jindai shuiquan jiufen jiejue jizhi yanjiu" 中国近代水权纠纷解决机制研究 (Study on the Mechanism of Solving Water

Rights Dispute in Modern China). PhD Dissertation, Zhongguo Zhengfa University, 2006.

Tilt, Bryan. *The Struggle for Sustainability in Rural China: Environmental Values and Civil Society*. New York: Columbia University Press, 2010.

Dams and Development in China: The Moral Economy of Water and Power. New York: Columbia University Press, 2015.

Tong, Jiandong 童建栋. *Zhongguo Xiaoshuidian* 中国小水电 (Small Hydropower in China). Beijing: Zhongguo shuili shuidian chubanshe, 2006.

Treiman, Donald J. and Andrew Walder. "The Impact of Class Labels on Life Chances in China." *American Journal of Sociology* 124, no. 4 (2019), 1125–1163.

Tucker, Richard P. and Russell Edmund (eds.). *Natural Enemy, Natural Ally: Toward an Environmental History of War*. Corvallis: Oregon State University Press, 2004.

Tuoteo (ed.), *Songshi, Hequ, Huanghe* 宋史,河渠一、黄河 (*The History of the Song Dynasty, Rivers, the Yellow River*), vol. 91. Beijing: Zhonghua shuju, 1985.

Turvey, Samuel. *Witness to Extinction: How We Failed to Save the Yangtze River Dolphins*. Oxford: Oxford University Press, 2008.

United States Public Health Service and Tennessee Valley Authority Health and Safety Department. *Malaria Control on Impounded Water*. Washington DC: US Government Printing Office, 1947.

Vassal, Gabrielle. *In and Round Yunnan Fou*. London: W. Heinemann, 1922.

van de Ven, Hans. *War and Nationalism in China, 1925–1945*. New York: Routledge, 2003.

China at War: Triumph and Tragedy in the Emergence of the New China. Cambridge, MA: Harvard University Press, 2018.

Wang, Guanglun 王光纶 (ed.). *Qingxishanhe: Zhang Guangdou zhuan* 情系山河：张光斗传 (*Biography of Zhang Guangdou*). Beijing: Zhongguo kexue-jishu chubanshe, 2014.

Wang, Huayun. 王化云. "Huangtu qiuling gouhequ shuitu baochi kaocha baogao" 黄土丘陵沟壑区水土保持考察报告 (Investigation Report of Water and Soil Conservation on the Loess Plateau). *Xin Huanghe* 新黄河 12 (1955), 26–31.

Wode Zhihe Shijian 我的治河实践 (*My Practice of River Management*). Zhengzhou: Henan kexue jishu chubanshe, 1989.

Wang, Jingya 王静雅. "Nanjing Guomin zhengfu jianshe weiyuanhui dianye guihua yu shijian yanji" 南京国民政府建设委员会电业规划与实践研究:以20世纪30年代长江中下游地区为例 (A Study of the Nanjing Nationalist Construction Commission's Electricity Industry Planning and Practice). *Huabei dianli daxue xuebao* 华北电力大学学报 6 (2012), 19–26.

Wang, Lihua 王利华. "Gudai huabei shuilijiagong xingshuai de shuihuanjing Beijing" 古代华北水力加工兴衰的水环境 (The Water Environment and the Evolution of Waterpower Use in Ancient North China). *Zhongguo jingji shi yanjiu* 中国经济史研究 (Chinese Economic History Studies) 1(2005), 30–39.

Wang, Ning. *Banished to the Great Northern Wilderness: Political Exile and Re-education in Mao's China*. Ithaca: Cornell University Press, 2018.

Wang, Pu, Shikui Dong, and James P. Lassoie. *The Large Dam Dilemma: An Exploration of the Impacts of Hydro Projects on People and the Environment in China*. New York: Springer, 2014.

Wang, Shuhuai 王树槐. "Zhongguo zaoqi de dianqi shiye 1882–1928" 中国早期的电气事业, 1882–1928 (The Early Electricity Industry in China, 1882–1928), in Zhongyangyanjiuyuan jinshisuo (ed.), *Zhongguo xiandaihua lunwenji* 中国现代化论文集 (*Essays on the Modernization of China*). Taipei: Institute of Modern History of Academia Sinica, 1991, 443–472.

Wang, Yanmou 王燕谋. *Zhongguo Shuini Fazhanshi* 中国水泥发展史 (*The History of Cment in China*). Beijing: Zhongguo jiancaigongye chubanshe, 2005.

Wang, Yong 王永. "Shoudu Beihai gongyuan shaonian xianfengdui shuidianzhan luocheng" 首都北海公园少年先锋队水电站落成 (The Young Pioneer Hydroelectric Station at Beihai Park in the Capital Has Been Completed). *Shuili Fadian* 水力发电 (*Hydropower*) 11 (1956), 38.

Wang, Yuan 王渊. *Dunhuang Yishi* 敦煌轶事 (*Dunhuang Anecdotes*). Lanzhou: Gansu renmin chubanshe, 2005.

Weller, Robert P. *Discovering Nature: Globalization and Environmental Culture in China and Taiwan*. Cambridge: Cambridge University Press, 2006.

Wemheuer, Felix. *Famine Politics in Maoist China and the Soviet Union*. New Haven: Yale University Press, 2014.

Westad, Odd Arne (ed.). *Brothers in Arms: The Rise and Fall of the Sino-Soviet Alliance, 1945–1963*. Washington DC: Woodrow Wilson Center Press, 1998.

White Jr., Lynn, *Medieval Technology and Social Change*. Oxford: Oxford University Press, 1962.

White, Richard, *The Organic Machine: The Remaking of the Columbia River*. New York: Hill and Wang, 1995.

Will, Pierre-Etienne. "State Intervention in the Administration of a Hydraulic Infrastructure: The Example of Hubei Province in Late Imperial Times," in Stuart Schram (ed.), *The Scope of State Power in China*. Hong Kong: Chinese University Press, 1985, 295–347.

Wittfogel, Karl A. *Oriental Despotism: A Comparative Study of Total Power*. New Haven: Yale University Press, 1957.

"Woguo kaishi xiang dianqihua maijin" 我国开始向电气化迈进 (Our Country Is Marching toward Electrification). *Renmin ribao*, May 15, 1958.

Worster, Donald. *Rivers of Empire: Water, Aridity, and the Growth of the American West*. Oxford: Oxford University Press, 1985.

"The Flow of Empire: Comparing Water Control in the United States and China." *RCC Perspectives* no. 5 (2011), 1–23.

"Woshi baitian'e baohu gongzuo zhashi youxiao" 我市白天鹅保护工作扎实有效 (Our City's Whooper Swan Protection Work Is Solid and Effective). *Sanmenxia Huanghe shidi guojiaji ziran baohuqu guanwang* 三门峡黄河湿地国家级自然保护区官网, November 17, 2011.

Wright, Tim. *Coal Mining in China's Economy and Society, 1895–1937*. Cambridge: Cambridge University Press, 1984.

"Electric Power Production in Pre-1937 China." *The China Quarterly* 126 (1991), 356–363.

Wu, Odoric Y. K. *Mobilizing the Masses: Building Revolution in Henan*. California: Stanford University Press, 1994.

Wu, Shellen X. *Empires of Coal: Fueling China's Entry into the Modern World Order, 1860–1920*. Stanford: Stanford University Press, 2015.

Wu, Xingzhi 吴兴帜. *Yanshen de pingxingxian: Dian-Yue tielu yu bianmin shehui* 延伸的平行线：滇越铁路与边民社会 (*Extended Parallel Lines: The Dian-Vietnam Railway and the Borderland Community*). Beijing: Beijing daxue chubanshe, 2012.

"Xiang nongcun dianqihua de daolu maijin" 向农村电气化的道路迈进 (Marching toward Rural Electrification). *Fujian ribao*, April 5, 1958.

Xiao, Lingyun, Zhi Lu, Xueyang Li, Xiang Zhao, and Binbin V. Li. "Why Do We Need a Wildlife Consumption Ban in China?" *Current Biology* 31 (2021), R161–R185.

"Xiaoshuidian zhixiang: Fujian Yongchuxian" 小水电之乡 – 福建省永春县 (Hometown of Small Hydropower: Yongchun, Fujian Province). *Fujian shuili shizhi ziliao* 福建水利史志资料 2 (1984), 13–17.

Xie, Zhaoping 谢朝平. *Da Qianxi* 大迁徙 (*Exodus*). Beijing: Huohua zazhishe, 2010.

Xu, Huaiyun 徐怀云, "Yangzi jiang sanxia gaoba sheji jishi" 扬子江三峡高坝设计纪实 (Yangzi River Three Gorges High Dam Design Document), in Zhongguo Changjiang Sanxia gongcheng lishi wenxian huibian bianweihui (ed.), *Zhongguo Chang Jiang Sanxia Gongcheng lishi wenxian huibian, 1918–1949* 中国长江三峡工程历史文献汇编, 1918–1949 (*Collected Records of the Three Gorges Project, 1918–1949*). Beijing: Zhongguo Sanxia chubanshe, 2010.

Xu, Shoubo 徐寿波. "Shuili fadian yu huoli fadian" 水力发电与火力发电 (Hydroelectricity and Thermoelectricity). *Renmin ribao*, March 6, 1963.

Xu, Weihua, Xinyue Fan, Jungai Ma, et al. "Hidden Loss of Wetlands in China." *Current Biology* 29 (2019), 3065–3071.

Xue, Yi 薛毅. "Kangzhanshiqi de sanyi xueshe" 抗战时期的三一学社 (The Sanyi Association during the War of Resistance against Japan). *Kangri Zhanzheng yanjiu* 抗日战争研究 2 (2003), 87–107.

Guomin zhengfu ziyuan weiyuanhui yanjiu 国民政府资源委员会研究 (*Study on the Nationalist National Resource Commission*). Beijing: Shehui kexue wenxian chubanshe, 2005.

Yan, Qiu 燕秋. *Wo jialege lieshi yigu: ji Luo Xibei de shuidian shengya* 我嫁了个烈士遗孤：记罗西北的水电生涯 (*I Married a Martyr's Orphan: Luo Xibei's Hydro Career*). Beijing: Zhongguo dianli chubanshe, 2002.

Yan, Yangchu 晏阳初. *Pingmin jianyu yu xiangcun jianshe yundong* 平民教育与乡村建设运动 (*Civilian Education and Rural Construction Movement*). Beijing: Shangwu yinshuguan, 2014.

Yang, Anqing 杨庆安 and Luo Qimin 罗启民. "Huanghe Sanmenxia shuikuqu de zhili jiqi jingyan" 黄河三门峡水库区的治理及其经验 (The Yellow River Sanmenxia Reservoir Area Management and Experience). *Renmin Huanghe* 人民黄河 5 (1986), 29–32.

Yang, Dali, Huayu Xu, and Ran Tao. "The Tragedy of the Nomenklatura? Career Incentives, Political Loyalty and Political Radicalism during China's Great Leap Forward." *Journal of Contemporary China* 23, no. 89 (2014), 864–883.

Yang, Jisheng. *Tombstone: The Great Chinese Famine, 1958–1962*. New York: Farrar, Straus and Giroux, 2013.

Yang, Kuisong. *Eight Outcasts: Social and Political Marginalization in China under Mao*. Berkeley: University of California Press, 2019.

Yang, Lijuan 杨利娟. "Shinian xinku buxunchang: Sanmenxia kuqu yimin yiliu wenti chuli gongzuo chengji feiran" 十年辛苦不寻常－三门峡库区移民遗留问题处理工作成绩斐然 (A Decade of Hard Work: Achievements in Solving Remaining Issues in the Sanmenxia Reservoir Displacement). Henan qingnianbao 河南青年报 (May 21, 1997).

Yang, Xianhui 杨显惠. *Jiabiangou Jishi* 夹边沟纪事 (*Jiabiangou Chronicle*). Guangzhou: Huacheng chubanshe, 2008.

Woman from Shanghai: Tales of Survival from a Chinese Labor Camp. New York: Anchor, 2009.

Yang, Yongnian 杨永年. "Jianguoqian Sichuan de shuidian jianshe" 建国前四川的水电建设 (Hydropower development in Sichuan before 1949). *Zhongguo shuilifadian shiliao* 2 (1987), 44–45.

Yang, Yuqiu 杨玉秋, and Luo Song 罗松. "Henan Huanghe shidi guojia ziran baohuqu Sanmenxiaduan jiben qingkuang jianjie" 河南黄河湿地国家级自然保护区三门峡段基本情况简介 (Brief Introduction of the National Yellow River Wetland Conservation in Sanmenxia, Henan). *Sanmenxia wenshiziliao* 三门峡文史资料 17 (2007), 431–435.

Yao, Hanyuan 姚汉源. *Huanghe shuilishi yanjiu* 黄河水利史研究 (*Study of the Yellow River Conservancy*). Zhengzhou: Huanghe shuili chubanshe, 2003.

Ye, Jianyun. 'Sanmenxia de mingtian" (The Future of Sanmenxia). *Renmin ribao*, November 26, 1956.

Yi, Si. "The World's Most Catastrophic Dam Failures: The August 1975 Collapse of the Banqiao and Shimantan Dams," in Dai Qing (ed.), *The River Dragon Has Come: Three Gorges Dam and the Fate of the Yangtze River and Its People*. New York: M. E. Sharpe, 1998, 25–38.

Yin, Liangwu. "The Long Quest for Greatness: China's Decision to Launch the Three Gorges Project." PhD Dissertation, Washington University, St. Louis, 1996.

"Yongchun shuidian shiye dafazhan" 永春水电事业大发展 (The Great Achievement of the Yongchun Hydropower Enterprise), *Fujian ribao* 福建日报 (*Fujian Daily*), January 21, 1959.

Yongchun xianzhi bainzhuan weiyuanhui 永春县志编撰委员会. *Yongchun xianzhi* 永春县志 (*Yongchun Gazetteer*). Beijing: Yuwen chubanshe, 1990.

Yongchunxian shuiliju dianlike 永春县水利局电力科 (Yongchun County Water Control Bureau Electric Power Office). Yongchunxian nongcun shuili shuidian ziliao 永春县农村水利水电站资料 (*Materials on Rural Water Conservancy and Hydropower in Yongchun County*). 1964.

Yuan, Li, Li Xiaomin, Yu Hongxian, et al. "The Status and Conservation of Whooper Swans (*Cygnus cygnus*) in China." *Journal of Forestry Research* 8, no. 4 (1997), 235–239.

Yun, Zhen 恽震. "Guanyu Sanxia shuili shoucikance" 关于三峡水力首次勘测 (On the first survey of the hydropower of the Three Gorges), in Zhongguo Changjiang Sanxiagongcheng lishi wenxian huibian bianweihui (ed.),

Zhongguo Chang Jiang Sanxia Gongcheng lishi wenxian huibian, 1918-1949 中国长江三峡工程历史文献汇编, 1918-1949 (*Collected Record of the Three Gorges project, 1918-1949*). Beijing: Zhongguo Sanxia chubanshe, 2010.

"Yunnan fu, Zhongguo de diyige shuidianzhan" 云南府, 中国的第一个水电站, 西门子杂志 (China's First Hydropower Plant). *Zhongguo Shuilifadian Shiliao* 5 (1989), 70-72.

Yunnan Sheng difangzhi bianzuan weiyuanhui 云南省地方志编纂委员会. *Yunnan Sheng zhi. juan 1, Di li zhi* 云南省志-卷一地理志 (*Yunnan Province Magazine: Roll 1, Geography Section*). Kunming: Yunnan renmin chubanshe, 1998.

Zanasi, Margherita. *Saving the Nation: Economic Modernity in Republican China*. Chicago: The University of Chicago Press, 2006.

Zeisler-Vralsted, Dorothy. *Rivers, Memory and Nation-Building: A History of the Volga and Mississippi Rivers*. New York: Berghahn Books, 2015.

Zelin, Madeleine. *The Merchants of Zigong: Industrial Entrepreneurship in Early Modern China*. New York: Columbia University Press, 2006.

Zeng, Zhaojin and Joshua Eisenman. "The Price of Persecution: The Long-Term Effects of the Anti-Rightist Campaign on Economic Performance in Post-Mao China." *World Development* 109 (2018), 249-260.

Zhang, Baichun 张柏春. "Zhongguo chuantong shuilun jiqi qudong jixie" 中国传统水轮及其驱动机械 (China's Traditional Waterwheel and Its Motor Mechanism). *Ziran kexue shi yanjiu* 自然科学史研究 (Studies in the History of Natural Sciences) 13, no. 2 (1994), 155-163.

Zhang, Baichun 张柏春, Fang Yao 姚芳, Jiuchun Zhang 张久春 and Long Jiang 蒋龙, *Sulian jishu xiang Zhongguo de zhuanyi, 1949-1966* 苏联技术向中国的转移, 1949-1966 (*Technology Transfer from the Soviet Union to the P. R. China, 1949-1966*). Jinan: Shandong jiaoyu chubanshe, 2004.

Zhang, Genfu 张根福. *Kangzhan shiqi de renkou qianyi* 抗战时期的人口迁移. Beijing: Guangming ribao chubanshe, 2006.

Zhang, Guangdou 张光斗. *Wode rensheng zhilu* 我的人生之路 (*The Road of My Life*). Beijing: Qinghua daxue chubanshe, 2002.

Zhang, Guogang 张国钢. Huanghe Sanmenxia kuqu yuedong datian'e de zhongqun xianzhuang" 黄河三门峡库区越冬大天鹅的种群现状 (The Current Status of Wintering Population of Whooper Swans at Sanmenxia Reservoir Region). *Dongwuxue zazhi* 动物学杂志 51, no.2 (2016), 190-197.

Zhang, Hanying. "Sidalin pailaideren zenyang zai Zhongguode heliushang gongzuozhe" (How People Sent by Stalin Work on Chinese Rivers). *Renmin ribao*, November 25, 1952.

"Xuexi Sulian xianjin jingyan, tigao womende sixiang yu jinshu shuiping" 学习苏联先进经验, 提高我们的思想与技术水平 (Learn from the Soviet Experience, Improve Our Thought and Technology Levels). *Renmin Shuili* 人民水利 2 (1952), 6-9.

Zhang, Jiaqin 张加芹. "Yongchun xiaoshuidian fazhan jianjie" 永春小水电发展简介 (Brief Introduction to Small Hydropower Development in Yongchun). *Yongchun wenshi ziliao* 永春文史资料 24 (1994), 60.

Zhang, Jin 张瑾. *Quanli, Chongtu yu Bianqe: 1926-1937 nian Chongqing Chengshi Xiandaihua Yanjiu* 权力、冲突与变革：1926-1937 年重庆城市现代化研究 (*Power, Conflict and Reform: A Study of the Modernization of Chongqing City, 1926-1937*). Chongqing: Chongqing chubanshe, 2003.

Zhang, Junfeng 张俊峰. *Shuili Shehui de Leixing: Ming Qing yilai Hongdong Shuili yu Xiangcun Shehui Bianqian* 水利社会的类型：明清以来洪洞水利与乡村社会变迁 (*The Pattern of Hydraulic Society: Water Conservancy and Rural Social Changes in Hongdong since the Ming and Qing dynasties*). Beijing: Beijing Daxue chubanshe, 2012.

Zhang, Ling. *The River, The Plain, and the State: An Environmental Drama in Northern Song China, 1048-1128*. Cambridge: Cambridge University Press, 2016.

Zhang, Peiji. "Sanmenxia jianshe zhongde Sulian zhuanjia" (Soviet Experts in the Sanmenxia Project), in *Wanli Huanghe diyiba*. Zhengzhou: Henan renmin chubanshe, 1992, 309.

Zhang, Zhihui 张志会. "Liujiaxia shuidianzhan gongchengjianshe de ruogan lishifansi" 刘家峡水电站工程建设的若干历史反思 (Historical Reflections on the Construction of the Liujiaxia Hydropower Project). *Gongcheng yanjiu* 工程研究 5, no. 1 (2013), 58–70.

Zhao, Cheng 赵诚. *Huang Wanli de Changhe gulu* 黄万里的长河孤旅 (*Huang Wanli's Lonely Journey down the Long River*). Xi'an: Shanxi renmin chubanshe, 2013.

Zhao, Hai. "Manchurian Atlas: Competitive Geopolitics, Planned Industrialization, and the Rise of Heavy Industrial State in Northeast China, 1918-1954." PhD Dissertation, University of Chicago, 2015.

Zhao, Rukun 赵入坤. "Ershi shiji wuliushinian daide zhongguo bianjiang yimin" 二十世纪五六十年代的中国边疆移民 (Chinese Frontier Migration in the 1950s and 1960s). *Zhonggong dangshi yanjiu* 中共党史研究 2 (2012), 52–64.

Zhao, Xingsheng 赵兴胜. "Zhang Jingjiang in 1928-1937" 1928-1937年的张静江. *Jindaishi yanjiu* 近代史研究 1 (1997), 237–251.

Zhao, Zhilin. "Jianguoqian Sanmenxia gongcheng yanjiu" (Studies of the Sanmenxia Project before 1949). *Zhongguo shuili fadian shiliao* 3 (1991), 13.

Zheng, Qun 郑群. *Zhongguo Diyisuo Shuidianzhan Shilongba Chuanqi* 中国第一座水电站石龙坝传奇 (*The Legend of China's First Hydroelectric Plant: Shilongba*). Kunming: Yunnan jianyu chubanshe, 2012.

Zheng, Youkui 郑友揆, Linsun Cheng 程麟荪, and Chuanhong Zhang 张传洪. *Jiu Zhongguo Ziyuanweiyuanhui, 1932-1949: Shishi yu pingjia* 旧中国的资源委员会，1932-1949 – 史实与评价 (*The National Resource Commission in Old China, 1932-1949, Historical Fact and Evaluation*). Shanghai: Shanghai shehui kexueyuan chubanshe, 1991.

"Zhiyuan Sanmenxia" (Support Sanmenxia). *Renmin ribao*, April 18, 1957.

Zhonggong Zhongyang wenxian yanjiushi 中共中央文献研究室 (ed.). *Zhou Enlai nianpu, erjuan, 1949-1976* 周恩来年谱1949-1976 (*A Chronicle of Zhou Enlai, vol. 2 (1949-1976)*). Beijing: Zhongyang wenxian chubanshe, 1997.

Zhongguo Changjiang Sanxia gongcheng lishi wenxian huibian bianweihui (ed.). *Zhongguo Chang Jiang Sanxia Gongcheng lishi wenxian huibian, 1918-1949* 中国长江三峡工程历史文献汇编, 1918-1949 (*Collected Record of the Three Gorges Project, 1918-1949*). Beijing: Zhongguo Sanxia chubanshe, 2010.

Zhongguo Kexueyuan Kaogu Yanjiusuo 中国科学院考古研究所. *Sanmenxia caoyu yiji* 三门峡漕运遗迹 (*Sanmenxia Canal Transportation Relic*). Beijing: Kexue chubanshe, 1959.

Zhongguo Nongcun jingji tongji daquan (1949-1986) 中国农村经济统计大全 (*Economic Statistics for Rural China*). Beijing: Nongye chubanshe, 1989.

Zhongguo renmin zhengzhixieshanghuiyi 中国人民政治协商会议三门峡市委员会, in Zhongguo shuilishuidian dishiyi gongchengju 中国水利水电第十一工程局 (eds.). *Wanli Huanghe Diyiba* 万里黄河第一坝 (*The First Dam on the Yellow River*). Zhengzhou: Henan renmin chubanshe, 1992.

Zhongguo shuili fadianshi bianji weiyuanhui 中国水力发电史编辑委员会. *Zhongguo Shuilifadianshi* 中国水力发电史 (*The History of Hydroelectric Power in China*), Vol. 1, 2. Beijing: Zhongguo dianli chubanshe, 2005.

Zhongguo Shuilifadian Shiliao Zhengji Bianji Weiyuanhui 中国水力发电史料征集编辑委员会. *Zhongguo Shuilifadian Shiliao* 中国水力发电史料 (*China Historical Materials of Water Power*), 28 volumes. Beijing, 1987-1997.

Zhonghua renmin gongheguo shuilibu 中华人民共和国水利部 and Guojia tongjiju 国家统计局. *Diyici quanguo shuilipucha gongbao* 第一次全国水利普查公报 (*Bulletin of First National Census for Water*). Beijing: Zhongguo shuili shuidian chubanshe, 2011.

Zhou, Xun (ed.). *The Great Famine in China, 1958-1962: A Documentary History*. New Haven, CT: Yale University Press, 2012.

"Zhoumi bushu, tianqian zuohao shidi baohu gongzuo anpai" 周密部署, 提前做好湿地保护工作安排 (Be Prepared for the Wetland Protection Work Arrangement). *Sanmenxia Huanghe shidi guojiaji ziran baohuqu guanwang*, October 30, 2012.

Zhu, Chengzhang 朱成章. "Longxihe tiji shuidianzhan kaifa jishi" 龙溪河梯级水电站开发纪实 (Records of the Longxi River Cascade Hydropower Exploitation). *Zhongguo shuilifadian shiliao* 中国水力发电史料 1 (1987), 37.

Zhu, Sulian 朱淑莲. "Liujiaxia shuidianzhan guangyao jiuju Huanghe sishichun" 刘家峡水电站光耀九曲黄河四十春 (Liujiaxia Hydropower Project Shines over the Yellow River for Four Decades). *Guojia dianwangbao* 国家电网报, September 3, 2009.

Index

agricultural mechanization, 135
Agriculture Ministry, 122
Allen, Yong J., 29
alluvial lands, 206, 214, 220, 228
   collapse of, 214
An, Zhen, 195
An, Ziwen, 105
analogy of hair and skin, 102
Andreas, Joel, 85
   red engineers, 85, 98, 105
Anik, Ferruh, 144
Anthropocene, 21, 205, 235
anti-party clique, 193
Anti-Rightist Movement, 91, 193
appropriation, adaptive, 28
Australia, 61
Austria, 127
Aydin, Mansour, 144

backyard furnaces, 130
Beihai Park, 88
Bennett, Gordon, 119, 122
Bi Mountain, 40
biodiversity, 204, 218, 232, 234
Biwa Lake, 155
Bo, Yibo, 87
Bolshevik Revolution, 113
Burdin, Claude, 28
Bureau of Business Promotion in Yunnan, 35
Bureau of Reclamation, 64, 69, 72, 232

Canada, 30–31
cannibalism, 195

Caohai, 222
carbofuran, 204
carbon footprint, 233
carbon neutrality, 236
carbon technocracy, 11
Carlowitz & Company, 36
cave houses, 201
CCP, *See* Chinese Communist Party
cement, shortage of, 112
censorship, 203
Central News Documentary Film Factory, 172
Central Political Bureau, 161
Chang, Kia-Ngua, 68
Changchun Film Studio, 115
check dams, 164
Chen, Guofu, 33
Chen, Huangmei, 115
Chen, Liangfu, 63–65, 69
Chen, Yun, 86–87
Chen, Zhang, 52
Chen, Zhaoxiang, 193
Chen, Zudong, 33–34
Chi, Ch'ao-Ting, 3, 97
Chiang, Kai-shek, 68
Chiang, Tingfu, 68
Chief Hydroelectric Bureau, 88
Chief Hydroelectric Construction Bureau,
   See Hydroelectric Engineering Bureau
Chile, 61
China Central Television, 227
China Rural Construction College, 40
Chinese Changchun Railway, 100
Chinese Civil War, 79

Index

Chinese Communist Party, 84, 93, 98
    leadership, 170
Chongqing, 11, 15, 48, 54, 136
    Beibei, 39, 45
    Changshou, 54, 56
Churchill, Winston, 235
class identity, 187
class struggle, 119
Cleansing of the Class Ranks, 103
climate change, 229, 233
coercive methods, 187
Cold War, 111
collectivization, 116, 136
Colorado River, 69
communism, 93
concrete, effort to revolutionize, 110
concrete revolution, 7, 13, 109, 117
Cotton, John, 107, 156
counterrevolutionary, 100, 106–107, 198
COVID-19, 237
Crook, Isabel, 57
Cultural Revolution, 92, 104, 106–107, 133
    revolution in higher education, 107
culture of accommodation, 105

Dadu River, 54
Danjiangkou, 110
declensionist narrative, 229
decommissioning, 236
delegation, electrical industry, 96
demographic engineering, 180, 183, 195
detaining water and flushing sediments, 173
detaining water and sediment, 173
development refugees, 203
Dian Lake, 35
dikes, 151, 174, 208
disaster, man-made, 46
displacement, 180, 184, 211, 230, 232
    education of the masses, 185
    scale of, 178
Dneprostroi project, 95
door blockade, 207

ecological civilization, 221, 227, 229
eco-tourism, 221, 223, 225
eight essentials, 190
Eighth National Plenum, 108
Electric Power Ministry, 151
electricity, conservation of, 51
Electricity Ministry, 89
electrification, 33–34, 38

Elisson, S., 153
embezzlement, 200
energy, 30–33, 39
    manual labor, 49
    nonrenewable source of, 49
    self-sufficiency, 32
England, 61
environmental consciousness, 232
environmental hazard, 209
environmental justice, 204
environmentalists, 233
Espy, Willard, 78
Executive Ministry, 68
exhibition, 89
experts from the old society, 97, 99

Fan, Jinfu, 126
Fan, Shouren, 194
farmlands, 42
    flooding of, 43
    scarcity of, 58
Fen River, 205
Fengman hydropower project, 94, 99, 154
first Sino-Japanese War, 32
fish ladders, 218
fish population, decline of, 218
Five-Year Plan, 90
fossil fuels, 26, 28, 56, 233
four old heads, 100
Fourneyron, Benoit, *See* Fourneyron turbine
Fourneyron turbine, 5, 28–29
fragmentation of rivers, 137
France, 30, 35
frontier reclamation, 180, 202
Fu, Zuoyi, 162, 170
Fuel Industry Ministry, 86–87, 160
Fujian, 15, 117–118, 128, 137
    Nanping, 38
full stomach environmentalism, 232
Fushun coal mine, 49
Fuyuan Company, 40

game meat industry, 220
Gansu, 17, 20, 179
    Dunhuang, 180, 185–199
    Tianshui, 158, 164
    Zhangye, 189, 193
Gauss, Clarence, 73
Germany, 30, 33, 38, 46
gigantomania, 95
global North, 235

Index

global South, 232
Gobi Desert, 191, 194, 196
grain as the key line, 215
Grand Canal, 151, 153
grass on the wall, 102
great acceleration, 6, 13, 113, 140
Great Famine, 195
Great Leap Forward, 92, 108, 119, 142, 165, 185
   hydropower projects, 109
   ideology of, 176
   small hydropower, 120
   war against nature, 102
Greater East Asia Co-Prosperity Sphere, 154
green revolution, 143
groundwater, 210–211
Guangning, 137
Guanzhong Plain, 166, 173, 205, 207

H5N1, 219, 224, 237
Hao, Yujiang, 182
He, Jingzhi, 175–176
Heilongjiang River, 107
Heiriku, Ogashi, 153
Henan, 162, 181, 188
   archives, 17
   displacement of inhabitants, 178
   failed river management in history, 4
   labors from, 110
   Lingbao, 180–181, 199, 214
   Mianchi, 217
   Sanmenxia, 191
   Shaanxian, 180–182, 201, 204
Hershatter, Gail, 135
high modernism, 9, 183
Hongqi Canal, 132
Hoover Dam, 69, 231
Hu, Jintao, 114
Hu, Shi, 62
Hu, Yingze, 179, 206
Huai River, 236
Huang, Kecheng, 86
Huang, Nanqiao, 59
Huang, Wanli, 165, 178
Huang, Wenxi, 49
Huang, Yuxian, 54, 84, 99–100
   criticism of the party, 107
   director of the Hydroelectric Bureau, 97
   former director, 86
   reactionary academic authority, 103
   survey team, 54

Huang, Ziduan, 127
Hubei, 50, 110
Huilongzhai project, 58
hukou, 198
Hunan, 50, 86
Hundred Flowers campaign, 102
Hutchins, Francis, 61
hydraulic community, 3–4
hydraulic mode of consumption, 8–9
hydraulic mode of production, 8
hydraulic society, 3–4, 44
Hydroelectric Bureau, 87
Hydroelectric Engineering Bureau, 86–87
hydroelectricity, 27–29, 32
   modernist discourse, 43
   new method of producing an inexhaustible source of power, 25
   potential of, 41
   promotion of, 88
   Western scientific theories of, 34
   widespread use of, 29
hydrology, 31
hydropower, See hydroelectricity
hydropower engineering education, 100
hydropower engineering, internationalization of, 60
hydropower nation, 6–12, 14–15, 17–21, 28, 79, 146, 233
   all encompassing, 173
   anthropocentric system, 235
   entanglement with Maoist politics, 113
   environmental aspect of, 205
   human cost, 178
   incipient, 51
   international cooperation, 60
   mass participation, 84
   might of, 150
   productivist, 203
   role of small hydropower, 117
   social and environmental aspects of, 150
   social limit of, 181
   technostructure of, 113
   transnational dimentions of, 74

illegal hunting, 221
   crackdown on, 223
India, 144
Industrial Revolution, 234
industrialization, 94
intellectuals, problem of, 101

intelligentsia, 91, 100
Italy, 30

Japan, 30, 32, 49
Ji,Tinghong, 38
Jialing River, 39–40
Jiang, Gengqiao, 42
Jiang, Guiyuan, 63
Jiangdu, 63
Jiangsu, 49
Jiaotong University, 64
Jihe project, 38
Jinsha River, 35
Jiuyuangou, 164

Keh, Chi-yang, 74–75
Khrushchev, Nikita, 119
killing two birds with one stone, 183
Knoxville, 65–66
Kongjiazhuang, 115
Korean War, 92–93
Kunming, 36
Kunming Chamber of Commerce, 36
Kuomintang, 39, 44, 48, 52

labor mobilization, mass, 110
land reclamation, 183
leaning to one side policy, 92, 106
learning from Dazhai, 132
Lend-Lease Act, 62, 65
Leng, Meng, 179
Lenin, 90, 93, 113
Leningrad Design Academy, 162
Li, Bai, 175
Li, Hongzhang, 25
Li, Jimao, 196
Li, Peng, 96, 114, 231
Li, Rui, 7, 86–88, 99
  background in propaganda, 88
  delegation to the Soviet Union, 96
  passion for hydropower, 87
  pragmatism, 100
  revolutionary cadre, 84
  right opportunist, 91
Li, Yizhi, 153, 158
Liangtan River, 40–41, 44, 46
Lilienthal, David, 61–62
Lin, Miaoqing, 125
Lin, Yishan, 89
Liu, Lanbo, 90, 96
Liu, Lingfang, 35

Liu, Shaoqi, 126
Liu, Shuanzhou, 192
Liu, Yingzhou, 42
Liu, Yongzhi, 191
load equilibrium theory, 90
local community, conflict with, 42
local community, well-being of, 60
Loess Plateau, 13, 157–159, 181, 205, 208
  heavy rainfall in summer, 213
  water and soil conservation project, 164
Longhai Railway, 199
Longxi River, 38, 54
Longxi River project, 50, 55
Lowdermilk, Walter C., 157–159, 164
Lowell, 26, 30
Lower Qingyan Cave project, *See* Xiadong hydropower plant
Lu, Shiqian, 31
Lü, Siqi, 42
Lu, Zuofu, 40–41
Lüda, 121
Luo, Xibei, 91
Lushan Plenum, 91, 198

Ma, Jiying, 192
Manchukuo, 94, 99
Manchuria, 32
Mao, Zedong, 89, 166, 198
Mao Zedong Thought Propaganda Team, 106
Maoism, radical, 103, 113
  class struggle, 111
  politics in command, 103
Maolin, 115
Martin, W. A. P., 29
mass campaign, 117
mass line, 119
mass mobilization, 130
materials, alternative, 109
McCully, Patrick, 174
mechanization, 110
Mengjin, 153
Mexico, 61
migrant cadres, 192
migratory birds, 217
militarization, 77
Ministry of Water Conservancy, 126, 144
Minle county, 191
Minsheng, 40–41
mobilization tactics, 185
Morita, Akira, 3

multipurpose exploitation, 150
multipurpose hydro-technical innovation, 75

Nanjing, 49, 104
Nanning Conference, 89
National Defense Commission, 56, 70
National Defense Strategy Commission, 53, 63
National People's Congress, 162, 231
National Planning Commission, 88
National Resources Commission, 17–18, 50, 52–55, 58, 60, 66, 97
  estimation of hydropower capacity, 31
  key role in harnessing hydropower reserves, 50
  opposition from landowners, 58
  practical training requirement, 64
  shortage of engineers, 63
  survey team, 54
  transnational exchange, 61
National Wetland Conservation Action Plan, 216
National Wetland Conservation Program, 216
National Writers Association, 172
nationalism, 32
Nationalist, *See* Kuomingtang
Nehru, Jawaharlal, 150
Nelson, Donald, 69
Niagara Falls, 28–29, 34
Niekoop, Raymond, 145
Nigeria, 144
Ningxia, 103, 179, 190, 196
nitrogen fertilizer, 71
non-Communist intellectuals, 98
nonhuman species, 206
North China Plain, 151, 157, 205, 213
  farmland, 162
Northeastern Hydropower Engineering Company, 94
northern wilderness, 91
Norway, 30

Office of the National Electric Industry, 53
Organization Department, 105, 123
Organization Department of the Kuomintang, 33
oriental despotism, 3
Outline of National Agricultural Development for 1956 to 1967, 126
overfishing, 218

Pacific War, 156
Palchinsky, Peter, 92
Pan, Jiazheng, 232
Pan-Asianism, 153
Paniushkin, A. S., 100
party technocrats, 97
party-state, 85, 114
Paschal, G. R., 69
Pearl on the Yellow River, 230
Peng, Dehuai, 91, 168, 197
Peng, Zhen, 100
People's Commune, 192
People's Daily, 170
People's Liberation Army, 104
Pietz, David, 8, 84, 96, 133, 150, 179
Pope, James, 67
poverty alleviation, 215
powershed, 12
principle of small, indigenous, mass, 117
Pritchard, Sara, 12
proletarian engineers, 124
proletarian hydropower technology, 95
proletarianization of higher education, 107

Qin, Xiudian, 100
Qinglong, 131–132
Qinling Mountains, 206

radical Maoism, 178, 200, 215
  shortcomings of, 119
Ramsar Convention on Wetlands, 215
reactionary academic authority, 103
reconstruction, 52–53, 74
red and expert, 114
Red Guards, 106
red revolution, 143
reeducation, 102, 106
reeducation through labor, 91
Ren, S. D., 66
renewable energy, 233
Reservoir Management Bureau, 213
reservoir migrants, 178–180
  access to farmland, 209
  agency of, 180
  anxiety, 188
  avaiability of arable land, 213
  inspection tour, 185
  invisibility of, 230
  malnutrition, 195
  quota of, 183
  repatriation of, 197

reservoir migrants (cont.)
  returning, 197
  sense of honor, 202
  subsistence needs of, 215
reservoir resettlement, standard process of, 184
resettlement, 20, 44, 75, 180–181, 184, 186–187, 214, 223
  as political task, 185
  assistance from local residents, 190
  attitudes of farmers toward, 76
  compensation policy, 200
  higher altitude, 182
  lack of permanent shelter, 194
  long distance, 187, 197
  politicization of, 193
  short-distance, 200
  stress of, 202
  sustainability of, 188
resources, total development of, 76
retained engineers, 98, 100, 104
revolution in electricity, 121
revolution in technique, 110
revolutionary cadres, 97–99
right opportunist, 91
rightists, 91, 102, 123, 129, 193
right-leaning conservatives, 111
river development, multipurpose, 96–97
river ecologies, destruction of, 137
riverbank collapse, 209–210, 214
  investigation of, 210
Rockefeller Foundation, 61
Roosevelt, Franklin, 60
Roshchin, N. V., 93
run-of-the-river pattern, 174
rural collectivization, 164, 180, 185
rural electrification, 115, 118, 133, 143
  improvement of living standard, 135
  major principles, 122
  political significance of, 140
  targets, 123

salinization, 173, 211
sanbian, 109
Sanmenxia, 11–13, 15, 17, 110
  flood control, 152
  Japanese plan, 154
  multipurpose development, 161
  passage, 152
Sanmenxia City
  Hubin, 217

Sanmenxia Dam, 163, 169, 175–176, 205–206, 236
Sanmenxia project
  controversy, 178
  debate of, 165
  opening ceremony, 170
  operating pattern, 174
  silting problem, 165
  social engineering, 172
Sanmenxia Radio and Television Station, 227
Sanmenxia reservoir, 182, 205, 207, 215, 219
  mode of operation, 208
Sanmenxia wetland, 217
Sanmenxia Yellow River and Migrants Management Bureau, 214
Savage, John L., 69–74, 103, 107
Schmalzer, Sigrid, 116, 143
Schumacher, E. F., 14, 138
Scott, James, 9, 184, 202
seasonal flooding, 213
Second Sino-Japanese war, *See* War of Resistance against Japan
sediment, 206, 212
  retention of, 207
self-reliance, 115
self-strengthening movement, 32
Seow, Victor, 11
Shaanxi, 20, 164, 178–179, 196, 205, 208
  concerns about the loss of arable land, 169
  Hua County, 174
  Huaxian, 211
  Huayin, 211
  power grid, 162
  Suide, 164
  Tongguan, 174
  Xi'an, 207
Shaanzhou, 149
Shan, Yubin, 32
Shandong, 177
Shanghai, 31, 36, 49
Shanxi, 20, 162, 164, 178, 205
  Liulin, 164
  Pinglu, 223
Shapiro, Judith, 179, 202
Shi, Hongxi, 63
Shi, Jiayang, 100
Shijiatan, 177, 201
Shilongba Hydropower Plant, *See* Yaolong Light Company
Shizitan, 54

shoaly lands, 179
shore lands, *See* alluvial lands
Shuanglong Commune, 124
Shuanglonghu wetlands, 221
Shui, Xiheng, 38
Shuzhuangtai, 175
Sichuan, 40, 50, 52–53, 109
    Ba County, 44
    Luzhou, 38
Siemens, 36, 38
sihua, 115
silt trap dams, 164
Sino-Soviet split, 207
Sino-Soviet Treaty of Alliance, Friendship, and Mutual Assistance, 159
sluice gate, 42, 45, 225
small hydropower, 116–117, 235
    agricultural mechanization, 118
    benefits of, 116
    definition of, 116
    impact on the environment, 137
    local materials and native knowledge, 124
    national security, 143
    role of women, 135
    top-down initiative, 120
    waste of human labor and local resources, 130
Sneddon, Christopher, 7, 69, 109
    concrete revolution, 8, 69
socialism, 141
socialist industrialization, 102
soil conservation, 157
soil erosion prevention, 159
Songhua River, 94, 154
Southern Sichuan Construction Bureau, 38
Soviet Union, 30, 33, 65, 92
    experts, 94–96, 160
    hydropower advisors, 94
    imperialist power, 100
    Moscow, 96
    revisionism, 111
soybeans, 223
Stalin, 93, 95
Stalinist, 93
starvation, 192
State Council, 160, 170, 182
statist, 178
struggle session, 103, 105
Su, Li, 115
Su, Yuren, 187

Suining, 123, 136, 142
Sun, Yat-sen, 25, 27
Sun, Yunxuan, 63
Sup'ung, 154
Suriname, 144
suspended river, 157
sustainability, 234
Suyun, 227
Swan City, 228, 230
Sweden, 30
Switzerland, 30
sworn kinships, 190
Syria, 144

Taiwan, 32, 79, 97
Tanglang River, 35
Taohuaxi plant, 55, 59
Taohuaxi River, 55, 59
technocracy, 113
technocrats, 18–19, 85, 114, 160
technological hubris, 176, 233
technological lock-in, 15
technological revolution, 111, 121
technologies, brute force, 13
technology, appropriate, 14, 138, 145
technology, brute force, 176, 234
technology, indigenous, 126
technology, intermediate, 138
technology, international, 92
technostructure, 84–85, 92, 233
Tekhintern, 92, 160
temple of modernity, 150
Tennessee Valley Authority, 5, 60–69, 72, 231
Tesla, Nikola, 29
Third Front, 110, 142
thought reform, 101, 103, 105
Thousand Islands Lake, 237
Three Gates Gorge, *See* Sanmenxia
Three Gorges, 1–2, 14, 70
Three Gorges project, 50, 89
Tianfu coal mine, 51
Tilt, Bryan, 234
Tsinghua University, 33, 54, 105, 114
    hydraulic professor at, 165
    Hydraulics Department, 108
    revolution in education, 85
tu expert, 127
Turkey, 144
tu-yang binary, 116
TVA, *See* Tennessee Valley Authority

UNESCO, 226
unimagined communities, 177, 237
United Kingdom, 108
United Nations Industrial Development
    Organization, 133, 144
United Nations Relief and Rehabilitation
    Administration, 68
United States, 31, 50
    hydraulic education, 49
    hydropower capacity, 30
    large generators and turbines from, 33
    prime mover of large hydropower
        projects, 231
Universal Engineering Digest, 67
Universal Trading Corporation, 66
Upper and Lower Qingyan caves,
    54
US State Department, 63
USSR, *See* Soviet Union

vernacular industrialism, 38
Vietnam, 35
    Haiphong, 35
Voith, 36

walk on two legs, 141
Wallace, Henry, 68
Wang, Huayun, 162, 164, 171
Wang, Pingyang, 63
Wang, Xiaozhai, 36
Wang, Yukuan, 192
Wangguan, 221
Wangguan wetland, 222, 225
War of Resistance against Japan, 44, 78,
    107
wasteland, 215
water and soil conservation, 165
Water Conservancy and Electricity
    Ministry, 90, 131, 175
Water Conservancy Ministry, 89, 118, 160,
    162
water conservation movement,
    119
waterfalls, 25, 28–29, 31, 40, 54
waterfowl, 217, 222, 224
watermill horizontal wheel, *See* Fourneyron
    turbine
watermills, 26, 29, 42, 124
    impacted by the dam, 42
    role in processing grain, 27
    waterpower plant to earn funds, 128

waterpower, 25–26, 28–29
waterwheel, 28–29
Wei River, 158, 167, 205, 207
    estuary block, 173
    silting reduction, 174
welfare migrants, 186
Wen, Shanzhang, 167
Weng, Wenhao, 53
Wetland Management Office,
    217, 221
wetlands, 215
    buffer zones, 226
    key preserve areas, 226
    kidneys of the earth, 216
white coal, 5, *See* hydropower
whooper swans, 20, 204, 218–219, 224,
    226, 230, 237
    city cultural emblem, 227
    economic opportunity, 226
    habitat conservation, 225
    wintering ground, 219
wildlife protection, 204, 220–221, 223
    enforcement of, 222
    surveillance, 221
    tension with agricultural
        production, 223
William the Conqueror, 26
winter habitat, 229
Wittfogel, Karl, 3
wooden propeller, 126, 145
work points, 191
World War II, 156
Worster, Donald, 3, 234
Wu, Zhipu, 183
Wuhan, 49

Xi, Zhongxun, 168
Xiadong hydropower plant,
    56
Xie, Peihe, 63
Xie, Yusheng, 126
Xie, Zhaoping, 179
Xin'anjiang, 110, 237
Xinglongchang, 57
Xiong, Shuchen, 104
Xu, Shoubo, 140
Xu, Ying, 74

Yalu River, 154
Yan, Yangchu, 40
Yang, Sen, 56

Yangtze dolphins, 229
Yangtze Gorges, *See* Three Gorges
Yangtze paddlefish, 229
Yangtze River, 1–2, 4, 14, 16, 33, 48
  1998 flood, 216
  Gezhouba Dam, 229
  Three Gorges project, 231, 234
Yangtze River project, 70, 73
Yaolong Light Company, 36, 46
Yellow River, 3, 11, 20, 160, 182
  Baili Hutong, 153
  China's Sorrow, 8, 150
  Hukou, 153
  Liujiaxia, 111
  Longmen, 179, 205
  Qingtongxia, 103
  Tongguan, 179, 205, 211
  V-shaped course, 212
  Xiaobeiganliu, 211
  Xiaolangdi, 153
  Yanguoxia, 199
Yellow River Conservancy Commission, 98, 153, 163, 168
Yellow River Planning Commission, 160
Yellow River Wetland Preserve, 228
Yellow River Wetland Provincial Level Natural Preserve, 216
Yen, Jimmy, *See* Yan, Yangchu
Yichang, 69
Yongchun, 15, 19, 117, 122, 125–129, 135–136, 138, 142
You, Yangzu, 127
Yu the Great, 3, 149, 176

Yu, Kaiquan, 99
Yu, Youcai, 200
Yunnan, 50, 52
  Kunming, 35–38

Zhang, Bing'ai, 1
Zhang, Changling, 100
Zhang, Desheng, 166
Zhang, Guangdou, 64, 69, 84, 104–108
  concern about the Sanmenxia project, 166
  model red engineer, 106
  target of the Red Guards, 106
Zhang, Hanying, 98, 163
Zhang, Jianye, 210
Zhang, Ling, 8, 212
Zhang, Tiezheng, 95
Zhang, Wenying, 182
Zhang, Xiaowa, 197
Zhao, Shishi, 204
Zhejiang, 49
Zheng, Shenglin, 127
Zhengguo Canal, 3
Zhongtiao Mountains, 206
Zhou, Enlai, 93, 105, 174
  concern over water and soil conservation, 164
  discussion of the Soviet design, 167
  on intellectuals, 102
  party leader, 85
  role in finding compromise, 169
  visiting hydropower exhibition, 89
  visiting Sanmenxia, 168
Zhu, De, 89
Zhu, Zhongping, 204

Other Books in the Series

David A. Bello *Across Forest, Steppe, and Mountain: Environment, Identity, and Empire in Qing China's Borderlands*
Erik Loomis *Empire of Timber: Labor Unions and the Pacific Northwest Forests*
Peter Thorsheim *Waste into Weapons: Recycling in Britain during the Second World War*
Kieko Matteson *Forests in Revolutionary France: Conservation, Community, and Conflict, 1669–1848*
Micah S. Muscolino *The Ecology of War in China: Henan Province, the Yellow River, and Beyond, 1938–1950*
George Colpitts *Pemmican Empire: Food, Trade, and the Last Bison Hunts in the North American Plains, 1780–1882*
John L. Brooke *Climate Change and the Course of Global History: A Rough Journey*
Paul Josephson et al. *An Environmental History of Russia*
Emmanuel Kreike *Environmental Infrastructure in African History: Examining the Myth of Natural Resource Management*
Gregory T. Cushman *Guano and the Opening of the Pacific World: A Global Ecological History*
Sam White *The Climate of Rebellion in the Early Modern Ottoman Empire*
Edmund Russell *Evolutionary History: Uniting History and Biology to Understand Life on Earth*
Alan Mikhail *Nature and Empire in Ottoman Egypt: An Environmental History*
Richard W. Judd *The Untilled Garden: Natural History and the Spirit of Conservation in America, 1740–1840*
James L. A. Webb, Jr. *Humanity's Burden: A Global History of Malaria*
Myrna I. Santiago *The Ecology of Oil: Environment, Labor, and the Mexican Revolution, 1900–1938*
Frank Uekoetter *The Green and the Brown: A History of Conservation in Nazi Germany*
Matthew D. Evenden *Fish versus Power: An Environmental History of the Fraser River*
Alfred W. Crosby *Ecological Imperialism: The Biological Expansion of Europe, 900–1900*, second edition
Nancy J. Jacobs *Environment, Power, and Injustice: A South African History*
Edmund Russell *War and Nature: Fighting Humans and Insects with Chemicals from World War I to Silent Spring*
Adam Rome *The Bulldozer in the Countryside: Suburban Sprawl and the Rise of American Environmentalism*
Judith Shapiro *Mao's War against Nature: Politics and the Environment in Revolutionary China*
Andrew Isenberg *The Destruction of the Bison: An Environmental History*
Thomas Dunlap *Nature and the English Diaspora*
Robert B. Marks *Tigers, Rice, Silk, and Silt: Environment and Economy in Late Imperial South China*
Mark Elvin and Tsui'jung Liu, *Sediments of Time: Environment and Society in Chinese History*
Richard H. Grove *Green Imperialism: Colonial Expansion, Tropical Island Edens and the Origins of Environmentalism, 1600–1860*

Thorkild Kjærgaard *The Danish Revolution, 1500–1800: An Ecohistorical Interpretation*
Donald Worster *Nature's Economy: A History of Ecological Ideas, second edition*
Elinor G. K. Melville *A Plague of Sheep: Environmental Consequences of the Conquest of Mexico*
J. R. McNeill *The Mountains of the Mediterranean World: An Environmental History*
Theodore Steinberg *Nature Incorporated: Industrialization and the Waters of New England*
Timothy Silver *A New Face on the Countryside: Indians, Colonists, and Slaves in the South Atlantic Forests, 1500–1800*
Michael Williams *Americans and Their Forests: A Historical Geography*
Donald Worster *The Ends of the Earth: Perspectives on Modern Environmental History*
Robert Harms *Games against Nature: An Eco-Cultural History of the Nunu of Equatorial Africa*
Warren Dean *Brazil and the Struggle for Rubber: A Study in Environmental History*
Samuel P. Hays *Beauty, Health, and Permanence: Environmental Politics in the United States, 1955–1985*
Arthur F. McEvoy *The Fisherman's Problem: Ecology and Law in the California Fisheries, 1850–1980*
Kenneth F. Kiple *The Caribbean Slave: A Biological History*
Ellen F. Arnold *Medieval Riverscapes: Environment and Memory in Northwest Europe, c.300–1100*

Milton Keynes UK
Ingram Content Group UK Ltd.
UKHW040149041124
450466UK00010B/2

9 781009 426565